CAMBRIDGE STUDIES IN
ADVANCED MATHEMATICS 4

Ultrametric calculus

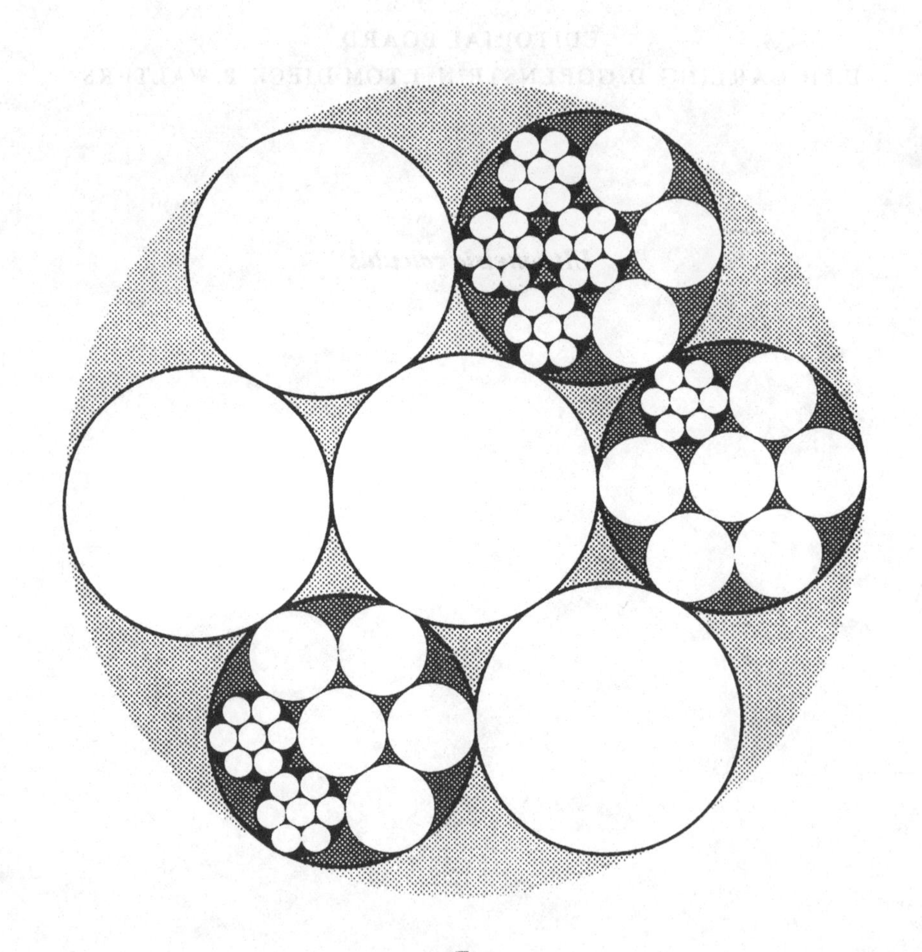

$\mathbb{Z}_7$

Ultrametric calculus

An introduction to p-adic analysis

W. H. SCHIKHOF

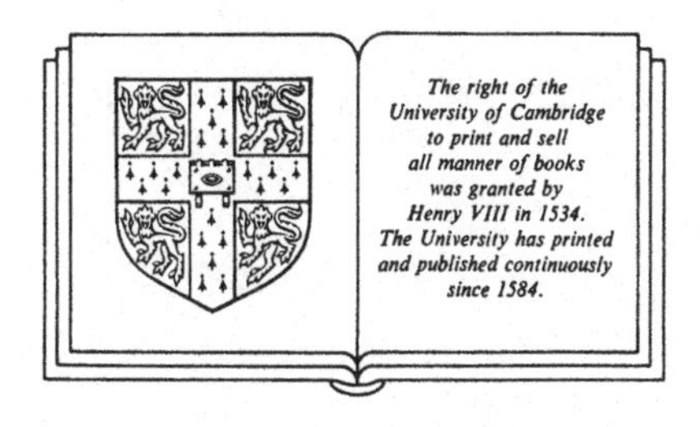
The right of the
University of Cambridge
to print and sell
all manner of books
was granted by
Henry VIII in 1534.
The University has printed
and published continuously
since 1584.

CAMBRIDGE UNIVERSITY PRESS

Cambridge

London New York New Rochelle

Melbourne Sydney

CAMBRIDGE UNIVERSITY PRESS
Cambridge, New York, Melbourne, Madrid, Cape Town, Singapore, São Paulo

Cambridge University Press
The Edinburgh Building, Cambridge CB2 2RU, UK

Published in the United States of America by Cambridge University Press, New York

www.cambridge.org
Information on this title: www.cambridge.org/9780521242349

First published 1984
This digitally printed first paperback version 2006

A catalogue record for this publication is available from the British Library

Library of Congress Catalogue Card Number: 83–7466

ISBN-13 978-0-521-24234-9 hardback
ISBN-10 0-521-24234-7 hardback

ISBN-13 978-0-521-03287-2 paperback
ISBN-10 0-521-03287-3 paperback

Contents

Preface

This elementary book is intended for advanced undergraduates or anyone on a higher level who wants to learn the basic facts of p-adic analysis. We only assume the reader to have some standard knowledge of analysis and algebra.

In analysis (and outside it) the fields $\mathbb{R}$ and $\mathbb{C}$ play a central role. For several reasons people started to study the implications of replacing $\mathbb{R}$ or $\mathbb{C}$ by a more general object, viz. a field K with a complete valuation $|\ |$ comparable to the absolute value function (see Definition 1.1). Many such fields other than $\mathbb{R}$ or $\mathbb{C}$ exist, their valuations are all 'non-archimedean', i.e. they satisfy the 'strong triangle inequality' $|x + y| \leqslant \max(|x|, |y|)$. The analysis in and over non-archimedean valued fields K is known as ultrametric (non-archimedean, p-adic) analysis.

In this book we shall treat the basic facts of ultrametric analysis together to form an alternative 'one variable calculus course'. Thus, in K we shall consider familiar concepts such as continuity, differentiability, (power) series, integration, etc. However, the strong triangle inequality causes fascinating deviations from the 'classical analysis' (over $\mathbb{R}$ or $\mathbb{C}$); let us mention a few of them.

(i) A series Σa_n in K converges if $\lim_{n \to \infty} a_n = 0$. The power series $\Sigma x^n/n!$ of exp (if it makes sense at all) converges only on a disc strictly contained in the closed unit disc $\{x : |x| \leqslant 1\}$. Hence $\Sigma 1/n!$ diverges (but $\Sigma n!$ converges in many K).
(ii) K is not ordered. Yet it is possible to define 'square roots of positive elements' in a natural way. It may happen that $\sqrt{16} = 4$ but $\sqrt{25} = -5$.
(iii) K is not connected. In fact, each disc is open and closed; each point of a disc is a centre; $\{x : |x| = 1\}$ is not the boundary of $\{x : |x| < 1\}$.
(iv) Liouville's theorem (a bounded analytic function $K \to K$ is constant) holds if and only if K is *not* locally compact.
(v) For each pair of continuous functions $f_1, f_2 : K^2 \to K$ there exists an $F : K^2 \to K$ such that $\partial F/\partial x = f_1$, $\partial F/\partial y = f_2$.

During the past few decades ultrametric analysis has grown from a relatively small and remote area maintained by a few enthusiastic pioneers to a widely recognized and mature discipline. However, since there are so many parts of mathematics for which a corresponding ultrametric theory exists or is at least conceivable, the term 'discipline' is somewhat misleading. In fact, in the 1980 classification scheme of the *Mathematical Reviews* explicit references to p-adic and non-archimedean theories are headed under Number theory; Algebraic number theory, field theory and polynomials; Algebraic geometry; Group theory and generalizations; Topological groups and Lie groups; Real functions; Functions of a complex variable; Several complex variables and analytic spaces and Functional analysis. It is clear that in an elementary book like this we cannot go into all these different branches, each one of which has its own origin, motivation, problems, language and applications. On the contrary, this book is written deliberately from no specific background at all but takes a 'naive' approach.

Depending on his or her point of view the beginner may use this book (at least the first three chapters of it) either as a 'main entrance' to the various specialistic theories or as an introduction to p-adic analysis in its own right. The specialist should not look for deep theorems. Yet I hope that he or she will appreciate having the elementary facts together with the proofs collected into a single volume.

For the reader's convenience many exercises of varying degree of difficulty have been included. They are meant for training purposes and also to indicate interesting by-paths. Exercises leading to results used in the main theory are marked*. In Appendix A we discuss some themes that have something to do with the subject but do not fit into the main text, mostly because of their functional analytic character. Appendix B contains several basic definitions and facts needed in the book and is meant as a refresher course, only to be consulted if the reader needs part of it.

A small list of books is given for further reading; in some of these one may find extensive lists of references. Perhaps closest to our book is Mahler (1980), whereas Bachman (1964) is also elementary but moves towards algebraic theory. In Amice (1975) one can find more about analytic functions. Monna (1970) and van Rooij (1978) treat functional analysis. Iwasawa (1972) considers special functions of interest for number theory. Koblitz (1977), after an elementary start, moves towards the p-adic zeta functions and Dwork's theory. More advanced studies and applications in algebraic number theory and algebraic geometry can be found in Koblitz (1980).

Only occasionally – in cases where a theorem is generally known as such – names of inventors are given. I have also added names to results that have not been published before and were pointed out to me by informal communication.

I am grateful to Lucien van Hamme and Arnoud van Rooij for their helpful comments and the stimulating effect they had on me.

October 1982

Wim Schikhof
Nijmegen, The Netherlands

1
Valuations

PART 1: VALUATIONS

1. Valuations

The absolute value function on the field of the real numbers $\mathbb{R}$ or the field of the complex numbers $\mathbb{C}$ has the following fundamental properties. For all elements x and y we have

(i) $|x| \geqslant 0$, $|x| = 0$ if and only if $x = 0$
(ii) $|x + y| \leqslant |x| + |y|$ (the *triangle inequality*)
(iii) $|xy| = |x|\,|y|$

In this book we shall be concerned with the following generalization.

DEFINITION 1.1. Let K be a field. A *valuation* on K is a map $|\ |: K \to \mathbb{R}$ satisfying the rules (i), (ii), (iii), for all $x, y \in K$. The pair $(K, |\ |)$ is a *valued field*. Often we will write simply K instead of $(K, |\ |)$.

There are many examples of valued fields other than subfields of $\mathbb{C}$ with the ordinary absolute value function. The most important one is the field of the p-adic numbers, which we shall introduce in Section 4. An obvious example is the *trivial valuation* that can be defined on any field by setting

$$|x| := \begin{cases} 0 \text{ if } x = 0 \\ 1 \text{ if } x \neq 0 \end{cases}$$

Let $(K, |\ |)$ be a valued field. The map $(x, y) \mapsto |x - y|$ is easily seen to be a metric on K which, in turn, yields a topology on K in the usual way (a set $U \subset K$ is called open if for each $a \in U$ there is $\delta > 0$ such that $\{x \in K: |x - a| < \delta\} \subset U$. Then the union of arbitrarily many open sets is open, the intersection of finitely many open sets is open). The above metric and topology are said to be *induced* by the valuation $|\ |$.

**Exercise* 1.A. Let 1_K denote the unit element of a valued field $(K, |\ |)$. Show that $|1_K| = 1$, $|-x| = |x|$ $(x \in K)$, $|x^{-1}| = |x|^{-1}$ $(x \in K, x \neq 0)$, $|x - y| \geqslant ||x| - |y||$ $(x, y \in K)$.

**Exercise* 1.B. The topology induced by a valuation $|\ |$ on a field K makes K into a *topological field*. That is, prove the following.
(i) Addition $(x, y) \mapsto x + y$ and multiplication $(x, y) \mapsto xy$ are continuous maps $K \times K \to K$. (Here $K \times K$ carries the product topology.)
(ii) The maps $x \mapsto -x$ $(x \in K)$ and $x \mapsto x^{-1}$ $(x \in K, x \neq 0)$ are continuous.

Exercise 1.C. Show that the trivial valuation on a field K induces the discrete topology on K (i.e. each subset of K is open).

Before starting with the actual calculus in Chapter 2 we first develop the necessary theory on valued fields, study the basic examples and consider some important metrical aspects. Those who want to reach the main subject as quickly as possible may skip over the hard proofs of Sections 14–18. Also observe that Appendixes A.1, A.9 and A.10 logically belong to Chapter 1.

2. The strong triangle inequality

In this book we shall mainly be interested in valued fields $(K, |\ |)$ whose valuation satisfies the *strong triangle inequality*

$$|x+y| \leqslant \max(|x|, |y|) \qquad (x, y \in K)$$

rather than the general, weaker, form

$$|x+y| \leqslant |x| + |y| \qquad (x, y \in K)$$

Such a field is constructed in Example 2.1 below. Although it is not itself going to play a central role in the sequel, it is quite easy to understand and illustrative for what follows.

For a nonzero polynomial $f \in \mathbb{R}[X]$ given by

$$f = a_0 + a_1 X + a_2 X^2 + \ldots + a_n X^n \qquad (a_0, \ldots, a_n \in \mathbb{R}, a_n \neq 0)$$

we set $d(f) := n$ (the *degree* of f). Let $d(f) := -\infty$ if f is the zero polynomial. Then we have the following rules for d.

$$\left.\begin{array}{l} d(f+g) \leqslant \max(d(f), d(g)) \\ d(fg) = d(f) + d(g) \end{array}\right\} \quad (f, g \in \mathbb{R}[X])$$

Let ρ be any real number, greater than 1. For $f \in \mathbb{R}[X]$ put

$$|f| := \begin{cases} 0 \text{ if } f = 0 \\ \rho^{d(f)} \text{ if } f \neq 0 \end{cases}$$

Then the above rules for d now look as follows.

$$\left.\begin{array}{l} |f+g| \leqslant \max(|f|, |g|) \\ |fg| = |f|\,|g| \end{array}\right\} \quad (f, g \in \mathbb{R}[X])$$

Also we have trivially

$$|f| \geqslant 0; \; |f| = 0 \text{ if and only } f = 0 \qquad (f \in \mathbb{R}[X])$$

So, forgetting for a moment that $\mathbb{R}[X]$ is not a field, we see that $|\ |$ behaves like a valuation. The strong triangle inequality here simply expresses the fact that *if two polynomials each have a degree $\leqslant n$ then so has their sum.*

We now extend $|\ |$ to the field $\mathbb{R}(X)$ of rational functions.

EXAMPLE 2.1. (A valued field) *Let $\rho > 1$. For $f \in \mathbb{R}[X]$ put*

$$|f| := \begin{cases} 0 \text{ if } f = 0 \\ \rho^{d(f)} \text{ if } f \neq 0 \end{cases}$$

For $s \in \mathbb{R}(X)$ set

$$|s| := |f|\,|g|^{-1} \quad (s = fg^{-1}; f, g \in \mathbb{R}[X], g \neq 0)$$

Then $|\ |$ is a valuation on $\mathbb{R}(X)$ satisfying the strong triangle inequality

$$|s + t| \leqslant \max(|s|, |t|) \qquad (s, t \in \mathbb{R}(X))$$

Remarks.

1. The above construction yields infinitely many valuations on $\mathbb{R}(X)$, one for each $\rho > 1$. See, however, Section 9.
2. The constant polynomials form a subfield of $\mathbb{R}(X)$, isomorphic to $\mathbb{R}$. If a is such a constant polynomial then

$$|a| = \begin{cases} 1 \text{ if } a \neq 0 \\ 0 \text{ if } a = 0 \end{cases}$$

so that $|\ |$ does not induce the ordinary absolute value on $\mathbb{R}$, but the trivial valuation.

3. In the above example we may without harm replace $\mathbb{R}$ by an arbitrary field K. It follows that there are non-trivial examples of valued fields of characteristic $p \neq 0$!

**Exercise* 2.A. Show that the definition of $|\ |$ given in Example 2.1 is meaningful and that $(\mathbb{R}(X), |\ |)$ is indeed a valued field. More generally, prove the following. Let D be an integral domain and let $|\ |$ be a map of D into $\mathbb{R}$ satisfying the conditions (i), (ii), (iii) (see Section 1) for a valuation. Then $|\ |$ can uniquely be extended to a valuation on the quotient field of D. If $|\ |$ satisfies the strong triangle inequality then so does the extension.

Exercise 2.B. Let σ: $\mathbb{R}(X) \to \mathbb{R}(X)$ be the automorphism of $\mathbb{R}(X)$ sending X into X^{-1}. Then, with $|\ |$ as in Example 2.1, set $|s|_1 := |\sigma(s)|$ $(s \in \mathbb{R}(X))$. Show that $|\ |_1$ is also a valuation on $\mathbb{R}(X)$ and that

$$|f|_1 = \rho^{-k} \qquad (f \in \mathbb{R}[X], f = a_k X^k + a_{k+1} X^{k+1} + \ldots + a_n X^n, a_k \neq 0)$$

**Exercise* 2.C. Let $|\ |$ be a valuation on a field K satisfying the strong triangle inequality. Let $x_1, x_2, \ldots, x_n \in K$, where $n \in \mathbb{N}$. Show that $|x_1 + x_2 + \ldots + x_n| \leqslant \max(|x_1|, |x_2|, \ldots, |x_n|)$.

3. The *p*-adic integers

In Sections 3 and 4 we shall construct the valued field $\mathbb{Q}_p$ of the p-adic numbers. This field, whose valuation satisfies the strong triangle inequality, is the fundamental example throughout this book. In this section we make a start by defining the subring $\{x: |x| \leqslant 1\}$ of this field.

In the decimal system we denote nonnegative integers by expressions such as 1028 $(8 + 2 \cdot 10 + 0 \cdot 10^2 + 1 \cdot 10^3)$. When we write down a sequence $a_n\, a_{n-1} \ldots a_0$ we mean $a_0 + a_1\, 10 + \ldots + a_n\, 10^n$. Here each a_i is one of the symbols 0, 1, ..., 9. Of course we may also write this as an infinite sequence

$$\ldots a_{n+2}\, a_{n+1}\, a_n \ldots a_0$$

where $a_i = 0$ for $i > n$. Further, instead of 10, we can choose any number $\in \{2, 3, \ldots\}$ as a base. These simple observations lead to the following.

DEFINITION 3.1. For any $n \in \{2, 3, \ldots\}$, let $\mathbb{Z}_n$ be the set of *all* infinite sequences

$$\ldots a_m\, a_{m-1} \ldots a_1\, a_0$$

where each a_m is one of the elements $0, 1, \ldots, n-1$. The elements of $\mathbb{Z}_n$ are *n-adic integers*. The sequences with $a_m = 0$ for sufficiently large m can be identified with the nonnegative integers. Thus we may write

$$\mathbb{N} \subset \mathbb{Z}_n$$

Remarks.

1. The elements of $\mathbb{Z}_n \backslash \{0, 1, 2, \ldots\}$ are sequences

$$\ldots a_2\, a_1\, a_0$$

for which $a_m \neq 0$ for infinitely many m. One may be tempted to think of these elements as being 'infinitely large' or 'supernatural' numbers. However, in the sequel we shall see that, according to a quite natural point of view, these elements are limits of sequences of natural numbers.

2. One may wonder why the elements of $\mathbb{Z}_n$ are called n-adic *integers* rather than n-adic natural numbers. For this, see Proposition 3.2.

We can define a natural addition and multiplication in $\mathbb{Z}_n$ that extend the operations on $\mathbb{N}$. The following examples (in $\mathbb{Z}_{10}$) will make this clear.

$$\begin{array}{r} \ldots.8427159 \\ \ldots.5478563 \\ \hline \ldots.3905722 \end{array} +$$

$$\begin{array}{r} \ldots.27159 \\ \ldots.78563 \\ \hline \ldots.81477 \\ \ldots.62954 \\ \ldots.35795 \\ \ldots.17272 \\ \ldots.90113 \\ \ldots\ldots\ldots \\ \vdots \\ \hline \ldots.92517 \end{array} \times$$

The ideas suggested in the above examples ('treat the elements of $\mathbb{Z}_n$ as ordinary integers') can of course be sharpened to correct definitions of addition and multiplication in $\mathbb{Z}_n$. (Let $x = \ldots . a_2 a_1 a_0$ and $y = \ldots . b_2 b_1 b_0$ be elements of $\mathbb{Z}_n$. Then $x + y = \ldots . c_2 c_1 c_0$, where the c_i are determined by

(i) $c_i \in \{0, 1, \ldots, n-1\}$ for each i

(ii) for each $m \in \{0, 1, 2, \ldots\}$

$$\sum_{i=0}^{m} c_i n^i \equiv \sum_{i=0}^{m} (a_i + b_i) n^i \pmod{n^{m+1}}$$

Similarly, $xy = \ldots . d_2 d_1 d_0$, where the d_i are determined by

(i) $d_i \in \{0, 1, \ldots, n-1\}$ for each i

(ii) for each $m \in \{0, 1, 2, \ldots\}$

$$\sum_{i=0}^{m} d_i n^i \equiv \left(\sum_{i=0}^{m} a_i n^i\right)\left(\sum_{i=0}^{m} b_i n^i\right) \pmod{n^{m+1}}$$

So, what we do is nothing but elementary school arithmetic.) Notice that in the second example we 'add' infinitely many elements of $\mathbb{Z}_{10}$. The 'sum' is well defined since every column has only finitely many (nonzero) entries. (See the preamble to Definition 3.4.)

A surprising fact is that one can subtract every element of $\mathbb{Z}_n$ from every other element of $\mathbb{Z}_n$. For example in $\mathbb{Z}_{10}$ we have

$$\begin{array}{r} \ldots.8427159 \\ \ldots.5478563 \\ \hline \ldots.2948596 \end{array} - \qquad \begin{array}{r} \ldots.5478563 \\ \ldots.8427159 \\ \hline \ldots.7051404 \end{array} -$$

Subtracting familiar numbers we may obtain a non-familiar result. For example, in $\mathbb{Z}_{10}$ we have $3-5 = \ldots.999998$. The same subtraction in $\mathbb{Z}_5$ yields $\ldots.444443$ as an answer. The following proposition is not hard to prove.

PROPOSITION 3.2. *With the above addition and multiplication, $\mathbb{Z}_n$ is a commutative ring with $0 := \ldots.000000$ as a zero element and $1 := \ldots.000001$ as a unit. $\mathbb{Z}$ can be identified with a subring of $\mathbb{Z}_n$.*

Remark. Let $\ldots.a_2\,a_1\,a_0$ and $\ldots.b_2\,b_1\,b_0$ be elements of $\mathbb{Z}_n$. Suppose we want to know, for a certain m, the last $m+1$ digits $c_m, c_{m-1}, \ldots, c_0$ of their product (sum) $\ldots.c_2\,c_1\,c_0$. Then we only need to compute the product (sum) $d_k\,d_{k-1}\ldots d_0$ of the nonnegative integers $a_m\,a_{m-1}\ldots a_0$ and $b_m\,b_{m-1}\ldots b_0$ in the ordinary way and we get $c_0 = d_0, \ldots, c_m = d_m$.

Exercise 3.A. (Extension of the ordering to $\mathbb{Z}_n$) Let $x = \ldots.a_2\,a_1\,a_0$ and $y = \ldots.b_2\,b_1\,b_0$ be elements of $\mathbb{Z}_n$. Consider the following two definitions (*) and (**).
(*) $x >_1 y$ if there is an $m \in \{0,1,2,\ldots\}$ for which $a_m > b_m$ and $a_j = b_j$ for $j > m$.
(**) $x >_2 y$ if there is an $m \in \{0,1,2,\ldots\}$ for which $a_m > b_m$ and $a_j \geqslant b_j$ for $j > m$.
Show that (*) and (**) define partial orderings $>_1$ and $>_2$ on $\mathbb{Z}_n$ extending the natural ordering on $\mathbb{N}$ and that $>_1 \neq >_2$. Prove, however, that in general $x >_1 y$ does not imply $x + 1 >_1 y + 1$ and that $x >_2 y$ does not imply $x + 1 >_2 y + 1$.

Division in $\mathbb{Z}_n$ is less simple. For example, in $\mathbb{Z}_{10}$ we can find nonzero elements x and y for which $xy = 0$, as suggested below.

```
          ....10112
          ....03125
          ----------- ×
          ....50560
         ....20224
        ....10112
       ....30336
      ....00000
          .
          .
          .
      -------------
          ....00000
```

(The reader is asked to show that one can indeed fill in the dots in such a way as to obtain the zero sequence.) Thus, $\mathbb{Z}_{10}$ is not an integral domain and there is no way to extend $\mathbb{Z}_{10}$ to a field. The situation becomes better if n is a prime number.

PROPOSITION 3.3. *Let p be a prime number. Then $\mathbb{Z}_p$ is an integral domain. An element $\ldots. a_2 a_1 a_0$ of $\mathbb{Z}_p$ has an inverse in $\mathbb{Z}_p$ if and only if $a_0 \neq 0$.*
Proof. As $\ldots. a_2a_1a_00 = p(\ldots. a_2a_1a_0)$, $\ldots. a_2a_1a_000 = p^2(\ldots. a_2a_1a_0)$, etc. it suffices to show the second statement. If $a_0 = 0$ then the product of $\ldots. a_0$ and any element of $\mathbb{Z}_p$ ends with a 0, so certainly $\ldots a_2a_1a_0$ has no inverse. Suppose $a_0 \neq 0$. We prove inductively that we can find $x_0, x_1, \ldots \in \{0,1,\ldots, p-1\}$ such that the product of $\ldots. x_2x_1x_0$ and $\ldots. a_2a_1a_0$ equals $\ldots. 001$. By looking at the 'long multiplication'

$$\begin{array}{r} \ldots\ldots a_2\,a_1\,a_0 \\ \ldots\ldots x_2\,x_1\,x_0 \\ \hline \ldots\ldots \\ \ldots\ldots \\ \cdot \\ \cdot \\ \hline \ldots 0\;0\;1 \end{array} \times$$

we see that we have to sole the following congruences.

$$x_0 a_0 \equiv 1 \pmod p$$
$$x_0 a_1 + x_1 a_0 + p^{-1}(a_0 x_0 - 1) \equiv 0 \pmod p, \text{ etc.}$$

The essential point is that for each $n \in \mathbb{N}$ it is required that

$$x_{n+1} a_0 \equiv c_{n+1} \pmod p$$

where c_{n+1} depends only on $x_0, x_1, \ldots, x_n$. Whatever c_{n+1} is, we always can solve this congruence since $a_0 \not\equiv 0 \pmod p$ and $\mathbb{Z}/p\mathbb{Z}$ is a field.

Exercise 3.B. Describe how at the p-adic elementary school one would carry out a division $(\ldots. b_2b_1b_0) \div (\ldots. a_2a_1a_0)$ in $\mathbb{Z}_p$. (Assume that $a_0 \neq 0$.)

Exercise 3.C. Let $n \in \{2,3,4,\ldots\}$. Show that an element $\ldots. a_2a_1a_0$ of $\mathbb{Z}_n$ has an inverse if and only if the greatest common divisor of a_0 and n equals 1.

We proceed to investigate $\mathbb{Z}_p$ where p is a prime number. In particular we want to introduce a 'valuation' on $\mathbb{Z}_p$ (that can be extended to a valuation on the quotient field of $\mathbb{Z}_p$, see Section 4). Let us return to the first example of a multiplication we have given for $\mathbb{Z}_{10}$. We mentioned there that the final 'addition' succeeds since every column contains only finitely many digits. More generally, if we are given a sequence $x_1, x_2, \ldots$ where $x_i \in \mathbb{Z}_p$ for every i we can formally define $\sum_{i=1}^{\infty} x_i$ if, from a certain i_0 on, the last digit of x_i is always 0, from a certain i_1 on the last two digits of x_i are 00, etc. If we want to consider $\sum_{i=1}^{\infty} x_i$ as the 'limit' of $\sum_{i=1}^{n} x_i$ for $n \to \infty$ it is intuitively clear that the x_i must 'tend to zero'. So we come to the somewhat unusual conclusion that an element of $\mathbb{Z}_p$ must be called 'small' if it ends in many zeros. This implies that an element of $\mathbb{N}$ is 'small' in the sense of $\mathbb{Z}_p$ if it is divisible by a large power of p. Thus, the sequence $1, p, p^2, \ldots$ will tend to zero in $\mathbb{Z}_p$! We formalize this as follows.

DEFINITION 3.4. Let p be a prime number and let $\ldots . a_2 a_1 a_0$ be an element of $\mathbb{Z}_p$. The *order* of $\ldots . a_2 a_1 a_0$ is the smallest m for which $a_m \neq 0$. More precisely,

$$\operatorname{ord}_p(\ldots . a_2 a_1 a_0) := \begin{cases} \infty \text{ if } a_i = 0 \text{ for all } i \\ \min\{s : a_s \neq 0\} \text{otherwise} \end{cases}$$

We set

$$|\ldots . a_2 a_1 a_0|_p := \begin{cases} 0 \text{ if } a_i = 0 \text{ for all } i \\ p^{-\operatorname{ord}_p(\ldots . a_2 a_1 a_0)} \text{ otherwise} \end{cases}$$

The function $| \; |_p$ is the *p-adic valuation* on $\mathbb{Z}_p$.

PROPOSITION 3.5. *Let p be a prime number and let* $x, y \in \mathbb{Z}_p$.
(i) $|x|_p \geqslant 0$; $|x|_p = 0$ *if and only if* $x = 0$.
(ii) $|x+y|_p \leqslant \max(|x|_p, |y|_p)$ (*the strong triangle inequality*).
(iii) $|xy|_p = |x|_p \, |y|_p$.

The easy proof is left to the reader. The strong triangle inequality for $| \; |_p$ reflects the fact that *if*, for some $s \in \{0,1,2,\ldots\}$, *two integers are divisible by* p^s *then so is their sum.* Observe that the set of values of $| \; |_p$ equals $\{0, 1, p^{-1}, p^{-2}, \ldots\}$.

Exercise 3.D. (Other valuations on $\mathbb{Z}_p$) Let $\rho \in \mathbb{R}$, $\rho > 1$. Define

$$|\ldots . a_2 a_1 a_0|_p' := \begin{cases} 0 \text{ if } a_i = 0 \text{ for all } i \\ \rho^{-\operatorname{ord}_p(\ldots . a_2 a_1 a_0)} \text{ otherwise} \end{cases}$$

Show that the properties (i), (ii), (iii) of Proposition 3.5 hold for $| \; |_p'$ in place of $| \; |_p$. Compare Remark 1 following Example 2.1 and Exercise 9.A.

Exercise 3.E. Find the '5-adic representation' $\ldots . a_2 a_1 a_0$ $(a_i \in \{0,1,2,3,4,\})$ of the numbers 15, -1 and -3. The numbers 2,3,4 have inverses in $\mathbb{Z}_5$. Find their 5-adic representations.

Exercise 3.F. Let p be an odd prime. Show that $2^{-1} = \ldots . a_2 a_1 a_0$ in $\mathbb{Z}_p$ where $a_0 = \frac{1}{2}(p+1)$ and $a_i = \frac{1}{2}(p-1)$ for $i \geqslant 1$.

Exercise 3.G. (On $\sqrt{-1}$) Show that the equation $x^2 + 1 = 0$ has no solutions in $\mathbb{Z}_3$ but has two solutions in $\mathbb{Z}_5$.

Exercise 3.H. Compute $\mathrm{ord}_p(p^n!)$ and $|p^n!|_p$ (p prime, $n \in \mathbb{N}$).

**Exercise* 3.I. (Basic facts on $\mathbb{Z}_p$. See the frontispiece for a 'picture' of $\mathbb{Z}_7$) Let p be a prime number. Prove the following.
(i) An element x of $\mathbb{Z}_p$ has an inverse in $\mathbb{Z}_p$ if and only if $|x|_p = 1$.
(ii) If x is a nonzero element of $\mathbb{Z}_p$ then $x = p^{\mathrm{ord}_p(x)} y$ where $y \in \mathbb{Z}_p$, $|y|_p = 1$.
(iii) Set

$$p\mathbb{Z}_p := \{py : y \in \mathbb{Z}_p\}$$

Then $p\mathbb{Z}_p$ is a maximal ideal of $\mathbb{Z}_p$ and $\mathbb{Z}_p/p\mathbb{Z}_p$ is a field of p elements. The additive cosets

$$p\mathbb{Z}_p,\ 1 + p\mathbb{Z}_p, \ldots , p-1+p\mathbb{Z}_p$$

form a partition of $\mathbb{Z}_p$. For each $j \in \{0,1,2,\ldots,p-1\}$ we have

$$j + p\mathbb{Z}_p = \{x \in \mathbb{Z}_p : |x-j|_p < 1\} = \{x \in \mathbb{Z}_p : |x-j|_p \leqslant p^{-1}\}$$

(iv) Let $p^n\mathbb{Z}_p := \{p^n y : y \in \mathbb{Z}_p\}$ $(n \in \mathbb{N})$. The cosets $p^n\mathbb{Z}_p, 1+p^n\mathbb{Z}_p, \ldots, p^n - 1 + p\mathbb{Z}_p$ form a partition of $\mathbb{Z}_p$. For each $j \in \{0,1,\ldots,p^n-1\}$ we have $j + p^n\mathbb{Z}_p = \{x \in \mathbb{Z}_p : |x-j|_p < p^{-n+1}\} = \{x \in \mathbb{Z}_p : |x-j|_p \leqslant p^{-n}\}$.

Exercise 3.J. (Valuation on $\mathbb{Z}_n$) Let $n \in \{2, 3, 4, \ldots\}$ be not a prime number. Define an 'n-adic valuation' $|\ |_n$ on $\mathbb{Z}_n$ in the spirit of Definition 3.4. Are the properties (i), (ii), (iii) of Proposition 3.5 true for p replaced by n and $x, y \in \mathbb{Z}_n$?

Exercise 3.K. (Some p-adic numerical analysis) Let p be a prime number, let $a \in \mathbb{Z}$, $|a|_p = 1$. We shall describe a method to approximate the inverse a^{-1} of a in $\mathbb{Z}_p$ by means of integers which is more efficient than the one that follows from the proof of Proposition 3.3.
(i) Choose $x_0 \in \mathbb{Z}$ such that $|1 - x_0 a|_p < 1$. The formula $1 - x_{n+1}a = (1-x_n a)^2$, i.e. $x_{n+1} = x_n(2 - x_n a)$ defines a sequence $x_0, x_1, \ldots$ of integers. Show that $|x_n - a^{-1}|_p \leqslant p^{-2^n}$ $(n \in \mathbb{N})$. In other words, if

$$x_n = \ldots . s_2 s_1 s_0,\quad a^{-1} = \ldots . a_2 a_1 a_0$$

then $a_j = s_j$ for $0 \leq j < 2^n$. Observe that this method is quadratic and that we do not have to worry about rounding errors as x_n gives us the exact values of a_j $(0 \leq j < 2^n)$.
(ii) Choose $p = 5$, $a = 23$ (twenty-three) and use (i) to find an integer s for which $|a^{-1} - s|_p \leq p^{-8}$.

4. The p-adic numbers

In this section we shall extend $|\ |_p$ to a valuation on the quotient field of $\mathbb{Z}_p$.
FROM NOW ON p IS A PRIME NUMBER
For a nonzero element x of $\mathbb{Z}_p$ we have, according to Exercise 3.I (ii)

$$x = p^n y$$

where $n = \text{ord}_p(x)$ and y is invertible in $\mathbb{Z}_p$. So, to find a concrete representation of 'the smallest field that contains $\mathbb{Z}_p$', we must find an inverse for p. Now the common notation in base p for p^{-1} is 0.1; p^{-2} is written 0.01 etc. This leads to the following definition.

DEFINITION 4.1. Let $\mathbb{Q}_p$ be the set of all two-sided sequences

$$\ldots . a_2 a_1 a_0 .\ a_{-1} a_{-2} \ldots$$

for which $a_i \in \{0,1,\ldots,p-1\}$ for each i and such that $a_{-n} = 0$ for large n. The elements of $\mathbb{Q}_p$ are *p-adic numbers*. The sequences $\ldots . a_2 a_1 a_0 .\ a_{-1} a_{-2} \ldots$ for which $a_{-1} = a_{-2} = \ldots = 0$ can be identified with the p-adic integers. So we may write

$$\mathbb{Z}_p \subset \mathbb{Q}_p$$

Addition and multiplication in $\mathbb{Z}_p$ can be extended to $\mathbb{Q}_p$ in a natural way. (Formally, let $x = \ldots . a_2 a_1 a_0 . a_{-1} a_{-2} \ldots$ and $y = \ldots . b_2 b_1 b_0 . b_{-1} b_{-2} \ldots$ be elements of $\mathbb{Q}_p$ and suppose that $a_{-n} = b_{-n} = 0$ for $n > N$. Then $x' := \ldots . a_0 a_{-1} \ldots a_{-N}$ and $y' := \ldots b_0 b_{-1} \ldots b_{-N}$ are p-adic integers. Let $x' + y' = \ldots . c_2 c_1 c_0$. Define $x + y$ to be $\ldots . c_N . c_{N-1} \ldots c_0 00 \ldots$ Similarly one defines the product xy of x and y.) Then the inverse of $p = 1.0$ becomes 0.1, the inverse of $p^2 = 10.0$ becomes 0.01, etc. It follows that every nonzero element of $\mathbb{Q}_p$ can be written as $p^n y$ where $n \in \mathbb{Z}$ and $y \in \mathbb{Z}_p$, $|y|_p = 1$. With this in mind, the following is not hard to prove.

PROPOSITION 4.2. *$\mathbb{Q}_p$ is a field containing $\mathbb{Q}$ as a subfield and $\mathbb{Z}_p$ as a subring. $\mathbb{Q}_p$ is (isomorphic to) the quotient field of $\mathbb{Z}_p$.*

Now we extend $|\ |_p$ to $\mathbb{Q}_p$.

DEFINITION 4.3. For a nonzero element

$$x = \ldots . a_2 a_1 a_0 . a_{-1} a_{-2} \ldots .$$

of $\mathbb{Q}_p$ the *order* of x is the following integer

$$\mathrm{ord}_p(x) := \min \{s : a_s \neq 0\}$$

The *p-adic value* of x is

$$|x|_p := p^{-\mathrm{ord}_p(x)}$$

Further, we define

$$|0|_p := 0.$$

$| \ |_p$ is the *p-adic valuation* on $\mathbb{Q}_p$.

THEOREM 4.4. *$| \ |_p$ is a valuation on $\mathbb{Q}_p$. It satisfies the strong triangle inequality. The 'closed unit disc' $\{x \in \mathbb{Q}_p : |x|_p \leqslant 1\}$ is equal to $\mathbb{Z}_p$. The set of values of $| \ |_p$ is $\{0\} \cup \{p^n : n \in \mathbb{Z}\}$.*

Exercise 4.A. Describe, in the spirit of the previous section, how to carry out additions, subtractions, multiplications and divisions in $\mathbb{Q}_p$.

**Exercise* 4.B. (The p-adic valuation on $\mathbb{Q}$) Let x be a nonzero rational number. Show that there is an $n \in \mathbb{Z}$ and that there are integers s, t not divisible by p such that

$$x = p^n \frac{s}{t}$$

and prove that

$$|x|_p = p^{-n}.$$

**Exercise* 4.C. ($\mathbb{Q}_p$ is not algebraically closed) Show that the equation $x^2 - p = 0$ has no roots in $\mathbb{Q}_p$.

Exercise 4.D. (Cosets of $\mathbb{Z}_p$) Let S be the set of all numbers of the form

$$a_1 p^{-1} + a_2 p^{-2} + \ldots + a_n p^{-n}$$

where $n \in \mathbb{N}$, $a_i \in \{0, 1, \ldots, p-1\}$ $(1 \leqslant i \leqslant n)$. Show that the sets $s + \mathbb{Z}_p$, where s runs through S, form a partition of $\mathbb{Q}_p$.

Exercise 4.E. (Squares in $\mathbb{Q}_p$) A p-adic number $x = \ldots . a_2 a_1 a_0 . a_{-1} a_{-2} \ldots .$ is a *square* (has a square root) if there is a p-adic number y such that $y^2 = x$.
(i) For the cases $p = 3$ and $p = 5$ give necessary and sufficient conditions on the digits a_i in order that $\ldots . a_2 a_1 a_0 . a_{-1} a_{-2} \ldots .$ is a square.
(ii) Do the same as in (i) but now for the case $p = 2$.

5. Topological properties of $\mathbb{Q}_p$

In this section we are concerned with the metric and the topology induced by the valuation $| \ |_p$.

THEOREM 5.1. *$\mathbb{Z}_p$ is compact.*

Proof. Let $x_1, x_2, \ldots$ be a sequence in $\mathbb{Z}_p$. We show that it has a convergent subsequence. Denote x_k by

$$x_k = \ldots a_2^k a_1^k a_0^k$$

Since there are only finitely many possibilities for a_0^k (namely $0,1,\ldots,p-1$) we can find $b_0 \in \{0,1,\ldots,p-1\}$ and a subsequence $x_{00}, x_{01}, \ldots$ of $x_1, x_2, \ldots$ such that the last digit of x_{0k} is always b_0. The same trick yields $b_1 \in \{0,1,\ldots,p-1\}$ and a subsequence $x_{10}, x_{11}, x_{12}, \ldots$ for which the last two digits are $b_1 b_0$. This procedure can be continued, and we obtain $b_0, b_1, \ldots$ and a sequence of sequences

$$\begin{array}{llll} x_{00} & x_{01} & x_{02} & \cdots \\ x_{10} & x_{11} & & \cdots \\ x_{20} & x_{21} & & \cdots \\ & \vdots & & \end{array}$$

such that each sequence is a subsequence of its predecessor and such that each element of the nth row ends in $b_n b_{n-1} \ldots b_0$. The diagonal sequence $x_{00}, x_{11}, x_{22}, \ldots$ is still a subsequence of the original sequence $x_1, x_2, \ldots$ and it obviously converges (to $\ldots b_2 b_1 b_0$).

COROLLARY 5.2. *$\mathbb{Z}_p$ is complete.*

THEOREM 5.3. *$\mathbb{Z}$ is dense in $\mathbb{Z}_p$.*

Proof. Let $x \in \mathbb{Z}_p, x = \ldots a_2 a_1 a_0$. For each $n \in \mathbb{N}$, set $x_n := \ldots 00 a_n a_{n-1} \ldots a_0 = \sum_{i=0}^{n} a_i p^i$. Then $x_n \in \mathbb{Z}$ and $|x - x_n|_p < p^{-n}$. The statement follows.

$\mathbb{Q}_p$ is not compact $(1, p^{-1}, \ldots$ does not have a convergent subsequence). But we have

THEOREM 5.4. *$\mathbb{Q}_p$ is locally compact. $\mathbb{Q}$ is dense in $\mathbb{Q}_p$.*

Proof. Because

$$\mathbb{Z}_p = \{x \in \mathbb{Q}_p : |x|_p \leqslant 1\} = \{x \in \mathbb{Q}_p : |x|_p < p\}$$

is both closed and open, $\mathbb{Z}_p$ is a compact neighbourhood of 0. It follows easily that for $a \in \mathbb{Q}_p$ the coset $a + \mathbb{Z}_p$ is a compact neighbourhood of a. To show that $\mathbb{Q}$ is dense in $\mathbb{Q}_p$, let $x = \ldots . a_2 a_1 a_0 . a_{-1} \ldots a_{-N} 000 \ldots .$ be a p-adic number. Then for each $n \in \mathbb{N}$ the element

$$x_n := \ldots . 00 a_n a_{n-1} \ldots a_1 a_0 . a_{-1} \ldots a_{-N} 00 \ldots . = \Sigma_{i=-N}^{n} a_i p^i$$

is in $\mathbb{Q}$ and $|x - x_n|_p < p^{-n}$. This finishes the proof.

COROLLARY 5.5. *$\mathbb{Q}_p$ is complete and separable.*

THEOREM 5.6. *The topology of $\mathbb{Q}_p$ is zerodimensional. $\mathbb{Q}_p$ is totally disconnected.*

Proof. For each $a \in \mathbb{Q}_p$ and $n \in \mathbb{Z}$ the set

$$\{x \in \mathbb{Q}_p : |x-a|_p \leqslant p^{-n}\} = \{x \in \mathbb{Q}_p : |x-a|_p < p^{-n+1}\}$$

is an open and closed neighbourhood of a. (For the terms 'zerodimensional' and 'totally disconnected' see Appendix B.3.)

The approximation of p-adic numbers by rational numbers in the proof of Theorem 5.4 leads to another way of describing p-adic numbers. The basic idea is the interpretation of a p-adic number $\ldots . a_2 a_1 a_0 . a_{-1} a_{-2} \ldots$ as the sum of the convergent series $\Sigma a_n p^n$.

THEOREM 5.7. *Each p-adic number can uniquely be written as the sum of a convergent series of the form*

$$(*) \qquad \sum_{n=-\infty}^{\infty} a_n p^n$$

where $a_n \in \{0,1,\ldots,p-1\}$ for each n and $a_{-n} = 0$ for large n. Conversely, if $a_n \in \{0,1,\ldots\ p-1\}$ for each $n \in \mathbb{Z}$ and $a_{-n} = 0$ for large n then (*) represents a p-adic number.

Proof. Let $\ldots . a_2 a_1 a_0 . a_{-1} a_{-2} \ldots .$ be a p-adic number and let $a_{-n} = 0$ for $n > N$. Then the sequence

$$n \mapsto \sum_{i=-N}^{n} a_i p^i = \sum_{i=-\infty}^{n} a_i p^i$$

converges to $\ldots . a_2 a_1 a_0 . a_{-1} a_{-2} \ldots . = \Sigma_{i=-\infty}^{\infty} a_i p^i$. The rest is easy.

The representation (*) is *the (standard) p-adic expansion of the element* $\ldots . a_2 a_1 a_0 . a_{-1} a_{-2} \ldots .$ If the a_n are not restricted to $\{0,1,\ldots,p-1\}$ then still the series $\Sigma a_n p^n$ may converge. For example, if $a_i \in \mathbb{Z}$ for each $i \in \{0,1,2,\ldots\}$ and if $m > n \geqslant 0$ we have (since $|a_i|_p \leqslant 1$ for all i)

$$\left| \sum_{i=n}^{m} a_i p^i \right|_p \leqslant \max(|p^n|_p, \ldots, |p^m|_p) = p^{-n}$$

so that the sequence $n \mapsto \sum_{i=0}^{n} a_i p^i$ is Cauchy, hence convergent. Thus,

$$\sum_{n=0}^{\infty} n!\, p^n, \qquad \sum_{n=-8}^{\infty} 3^n p^n$$

are legitimate p-adic numbers, yet the defining formulas do not represent the standard expansion. It is clear from the above that now we can write addition and multiplication in $\mathbb{Z}_p$ as follows.

$$\sum_{n=0}^{\infty} a_n p^n + \sum_{n=0}^{\infty} b_n p^n = \sum_{n=0}^{\infty} (a_n + b_n) p^n$$

$$\left(\sum_{n=0}^{\infty} a_n p^n\right)\left(\sum_{n=0}^{\infty} b_n p^n\right) = \sum_{n=0}^{\infty}\left(\sum_{i=0}^{n} a_i b_{n-i}\right) p^n$$

The left-hand expansions are standard but the right-hand ones, in general, are not.

Exercise 5.A. Show that each one of the following sets is dense in $\mathbb{Z}_p$. $\mathbb{N}$, $\mathbb{Z}_p \backslash \mathbb{Z}$, $\mathbb{Z}_p \backslash \mathbb{N}$, $\{n \in \mathbb{N} : n \geq 8\}$, the set $n\mathbb{N} := \{mn : m \in \mathbb{N}\}$ ($n \in \mathbb{N}$, n not divisible by p).

Exercise 5.B. (On other p-adic expansions) Let $\nu_1, \ldots, \nu_{p-1}$ be p-adic numbers satisfying $|\nu_i - i|_p < 1$ for $i = 1, \ldots, p-1$. Show that for each $x \in \mathbb{Q}_p$ there is a unique sequence $\ldots, b_2, b_1, b_0, b_{-1}, b_{-2}, \ldots$ such that
(i) for each $n \in \mathbb{Z}$ the elements b_n are in $\{0, \nu_1, \nu_2, \ldots, \nu_{p-1}\}$
(ii) $b_{-n} = 0$ for large n
(iii) $x = \sum_{n=-\infty}^{\infty} b_n p^n$. (For a natural choice of $\nu_1, \ldots, \nu_{p-1}$ other than $1, 2, \ldots p-1$, see Exercise 27.I.)

**Exercise* 5.C. Let K be a valued field whose valuation satisfies the strong triangle inequality. Show that the induced topology is zerodimensional and that K is totally disconnected. (See Theorem 5.6.)

6. $\mathbb{Q}_p$ as a completion of $\mathbb{Q}$

Theorem 5.4 gives us another way to define $\mathbb{Q}_p$, namely as the completion of $\mathbb{Q}$ with respect to the valuation $|\ |_p$ (see Exercise 4.B). Since we shall need it later on we shall describe the completion of a general valued field. As most readers will be familiar with similar constructions (e.g. the construction of $\mathbb{R}$ out of $\mathbb{Q}$) we will leave the technical details as an exercise.

DEFINITION 6.1. Let $(K, |\ |)$ be a valued field. A *completion* of $(K, |\ |)$ is a complete valued field $(L, |\ |)$, together with a dense isometrical field embedding $j : K \to L$. That is, a map $j : K \to L$ whose range is dense in L and such that for all $x,y \in K$

$$j(x+y) = j(x) + j(y)$$
$$j(xy) = j(x)j(y)$$
$$|j(x)| = |x|$$

LEMMA 6.2. *Let $(L, |\ |)$ and $(L', |\ |')$ be completions of a valued field $(K, |\ |)$ with embeddings $j : K \to L$ and $j' : K \to L'$ respectively. Then $(L, |\ |)$ and $(L', |\ |')$ are isomorphic in the following sense. There exists a bijective map $\sigma : L \to L'$ such that $\sigma \circ j = j'$ and such that for all $x, y \in L$ we have $\sigma(x+y) = \sigma(x) + \sigma(y)$, $\sigma(xy) = \sigma(x)\,\sigma(y)$, $|\sigma(x)|' = |x|$.*

Thus, in the sense of the above, a completion is unique. Together with the next theorem this makes it clear why, in the future, we shall speak about 'the' completion of a valued field.

THEOREM 6.3. *Each valued field has a completion.*
Proof. Let $(K, |\ |)$ be a valued field.
(i) The Cauchy sequences in K (i.e. sequences for which $\lim_{n,m\to\infty} |a_n - a_m| = 0$) form a ring C under the operations

$$(a_1, a_2, \ldots) + (b_1, b_2, \ldots) = (a_1 + b_1, a_2 + b_2, \ldots)$$
$$(a_1, a_2, \ldots) \cdot (b_1, b_2, \ldots) = (a_1 b_1, a_2 b_2, \ldots)$$

The unit element of C is $(1,1,1,\ldots)$.
(ii) Let N be the collection of null sequences, i.e.

$$N := \left\{(a_1, a_2, \ldots) \in C : \lim_{n\to\infty} |a_n| = 0\right\}$$

Then N is a maximal ideal in C so that the quotient C/N is a field.
(iii) The map

$$j : a \mapsto (a,a,a,\ldots) \bmod N$$

is an injection of K into C/N. For all $a,b \in K$ we have $j(a+b) = j(a) + j(b)$, $j(ab) = j(a)j(b)$.
(iv) The definition

$$\|(a_1, a_2, \ldots) \bmod N\| := \lim_{n\to\infty} |a_n| \quad ((a_1, a_2, \ldots) \in C)$$

makes sense. $\|\ \|$ is a valuation on C/N.
(v) $(C/N, \|\ \|)$ is complete.
(vi) $(C/N, \|\ \|)$, together with the embedding j of (iii) is a completion of $(K, |\ |)$.

THEOREM 6.4. (Alternative definition of $\mathbb{Q}_p$) *$\mathbb{Q}_p$ is the completion of $\mathbb{Q}$ with respect to the p-adic valuation on $\mathbb{Q}$.*
Proof. Theorem 5.4 and Corollary 5.5.

Exercise 6.A. Show that $\mathbb{Q}_p$ is uncountable. Deduce that $\mathbb{Q} \neq \mathbb{Q}_p$ and that $\mathbb{Q}$ is not complete with respect to any p-adic valuation.

Exercise 6.B. Is the field $\mathbb{R}(X)$ of Example 2.1 complete?

7. $\mathbb{Q}_p$ compared to $\mathbb{R}$

From a certain point of view the valued fields $\mathbb{R}$ and the various $\mathbb{Q}_p$ are similar since they are all completions of $\mathbb{Q}$ but with respect to different valuations. We have seen in Exercise 4.C and Theorem 5.4 that $\mathbb{Q}_p$ is not algebraically closed (neither is $\mathbb{R}$) and that $\mathbb{Q}_p$ is locally compact (so is $\mathbb{R}$). But there are also striking differences. First we recall that $\{|x|_p : x \in \mathbb{Q}_p\}$ equals the set $\{0\} \cup \{p^n : n \in \mathbb{Z}\}$. Another difference is that $\mathbb{Q}_p$ is not connected. (Theorem 5.6.) This is a simple consequence of the strong triangle inequality. (Exercise 5.C.)

A more superficial analogy between $\mathbb{R}$ and $\mathbb{Q}_p$ is the existence of 'expansions to the base p'. In fact, if $x \in \mathbb{Q}_p$ then there are $a_n \in \{0,1,2,\ldots, p-1\}$ such that $a_{-n} = 0$ for $n > N$ and

$$x = \sum_{n=-N}^{\infty} a_n p^n \qquad \text{('expansion to the right')}$$

If $y \in \mathbb{R}$ then there are $b_n \in \{0,1,2,\ldots,p-1\}$ such that $b_n = 0$ for $n > N$ and

$$y = \sum_{n=-\infty}^{N} b_n p^n \qquad \text{('expansion to the left')}$$

But the a_n are unique and the b_n are not. Further, in $\mathbb{R}$ we can choose *every* prime number as a base. The latter is impossible in $\mathbb{Q}_p$.

In the next two exercises we encounter other connections between $\mathbb{Q}_p$ and $\mathbb{R}$.

Exercise 7.A. (i) Let $x \in \mathbb{Q}$ have the standard p-adic expansion $\sum_{n=N}^{\infty} a_n p^n$ for certain $N \in \mathbb{Z}$. Show that the sequence $a_N, a_{N+1}, \ldots$ is periodic, i.e. there are $m \in \mathbb{Z}, n \in \mathbb{N}$ such that $a_i = a_{i+n}$ for all $i > m$.
(ii) Prove the converse of (i) : if $a_N, a_{N+1}, \ldots$ is periodic in the sense of (i) ($a_i \in \{0,1,\ldots,p-1\}$ for each $i > N$) then $\sum_{n=N}^{\infty} a_n p^n \in \mathbb{Q}$.

Exercise 7.B. (A continuous map of $\mathbb{Z}_p$ onto $[0,1]$) The formula

$$f(\ldots.a_2a_1a_0) := \sum_{n=0}^{\infty} a_n p^{-n-1}$$

defines a map $f: \mathbb{Z}_p \to \mathbb{R}$ (of course the convergence of the right-hand side is meant to be with respect to the absolute value function). Show that f is a continuous map of $\mathbb{Z}_p$ onto the closed unit interval $[0,1]$. Show that f is not bijective, so that f^{-1} does not exist. More generally, prove that every continuous map $[0,1] \to \mathbb{Z}_p$ is constant.

The algebraic closure $\mathbb{C}$ of $\mathbb{R}$ is a twodimensional space over $\mathbb{R}$. In Sections 16 and 17 we shall see that the algebraic closure of $\mathbb{Q}_p$ is infinite dimensional, from which it follows directly that as a field $\mathbb{Q}_p$ is not isomorphic to $\mathbb{R}$. Exercise 33.B shall prove that, for distinct primes p and q, $\mathbb{Q}_p$ and $\mathbb{Q}_q$ are not isomorphic either. In Section 24 we shall discuss the possibility of defining a structure on $\mathbb{Q}_p$ that resembles somewhat the ordering we have in $\mathbb{R}$. (See also Part 3 of Chapter 4 for a further elaboration on this theme.)

Exercise 7.C. Prove the following. A rational number is a square in $\mathbb{Q}$ if and only if it is a square in $\mathbb{R}$ and for all (primes) p is a square in $\mathbb{Q}_p$.

Exercise 7.D. Let $n \in \{2,3,4,\ldots\}$. Show that the series $\sum k!$ converges in $\mathbb{Z}_n$. It does not seem to be known whether $\sum_{k=0}^{\infty} k!$ is rational in $\mathbb{Z}_p$ for some prime p. Prove however that $\sum_{k=0}^{\infty} k!$ cannot be rational in every $\mathbb{Z}_n$. (Hint. Suppose the sum is rational in every $\mathbb{Z}_n$. Show first that the sum does not depend on n, next that the sum has to be an integer.)

Exercise 7.E. A famous (and by no means trivial) theorem of Dirichlet in number theory states that there exist infinitely many primes in any arithmetical progression $a, a+n, a+2n, \ldots$ (where $a \in \mathbb{Z}$, $n \in \mathbb{N}$ have no common factors). Use Dirichlet's theorem to prove the following. Let $P := \{2,3,5,\ldots\}$ be the collection of primes. For each $p \in P$ the set $P \backslash \{p\}$ is dense in $\{x \in \mathbb{Z}_p : |x|_p = 1\}$.

Exercise 7.F. (On $\mathbb{Q}_n$, see also Exercise 3.J) Let $n \in \{2,3,4,\ldots\}$. Define $\mathbb{Q}_n$ to be the set of all two-sided sequences $x = \ldots.a_2a_1a_0.a_{-1}a_{-2}\ldots.$ where $a_{-n} = 0$ for large n and each $a_i \in \{0,1,\ldots,n-1\}$. Define $\operatorname{ord}_n(x) := \min \{s : a_s \neq 0\}$ $(x \in \mathbb{Q}_n, x \neq 0)$ and

$$|x|_n := \begin{cases} 0 \text{ if } x = 0 \\ n^{-\operatorname{ord}_n(x)} & \text{if } x \neq 0 \end{cases}$$

Define addition and multiplication of elements of $\mathbb{Q}_n$ in the spirit of Section 4.
(i) Show that every nonzero integer has an inverse in $\mathbb{Q}_n$, so that we may write $\mathbb{Q} \subset \mathbb{Q}_n$.

(ii) Show that $\mathbb{Q}_n$ is a commutative ring with identity. Is it a field?
(iii) Let $x, y \in \mathbb{Q}_n$. Show that (1) $|x|_n \geq 0$; $|x|_n = 0$ if and only if $x = 0$, (2) $|x|_n = |-x|_n$, (3) $|x + y|_n \leq \max(|x|_n, |y|_n)$, (4) $|xy|_n \leq |x|_n\, |y|_n$, (5) $|xy|_n = |x|_n\, |y|_n$ if n is a prime number.
(iv) Show that $\mathbb{Q}_n$ is complete with respect to the metric $(x, y) \mapsto |x - y|_n$ and that $\mathbb{Q}$ is dense in $\mathbb{Q}_n$.

8. Archimedean and non-archimedean valuations

The axiom of Archimedes can be formulated as

$$\mathbb{N} \text{ is not bounded}$$

In this section we shall see that a valuation on a field K satisfies the strong triangle inequality if and only if the set $\{1_K, 1_K + 1_K, 1_K + 1_K + 1_K, \ldots\}$ (where 1_K denotes the unit element of K) is bounded. This explains the term 'non-archimedean' in Definition 8.1. In the sequel we shall follow a bad but widespread habit and omit the subscript K in 1_K and n_K (:= the sum of n times 1_K). One should realize that this can be dangerous in expressions like $|1| = 1$ and even more so in

'if the characteristic of K is $p \neq 0$ then $p = 0$'(!)

but one gets used to it after a while.

DEFINITION 8.1. A valuation on a field K is *archimedean* if $|2| > 1$, *non-archimedean* if $|2| \leq 1$.

The absolute value function on $\mathbb{C}$ is archimedean. The p-adic valuation on $\mathbb{Q}_p$ is non-archimedean.

LEMMA 8.2. *Let* $|\ |$ *be a non-archimedean valuation on a field* K.
(i) $|n| \leq 1$ *for all* $n \in \mathbb{N}$.
(ii) $|x + y| \leq \max(|x|, |y|)$ $(x, y \in K)$.

Proof. (i) Let $n \in \mathbb{N}$, $n \geq 2$. We shall prove that $|n| \leq 1$. Write n using the base 2. Then

$$n = a_0 + a_1 2 + \ldots + a_s 2^s$$

where $a_0, \ldots a_s \in \{0,1\}$ and $a_s = 1$. Then, since $|2| \leq 1$,

$$|n| \leq \sum_{i=0}^{s} |a_i|\, |2^i| \leq \sum_{i=0}^{s} |2|^i \leq s + 1$$

Now let $k \in \mathbb{N}$ and take the kth power of n. We have $n < 2^{s+1}$, so that

$n^k < 2^{k(s+1)}$ and therefore

$$n^k = b_0 + b_1 2 + \ldots + b_t 2^t$$

where $b_0, \ldots, b_t \in \{0, 1\}$, $b_t = 1$ and $t < (s+1)k$. Hence

$$|n^k| \leqslant t + 1 \leqslant (s+1)k$$

so that

$$|n| \leqslant \lim_{k \to \infty} \sqrt[k]{(s+1)k} = 1$$

(ii) Let $n \in \mathbb{N}$. Since $\binom{n}{k}$ is an integer we have, by (i), that $|\binom{n}{k}| \leqslant 1$ so that $|(x+y)^n| = |\sum_{k=0}^{n} \binom{n}{k} x^k y^{n-k}| \leqslant \sum_{k=0}^{n} |x|^k |y|^{n-k} \leqslant (n+1) \max(|x|, |y|)^n$.
It follows that

$$|x+y| \leqslant \lim_{n \to \infty} \sqrt[n]{n+1} \max(|x|, |y|) = \max(|x|, |y|)$$

As a corollary we get the following characterization.

THEOREM 8.3. *The conditions* (α), (β), (γ) *for a valuation on a field are equivalent.*
(α) *The valuation satisfies the strong triangle inequality.*
(β) *The valuation is non-archimedean.*
(γ) *The set* $\{1, 2, 3, \ldots\}$ *is bounded.*
Proof. To prove the implication (γ) $\to$ (α) (the only one that may need some explanation), modify the proof of part (ii) of Lemma 8.2.

Remark. (On the logarithm of a valuation) In Example 2.1 it appears more natural to work with the degree $d(f)$ of a polynomial f rather than $\rho^{d(f)}$. Similarly, in $\mathbb{Q}_p$ one might prefer $\text{ord}_p(x)$ instead of $p^{-\text{ord}_p(x)}$. Thus, in a non-archimedean valued field $(K, |\ |)$ we define the function v (called 'valuation' by many authors) as follows.

$$v(a) := \begin{cases} -\log |a| & \text{if } a \in K, a \neq 0 \\ \infty & \text{if } a = 0 \end{cases}$$

Then v is a mapping of K into the extended real numbers $\mathbb{R} \cup \{\infty\}$. For all $a, b \in K$ we have

(i) $v(a+b) \geqslant \min(v(a), v(b))$
(ii) $v(ab) = v(a) + v(b)$
(iii) $v(a) = \infty$ if and only if $a = 0$

Of course, the choice between $|\ |$ and v is just a matter of preference. People with an analysis background mostly like $|\ |$ because of its analogy with the absolute value.

Exercise 8.A. (On powers of valuations)
(i) Let $| \ |$ be the absolute value function on $\mathbb{C}$. Show that $x \mapsto \sqrt{|x|}$ is a valuation on $\mathbb{C}$ and that $x \mapsto |x|^2$ is not.
(ii) Let $(K, | \ |)$ be a valued field, $0 < s \leqslant 1$. Then $| \ |^s$ is a valuation on K. Prove this. (Hint. To prove the triangle inequality, show that $(a + 1)^s \leqslant a^s + 1$ for $a \in \mathbb{R}, a \geqslant 0$.)
(iii) Let $(K, | \ |)$ be a non-archimedean valued field, $s > 0$. Prove that $| \ |^s$ is a non-archimedean valuation on K.

Exercise 8.B. Prove the following.
(i) The trivial valuation on a field K is non-archimedean.
(ii) On a finite field the only possible valuation is the trivial one.
(iii) If the characteristic of a field K is not zero, then any valuation on K is non-archimedean.

**Exercise* 8.C. (Important consequences of the strong triangle inequality) Show the following facts (compare the properties of the degree of a polynomial, see Section 2). If x, y are elements of a non-archimedean valued field $(K, | \ |)$ then

$$\text{(i) } |x| \neq |y| \text{ implies } |x + y| = \max(|x|, |y|)$$
$$\text{(ii) } |x + y| \geqslant |x| \text{ implies } |x + y| \geqslant |y|$$

**Exercise* 8.D. (No new values of $| \ |$ after completion) Let K be a non-archimedean valued field with completion L. Show that valuation on L is also non-archimedean and prove the remarkable fact that $\{ |x| : x \in K \} = \{ |x| : x \in L \}$.

9. Equivalence of valuations

Valuations which are powers of one another (see Exercise 8.A) do not seem to be materially different from an analytic point of view.

DEFINITION 9.1. Two valuations on a field K are *equivalent* if they induce the same topology on K.

THEOREM 9.2. *Let $| \ |_1$ and $| \ |_2$ be equivalent valuations on a field K. Then there is a positive real number c such that $| \ |_2 = | \ |_1^c$.*

Proof. If $| \ |_1$ is trivial then $\{0\}$ is open with respect to $| \ |_1$ and $| \ |_2$ so there is $\delta > 0$ such that $\{x \in K : |x|_2 < \delta\} = \{0\}$. It follows easily that $| \ |_2$ is trivial so that for this case the theorem is established. Now let $| \ |_1$ be not trivial and choose $\pi \in K$ such that $|\pi|_1 > 1$. Let $s \in K, s \neq 0$. Then there is an $a \in \mathbb{R}$ such that $|s|_1 = |\pi|_1^a$. Let $r = mn^{-1}$ ($m \in \mathbb{Z}$, $n \in \mathbb{N}$) be rational and $r > a$. Then $|s|_1 < |\pi|_1^r$. We prove that $|s|_2 < |\pi|_2^r$. In fact, we have $|s^n \pi^{-m}|_1 < 1$. Now in any valued field $(K, |\ |)$, a sequence $x, x^2, x^3, \ldots$ has limit 0 if and only if $|x| < 1$. By equivalence, $|s^n \pi^{-m}|_2 < 1$, i.e. $|s|_2 < |\pi|_2^r$. Similarly one proves that if $r \in \mathbb{Q}$, $r < a$ then $|s|_2 > |\pi|_2^r$. We may conclude that $|s|_2 = |\pi|_2^a$. It follows that $\log |s|_2 = c \log |s|_1$, where $c = \log |\pi|_2 (\log |\pi|_1)^{-1}$. The theorem follows.

**Exercise* 9.A. (i) Show that the function $|\ |_p'$ defined on $\mathbb{Z}_p$ in Exercise 3.D can be extended to a valuation on $\mathbb{Q}_p$ and that this valuation is equivalent to the p-adic valuation.
(ii) Prove that for distinct primes p and q the valuations $|\ |_p$, $|\ |_q$ on $\mathbb{Q}$ are not equivalent and also that no p-adic valuation on $\mathbb{Q}$ is equivalent to the absolute value function.

Exercise 9.B. Show that the following conditions for valuations $|\ |_1$ and $|\ |_2$ on a field K are equivalent.
(α) $|x|_1 < 1$ if and only if $|x|_2 < 1$ ($x \in K$).
(β) $|x|_1 \leqslant 1$ if and only if $|x|_2 \leqslant 1$ ($x \in K$).
(γ) $|\ |_1$ and $|\ |_2$ are equivalent.
(δ) $|\ |_1$ is a positive power of $|\ |_2$.

**Exercise* 9.C. Let $|\ |_1$ and $|\ |_2$ be valuations on a field K and let $|\ |_1$ be non-trivial. Suppose that $x \in K$, $|x|_1 < 1$ implies $|x|_2 < 1$. Prove that from this it follows already that $|\ |_1$ and $|\ |_2$ are equivalent.

Exercise 9.D. (Conclusion from the previous exercise) Prove that the inverse of a continuous automorphism of a valued field is again continuous.

Exercise 9.E. (Comparison with equivalence of norms) Let $|\ |_1$ and $|\ |_2$ be valuations on a field K and let c_1 and c_2 be positive constants such that $c_1 |\ |_1 \leqslant |\ |_2 \leqslant c_2 |\ |_1$. Prove that $|\ |_1 = |\ |_2$.

Exercise 9.F. If two valuations are equivalent then they are both archimedean or both non-archimedean. Show this.

Exercise 9.G. Let $|\ |_1$ and $|\ |_2$ be two equivalent valuations on a field K. Let $(L_1, \|\ \|_1)$, $(L_2, \|\ \|_2)$ be the completions of $(K, |\ |_1)$ and $(K, |\ |_2)$ respectively. Show that $L_1 = L_2$ and that $\|\ \|_1$ is equivalent to $\|\ \|_2$.

10. All valuations on $\mathbb{Q}$

In the preceding sections we have met the p-adic valuations and the absolute value function on $\mathbb{Q}$. The next theorem states that, essentially, there are no others.

THEOREM 10.1. (Ostrowski) *Each non-trivial valuation on the field of the rational numbers is equivalent either to the absolute value function or to some p-adic valuation.*

Proof. (i) Let $|\ |$ be an archimedean valuation on $\mathbb{Q}$. Since $1 < |2| \leqslant |1| + |1| = 2$ there is a number $c \in \mathbb{R}, 0 < c \leqslant 1$ for which

$$|2| = 2^c$$

We shall prove that

$$|n| = n^c$$

for $n \in \mathbb{N}$ (this obviously implies that, on $\mathbb{Q}$, $|\ |$ is the cth power of the absolute value function). Thus, let $n \in \mathbb{N}$, $n \geqslant 2$, and write n using the base 2

$$n = a_0 + a_1 2 + \ldots + a_s 2^s \quad (a_0, a_1, \ldots \in \{0, 1\},\ a_s = 1)$$

Then $2^s \leqslant n < 2^{s+1}$ so that

$$(*) \qquad 2^{sc} \leqslant n^c < 2^{c(s+1)}$$

We first prove that $|n| \leqslant n^c$ as follows. Applying (*) we get

$$|n| \leqslant \sum_{i=0}^{s} |a_i|\ |2|^i \leqslant \sum_{i=0}^{s} 2^{ic} = 2^{sc}(1 + 2^{-c} + \ldots + 2^{-sc}) \leqslant n^c M$$

where $M := \sum_{i=0}^{\infty} 2^{-ic}$ does not depend on n. Since n was arbitrary we have also

$$|n^k| \leqslant n^{kc} M \quad (k \in \mathbb{N})$$

so that

$$(**) \qquad |n| \leqslant \lim_{k \to \infty} n^c \sqrt[k]{M} = n^c$$

To prove the opposite inequality observe that

$$|n| = |2^{s+1} - (2^{s+1} - n)| \geqslant |2^{s+1}| - |2^{s+1} - n|$$

Now $|2^{s+1}| = |2|^{s+1} = 2^{c(s+1)}$. By (*) and (**)

$$|2^{s+1} - n| \leqslant (2^{s+1} - n)^c \leqslant (2^{s+1} - 2^s)^c = 2^{sc}$$

so that

$$|n| \geqslant 2^{c(s+1)} - 2^{cs} = 2^{c(s+1)}(1 - 2^{-c})$$

Again by (*), $2^{c(s+1)} > n^c$; with $M' := 1 - 2^{-c}$ we obtain

$$|n| \geqslant n^c M'$$

The kth power trick yields

$$|n| \geqslant \lim_{k\to\infty} n^c \sqrt[k]{M'} = n^c$$

which, together with (**), finishes this part of the proof.

(ii) Now let $|\ |$ be a non-archimedean valuation on $\mathbb{Q}$; we prove it to be a power of some p-adic valuation. The valuation is assumed to be non-trivial so the set $\{n \in \mathbb{N}: |n| < 1\}$ is not empty, let p be its minimal element. We claim that p is a prime number. In fact, $p \neq 1$. If $p = ab$ for some $a, b \in \mathbb{N}$, $a < p$, $b < p$, then $|a| = |b| = 1$, so $|p| = |ab| = 1$, a contradiction. Next we show that $|q| = 1$ for any $q \in \mathbb{N}$ that is not divisible by p. Write $q = ap + r$ where $a \in \{0, 1, 2, \ldots\}$ and $1 \leqslant r < p$. Then $|r| = 1$ and $|ap| = |a|\ |p| \leqslant |p| < 1$. By the strong triangle inequality $1 = |r| \leqslant \max(|ap + r|, |-ap|) = \max(|q|, |ap|) = |q|$. So $|q| \geqslant 1$, i.e. $|q| = 1$. It follows that for each $n \in \mathbb{N}$

$$|n| = |p|^k$$

where k is the number of factors p of n. We see that for each $n \in \mathbb{N}$

$$|n| = |n|_p^c$$

where $c = -\log |p| (\log p)^{-1}$. It follows easily that

$$|x| = |x|_p^c \quad (x \in \mathbb{Q})$$

which completes the proof of Theorem 10.1.

Together with Exercises 9.A. (ii) and 9.G Ostrowski's theorem furnishes a complete list of the possible completions of $\mathbb{Q}$. The absolute value on $\mathbb{Q}$ leads via completion to the ordinary analysis over $\mathbb{R}$ and $\mathbb{C}$. The trivial valuation induces the discrete topology on $\mathbb{Q}$. From an analyst's point of view this is not an interesting object to consider, so we shall exclude it.

FROM NOW ON, UNLESS EXPLICITLY STATED OTHERWISE,
THE TERM 'VALUATION' INDICATES A NON-TRIVIAL VALUATION

What interest us here are the p-adic valuations on $\mathbb{Q}$ and the corresponding completions $\mathbb{Q}_p$. More generally, from now on we shall focus our attention mainly on non-archimedean valued fields.

Exercise 10.A. It follows from Ostrowski's theorem that for an archimedean valuation $|\ |$ on a field we have $|n| > 1$ for all $n \in \mathbb{N}$, $n \geqslant 2$. Prove this fact directly. (Hint. Suppose $|m| \leqslant 1$ for some $m > 2$. Write $n \in \mathbb{N}$ using the base m.)

**Exercise* 10.B. (The product formula for valuations) Let $| \ |_\infty$ be the absolute value function on $\mathbb{Q}$. Let $x \in \mathbb{Q}, x \neq 0$.
(i) Show that $|x|_p \neq 1$ only for finitely many p so that $\Pi_p \ |x|_p$, where the product is taken over all primes p, is well defined.
(ii) Prove the *product formula*

$$|x|_\infty \prod_p |x|_p = 1$$

(iii) Conclude from (ii) or prove directly that

$$|n|_p \geq \frac{1}{n} \qquad (n \in \mathbb{N})$$

Exercise 10.C. (Approximation by inequivalent valuations) Let p, q be primes, $p < q$, and let $| \ |_\infty$ be as in the previous exercise. Show that

$$\lim_{n \to \infty} \frac{p^n}{p^n + q^n} = \begin{cases} 0 \text{ in } (\mathbb{Q}, | \ |_p) \text{ and } (\mathbb{Q}, | \ |_\infty) \\ \\ 1 \text{ in } (\mathbb{Q}, | \ |_q) \end{cases}$$

Use this to prove the following remarkable fact. *If* $| \ |$ and $| \ |'$ *are inequivalent valuations on* $\mathbb{Q}$ *then for each pair of rational numbers s, t there is a sequence* $a_1, a_2, \ldots$ *of rational numbers such that* $\lim_{n \to \infty} |s - a_n| = 0$ *and* $\lim_{n \to \infty} |t - a_n|' = 0$.

Exercise 10.D. (Sequel to Exercise 10.C) Let L_k be the product of the first k primes, i.e. $L_1 = 2, L_2 = 2 \cdot 3, L_3 = 2 \cdot 3 \cdot 5$, etc. Show that

$$\lim_{n \to \infty} L_k^k L_{k^2}^{-1} = 0$$

with respect to every valuation on $\mathbb{Q}$. Does there exist a sequence $a_1, a_2, \ldots$ of rational numbers for which $\lim_{n \to \infty} a_n = 1$ in $(\mathbb{Q}, | \ |_p)$ for some prime p, $\lim_{n \to \infty} a_n = 0$ in $(\mathbb{Q}, | \ |_\infty)$ and $(\mathbb{Q}, | \ |_q)$ for each prime q different from p?

11. The residue class field and the value group

In the remaining sections of this chapter we shall leave $\mathbb{Q}_p$ temporarily and turn to general non-archimedean valued fields. This is for two reasons. The most important one is that we shall use the results of Sections 11–16 to construct, in Section 17, the p-adic analogue $\mathbb{C}_p$ of the field $\mathbb{C}$ of the complex numbers, i.e. the – in a certain sense – smallest non-archimedean valued field that is complete, algebraically closed, and that contains $\mathbb{Q}_p$. Secondly, the results of Sections 11–17 bring about many examples of non-archimedean valued fields other than $\mathbb{Q}_p$.

FROM NOW ON K IS A NON-ARCHIMEDEAN VALUED FIELD

In this section we introduce some basic concepts.
Let $a \in K, r > 0$. The *'open' disc* (or *ball*) *of radius r with centre a* is

$$B_a(r^-) := \{x \in K : |x-a| < r\}$$

The *'closed' disc* (*ball*) *of radius r with centre a* is

$$B_a(r) := \{x \in K : |x-a| \leqslant r\}$$

(Warning. $B_a(r^-)$ is also closed, $B_a(r)$ is open. See Section 18 for further elaborations on this theme.) The discs $B_0(1)$ and $B_0(1^-)$ are the *'closed' unit disc* and the *'open' unit disc* respectively. The set

$$B_0(1)\backslash B_0(1^-) = \{x \in K : |x| = 1\}$$

is the *unit circle* or *unit sphere*. All these names are directly borrowed from the analysis in $\mathbb{R}$ and $\mathbb{C}$. The next proposition shows that there are striking differences between K and $\mathbb{C}$.

PROPOSITION 11.1. *$B_0(1)$ is a subring of K. $B_0(1^-)$ is a maximal ideal in $B_0(1)$.*

Proof. If $x, y \in B_0(1)$ then $|x+y| \leqslant \max(|x|, |y|) \leqslant 1$ so $x+y \in B_0(1)$. Obviously, $-x$ and xy are in $B_0(1)$. It follows that $B_0(1)$ is a ring. By the same token, $B_0(1^-)$ is a ring. If $x \in B_0(1), y \in B_0(1^-)$ then $|xy| = |x|\,|y| < 1$, hence $xy \in B_0(1^-)$. This proves that $B_0(1^-)$ is an ideal. It is maximal because $B_0(1)\backslash B_0(1^-)$ consists of invertible elements of $B_0(1)$. ($B_0(1^-)$ is the unique maximal ideal of $B_0(1)$.)

It follows from elementary algebra that $B_0(1)/B_0(1^-)$ is a field. We consider this field as an algebraic object (i.e. we are not aiming at a topology or a valuation).

DEFINITION 11.2. The *residue class field* of K is the field

$$k := B_0(1)/B_0(1^-) = \{x \in K : |x| \leqslant 1\} / \{x \in K : |x| < 1\}$$

The natural quotient map $B_0(1) \to k$ is often denoted by

$$x \mapsto \bar{x} \qquad (x \in B_0(1))$$

The map $x \mapsto |x|$ is a homomorphism of the multiplicative group $K\backslash\{0\}$ into the multiplicative group $\mathbb{R}\backslash\{0\}$. Its range, therefore, is a subgroup of $\mathbb{R}\backslash\{0\}$. This leads to the following definition.

DEFINITION 11.3. For a field L, let $L^\times := \{x \in L : x \neq 0\}$. For a subset $X \subset K$, let $|X| := \{|x| : x \in X\}$. The *value group* of K is the subgroup $|K^\times|$ of the multiplicative group of the positive real numbers.

Examples and a few immediate observations are collected in the following exercises.

**Exercise* 11.A. Determine the residue class field and the value group of $\mathbb{Q}_p$. Answer the same question for the field $\mathbb{R}(X)$ of Example 2.1.

**Exercise* 11.B. (Value group and residue class field of the completion) Let L be the completion of K. Show that the value groups of L and K are the same and that the residue class fields of L and K are isomorphic. (See Exercise 8.D.)

**Exercise* 11.C. (Value groups and residue class fields for equivalent valuations) Let $|\ |'$ be a valuation on K equivalent to $|\ |$, and let $L = (K, |\ |')$. Show that the value groups of L and K are isomorphic and that the residue class fields of L and K are the same.

**Exercise* 11.D. (The characteristics of K and its residue class field) Let k be the residue class field of K. For a field L, denote its characteristic by $\mathrm{char}(L)$. Give examples of each of the following cases.
(i) ('Equal characteristics') (1) $\mathrm{char}(K) = \mathrm{char}(k) = 0$, (2) $\mathrm{char}(K) = \mathrm{char}(k) = p$.
(ii) ('Mixed characteristics') $\mathrm{char}(K) = 0$, $\mathrm{char}(k) = p$.
Show that $\mathrm{char}(K) = p$ implies $\mathrm{char}(k) = p$, so that (i) and (ii) above describe all possible situations. See also Remark 2 below.

**Exercise* 11.E. (Additive and multiplicative groups in K) Prove the following.
(i) For each $r > 0$ the discs $B_0(r)$ and $B_0(r^-)$ are additive subgroups of K.
(ii) The unit sphere is a multiplicative subgroup of $K^\times$.
(iii) $B_1(1^-) = \{x \in K : |1 - x| < 1\}$ is a multiplicative subgroup of the unit sphere (!)
(iv) For each $r \in \mathbb{R}$, $0 < r < 1$ the discs $B_1(r)$ and $B_1(r^-)$ are multiplicative subgroups of $B_1(1^-)$.

DEFINITION 11.4. The valuation on K is *discrete* if 1 is not an accumulation point of the value group $|K^\times|$. Otherwise, the valuation is *dense*.

An explanation for the use of the terms 'discrete' and 'dense' is contained in Exercise 11.F. See also Remark 1 below.

**Exercise* 11.F. Let G be a multiplicative subgroup of the group $(0, \infty)$ of the positive real numbers. Prove the following.
(i) If 1 is an accumulation point of G then G is dense in $(0, \infty)$.

(ii) If 1 is not an accumulation point of G then there is an $s \in G, 0 < s < 1$ such that $G = \{ s^n : n \in \mathbb{Z} \}$.
Deduce from (i) and (ii) that a valuation on K is discrete or dense according to whether $|K^\times|$ is a discrete or a dense subset of $(0, \infty)$.

Exercise 11.G. Show that the p-adic valuation on $\mathbb{Q}_p$ and the valuation on $\mathbb{R}(X)$ of Example 2.1 are discrete. If the valuation on K is discrete (dense) then so is any equivalent valuation and the valuation on the completion. (See Exercise 8.D.)

**Exercise* 11.H. Let K be locally compact. Prove that its residue class field is finite and its valuation is discrete. (Hint. Consider cosets of $B_0(1^-)$ in $B_0(1)$; if $0 < |x_1| < |x_2| < \ldots < 1$ then $x_1, x_2, \ldots$ has no convergent subsequence.)

Exercise 11.I. Find examples of complete valued fields K and L whose respective residue class fields and value groups are isomorphic but such that K and L are not isomorphic as fields.

Remarks.
1. In Sections 16 and 17 we shall see that there do exist dense valuations. In Appendix A.9 it is even proved that for each field k and each subgroup Γ of $(0, \infty)$ there exists a non-archimedean valued (complete) field whose residue class field is (isomorphic to) k and whose value group is Γ.
2. In the case of 'mixed characteristics' (K has zero characteristic, its residue class field has characteristic p, see Exercise 11.D) we may consider $\mathbb{Q}$ as a valued subfield of K. It is easy to see that this valuation must be equivalent to the p-adic valuation. So by taking a suitable power of the valuation on K we can arrange that the valuation on $\mathbb{Q}$ is the p-adic one and that, in consequence, we may assume that $\mathbb{Q}_p \subset K$. *Henceforth we shall assume that if K is a valued field of characteristic* 0 *and if its residue class field has characteristic p, then the valuation on K is chosen in such a way that $\mathbb{Q}_p$ is a valued subfield of K.*

12. Series expansions of elements of K

We shall show that elements of an arbitrary (non-archimedean) valued field admit expansions similar to the p-adic expansions we have met in Theorem 5.7. Apart from their theoretical value these expansions shall be of use to us in the future when constructing (counter)examples. We consider discrete valuation first.

Let K have a discrete valuation. Then there is an element $\pi \in K$ such that $|\pi| = \max |K^\times| \cap (0, 1)$ and $|K^\times| = \{|\pi|^n : n \in \mathbb{Z}\}$, see Exercise 11.F. The element π shall take over the role played by p in $\mathbb{Q}_p$. Let R be a full set of representatives in $B_0(1)$ modulo $B_0(1^-)$. In other words, R is a subset of the 'closed' unit disc $B_0(1)$ of K satisfying
(i) if $r_1, r_2 \in R, r_1 \neq r_2$ then $|r_1 - r_2| = 1$
(ii) for each $x \in B_0(1)$ there is $r \in R$ such that $|x - r| < 1$.
The map $x \mapsto \bar{x}$ of Definition 11.2 sends R bijectively onto the residue class field of K. If $K = \mathbb{Q}_p$ the set $R := \{0, 1, 2, \ldots, p-1\}$ satisfies (i), (ii). In general, R may be an infinite set. For reasons of simplicity we assume that $0 \in R$. (See Exercise 12.B.)

THEOREM 12.1. (Series expansion, discrete case) *Let K have a discrete valuation and let π, R be as above. Then for each $x \in K$ there exists a unique two-sided sequence $\ldots a_{-1}, a_0, a_1, \ldots$ of elements of R such that $a_{-n} = 0$ for large n and*

$$(*) \qquad x = \sum_{j=-\infty}^{\infty} a_j \pi^j$$

If, in addition, K is complete then for any choice of $a_j \in R$ ($j \in \mathbb{Z}$) for which $a_{-n} = 0$ for large n formula (*) defines an element of K.

Proof. (i) (Existence) Let $x \in K$, $x \neq 0$. Then there is an $m \in \mathbb{Z}$ such that $|x| = |\pi|^m$ so that $|\pi^{-m}x| \leqslant 1$. There is $a_0 \in R$ such that $|\pi^{-m}x - a_0| < 1$, hence $|\pi^{-m}x - a_0| \leqslant |\pi|$. So there is $x_1 \in B_0(1)$ such that

$$\pi^{-m}x = a_0 + \pi x_1$$

Repeating the reasoning for x_1 in place of $\pi^{-m}x$ we obtain $a_1 \in R$ and $x_2 \in B_0(1)$ such that $x_1 = a_1 + \pi x_2$, so

$$\pi^{-m}x = a_0 + \pi a_1 + \pi^2 x_2$$

Let $n \in \mathbb{N}$. By induction we obtain $a_0, a_1, \ldots, a_n \in R$ and $x_{n+1} \in B_0(1)$ such that

$$\pi^{-m}x = a_0 + a_1\pi + \ldots + a_n\pi^n + x_{n+1}\pi^{n+1}$$

Now $\lim_{n\to\infty} x_{n+1}\pi^{n+1} = 0$. It follows that $\pi^{-m}x = \sum_{j=0}^{\infty} a_j\pi^j$, and that x has an expansion of the desired form.
(ii) (Uniqueness) Let, for some $m \in \mathbb{Z}$, $\sum_{j=m}^{\infty} a_j\pi^j = \sum_{j=m}^{\infty} b_j\pi^j$ be two such expansions. Then

$$\sum_{j=m}^{\infty} (a_j - b_j)\,\pi^j = 0$$

Division by π^m yields

$$(a_m - b_m) + (a_{m+1} - b_{m+1})\,\pi + \ldots = 0$$

Since $\bar{\pi} = 0$ we get $\bar{a}_m - \bar{b}_m = \bar{0}$, i.e. $a_m - b_m \in B_0(1^-)$.
But, since a_m and b_m are in R we have $a_m = b_m$. So we obtain

$$\sum_{j=m+1}^{\infty} (a_j - b_j)\,\pi^j = 0$$

We can use induction to prove that $a_{m+1} = b_{m+1}, a_{m+2} = b_{m+2}, \ldots$

Finally, let K be complete and let $a_m, a_{m+1}, \ldots$ be in R for some $m \in \mathbb{Z}$. The sequence $n \mapsto \sum_{j=m}^{n} a_j\,\pi^j$ is Cauchy, hence convergent. The statement follows.

COROLLARY 12.2. *A non-archimedean complete valued field is locally compact if and only if its residue class field is finite and its value group is discrete.*

Proof. Let K be locally compact. Then by Exercise 11.H its residue class field is finite and $|K^\times|$ is discrete. For the converse observe that $B_0(1)$ consists of all elements of the form

$$\sum_{n=0}^{\infty} a_n\,\pi^n \qquad (a_n \in R)$$

To prove compactness of $B_0(1)$ observe that R is a finite set so that we can apply the technique used in Theorem 5.1 to prove the compactness of $\mathbb{Z}_p$.

Next we turn to the general case. For a dense valuation the set $|K^\times| \cap (0, 1)$ has no largest element, so we shall content ourselves with a less standard choice of π; let us fix *any* $\pi \in K$, $0 < |\pi| < 1$. Accordingly we have to modify the definition of R as follows. Let $J := \{x \in K : |x| \leqslant |\pi|\}$. Then J is an ideal in $B_0(1)$ so that $B_0(1)/J$ is a ring (that may have zero divisors). Let R be a full set of representatives in $B_0(1)$ modulo J. In other words, R is a subset of $B_0(1)$ satisfying

(i) if $r_1, r_2 \in R, r_1 \neq r_2$ then $|\pi| < |r_1 - r_2| \leqslant 1$

(ii) for each $x \in B_0(1)$ there is an $r \in R$ such that $|x - r| \leqslant |\pi|$

Again we assume $0 \in R$. Of course the above construction of R and π applies also to discretely valued fields, so we can formulate

THEOREM 12.3. (Series expansion, general case) *The conclusion of Theorem* 12.1 *holds for an arbitrary K provided the definitions of R and π are modified in the sense of the above.*

Proof. With the necessary alterations and changes, the proof of Theorem 12.1 applies.

Exercise 12.A. Show that the field $\mathbb{R}(X)$ of Example 2.1 is not locally compact.

Exercise 12.B. (i) Show that every element of $\mathbb{Z}_p$ has a unique expansion $\sum_{n=0}^{\infty} a_n p^n$ where $a_n \in \{1, 2, \ldots, p\}$ for each n. Find the expansion of 0.
(ii) Explain what (minor) modifications are to be made in Theorems 12.1 and 12.3 if we drop the condition $0 \in R$.

Remark. By choosing in the first exercise of Appendix A.9 for F a finite field we obtain an example of a locally compact field with a nonzero characteristic.

13. Normed spaces

In the sequel we shall need several concepts from functional analysis. Right away we would like to reassure the alarmed reader by stating that a specific attention to 'functional analysis' does not really belong here. In fact, we shall only introduce a few basic notions and check whether some well-known facts of classical functional analysis remain valid for $\mathbb{R}$ or $\mathbb{C}$ replaced by K, only in so far as is necessary for our theory. As a by-product we can take advantage of having a terminology at hand that may ease the formulation of certain results. This section may be disregarded by the reader until he or she needs part of it.

IN THIS SECTION K IS COMPLETE

DEFINITION 13.1. Let E be a vector space over K (K-vector space, K-linear space). A *seminorm* on E is a map $q : E \to \mathbb{R}$ such that for $x, y \in E$, $\lambda \in K$

(i) $q(x) \geqslant 0$
(ii) $q(\lambda x) = |\lambda| q(x)$
(iii) $q(x+y) \leqslant \max(q(x), q(y))$

If, in addition, q satisfies

(i)' $q(x) = 0$ if and only if $x = 0$

then q is a *norm* on E. One usually denotes a norm by $\| \ \|$ rather than q. A *normed (vector) space over* K is a pair $(E, \| \ \|)$ where E is a vector space over K and where $\| \ \|$ is a norm on E. Often we shall write simply E instead of $(E, \| \ \|)$. E is a *Banach space over* K (*K-Banach space*) if E is complete with respect to the *induced metric* $(x, y) \mapsto \|x - y\|$. A *locally convex space* over K is a K-vector space together with a collection of seminorms on it.

The induced metric on a normed space E over K induces a topology on E. So we can talk about open, closed subsets of E, boundary, closure, interior, etc. A closed K-linear subspace of a Banach space over K is, with the inherited norm, itself a Banach space over K. The reason for requiring the strong triangle inequality (iii) instead of $q(x + y) \leqslant q(x) + q(y)$ lies in the fact that most norms that 'occur in nature' have this property (see, however, Exercise 13.B). In this book normed spaces arise mainly in two colours.

(1) If $(L, |\ |)$ is a valued field containing K as a valued subfield then addition $L \times L \to L$ and multiplication $K \times L \to L$ make L into a K-vector space. $|\ |$ is a norm on L so that $(L, |\ |)$ becomes a normed space over K.

(2) Let X be a set. A function $f: X \to K$ is *bounded* if

$$\|f\|_\infty := \sup \{|f(x)| : x \in X\} < \infty$$

Let $B(X \to K)$ be the set of all bounded functions $X \to K$. With the operations defined (for $f, g \in B(X \to K)$) by

$$(f + g)(x) = f(x) + g(x) \qquad (x \in X)$$
$$(\lambda f)(x) = \lambda f(x) \qquad (x \in X, \lambda \in K)$$

and the norm $\|\ \|_\infty$, $B(X \to K)$ is a normed space over K. Most function spaces we shall encounter will turn out to be (closed) subspaces of some $B(X \to K)$.

PROPOSITION 13.2. *$B(X \to K)$ is a Banach space over K.*

Proof. (It runs exactly like the 'classical' proof for spaces over $\mathbb{R}$ or $\mathbb{C}$ but is included for completeness) Let $f_1, f_2, \ldots$ be a Cauchy sequence in $B(X \to K)$. For each $x \in X$ and $n, m \in \mathbb{N}$ we have

$$|f_n(x) - f_m(x)| \leqslant \|f_n - f_m\|_\infty$$

so that $f_1(x), f_2(x), \ldots$ is Cauchy in K, hence convergent. It follows that

$$f(x) := \lim_{n \to \infty} f_n(x) \qquad (x \in X)$$

defines a function $f: X \to K$. For each $x \in X$, $n, m, \in \mathbb{N}$ we have

$$|f(x) - f_n(x)| \leqslant |f(x) - f_m(x)| + |f_m(x) - f_n(x)|$$
$$\leqslant |f(x) - f_m(x)| + \|f_m - f_n\|_\infty$$

After taking $\overline{\lim}_{m \to \infty}$ of the right-hand side we arrive at

$$|f(x) - f_n(x)| \leqslant \overline{\lim_{m \to \infty}} \|f_m - f_n\|_\infty$$

from which it follows easily that f is bounded and $\lim_{n \to \infty} \|f - f_n\|_\infty = 0$.

**Exercise* 13.A. Show that the following spaces are Banach spaces by interpreting them as some $B(X \to K)$.
(i) For $n \in \mathbb{N}$, the space K^n consisting of all finite sequences $(\xi_1, \xi_2, \ldots, \xi_n)$ of elements of K normed by $\|(\xi_1, \xi_2, \ldots, \xi_n)\| = \max_j |\xi_j|$.
(ii) The space l^∞ consisting of all bounded sequences $(\xi_1, \xi_2, \ldots)$ of elements of K normed by $\|(\xi_1, \xi_2, \ldots)\| = \sup_j |\xi_j|$. Show that $c_0 := \{(\xi_1, \xi_2, \ldots) \in l^\infty : \lim_{n\to\infty} |\xi_n| = 0\}$ is a closed subspace of l^∞.

Exercise 13.B. Let $l^1 := \{(\xi_1, \xi_2, \ldots) \in l^\infty : \sum_{n=1}^\infty |\xi_n| < \infty\}$. Show that l^1 is a vector space over K and that $(\xi_1, \xi_2, \ldots) \mapsto \sum_{n=1}^\infty |\xi_n|$ satisfies (i)', (i) , (ii) of Definition 13.1 and

$$\|x + y\| \leqslant \|x\| + \|y\| \qquad (x, y \in l^1)$$

but not the strong triangle inequality (iii).

**Exercise* 13.C. (Equivalent norms, compare Section 9) Two norms $\|\ \|_1$ and $\|\ \|_2$ on a K-vector space E are *equivalent* if they induce the same topology on E. Show that $\|\ \|_1$ and $\|\ \|_2$ are equivalent if and only if there are positive constants c_1 and c_2 such that $c_1 \|x\|_1 \leqslant \|x\|_2 \leqslant c_2 \|x\|_1$ for all $x \in E$.

The following theorem will not come as a surprise. We need it in Section 15.

THEOREM 13.3. *All norms on a finite dimensional K-vector space E are equivalent. E is a Banach space with respect to each norm.*

As K may not be locally compact, for a proof of Theorem 13.3 we cannot rely on compactness arguments frequently used in textbooks to prove the 'archimedean' version (all norms on a finite dimensional space over $\mathbb{R}$ or $\mathbb{C}$ are equivalent). To prove Theorem 13.3 we start with a simple lemma that will be applied several times in the sequel, especially for the case $c = 1$ (compare Exercise 8.C).

LEMMA 13.4. *Let E be a normed space over K. Let $x, y \in E$. Suppose there is $c \in (0, 1]$ such that*

$$\|x + y\| \geqslant c\, \|x\|$$

Then also

$$\|x + y\| \geqslant c\, \|y\|$$

Proof. $\|y\| = \|x + y - x\| \leqslant \max(\|x + y\|, \|x\|) \leqslant c^{-1} \|x + y\|$.

Proof of Theorem 13.3. We prove the statement by induction on $n := \dim E$. For $n = 1$ it is obvious. Suppose the statement is true for $(n-1)$-dimensional spaces. Let $\dim E = n$. Choose a base $e_1, \ldots, e_n$ of E and define

$$\|x\|_\infty := \max_j |\xi_j| \qquad (x = \sum_{j=1}^n \xi_j e_j \in E)$$

It suffices to prove that any given norm $\| \ \|$ on E is equivalent to $\| \ \|_\infty$. We have for $x = \sum_{j=1}^n \xi_j e_j \in E$

$$\|x\| \leqslant \max_j |\xi_j| \, \|e_j\| \leqslant M \|x\|_\infty$$

where $M := \max_j \|e_j\|$. We proceed to prove the existence of a positive constant N such that

$$\|x\| \geqslant N \|x\|_\infty \qquad (x \in E)$$

Set $D := [\![e_1, \ldots, e_{n-1}]\!]$. By the induction hypothesis and Exercise 13.C there is $c > 0$ such that $\|x\| \geqslant c \|x\|_\infty$ $(x \in D)$. Further, D is complete, hence $\| \ \|$-closed in E so

$$c' := \|e_n\|^{-1} \inf \{\|e_n - y\| : y \in D\}$$

is positive. Now set $N := \min (c'c, c'\|e_n\|)$. Let $x \in E$, $x = \sum_{j=1}^n \xi_j e_j$. Then $x = y + \xi_n e_n$ where $y = \sum_{j=1}^{n-1} \xi_j e_j \in D$. We have, if $\xi_n \neq 0$, $\|x\| = |\xi_n| \, \|e_n + \xi_n^{-1} y\| \geqslant |\xi_n| \, \|e_n\| \, c' = c' \|\xi_n e_n\|$. So, by Lemma 13.4

$$\|x\| \geqslant c' \|y\|$$

(The latter inequality is also true if $\xi_n = 0$.) We get

$$\begin{aligned} \|x\| &\geqslant c' \max (\|y\|, \|\xi_n e_n\|) \\ &\geqslant c' \max (c \|y\|_\infty, \|e_n\| \, |\xi_n|) \\ &\geqslant N \max (\|y\|_\infty, |\xi_n|) \\ &= N \max (|\xi_1|, \ldots, |\xi_n|) = N \|x\|_\infty \end{aligned}$$

Remark. Although at first sight the above proof seems to be typically 'non-archimedean', it is possible to rephrase it so as to become a valid proof for the 'archimedean' case too.

Let E, F be normed spaces over K. A K-linear map $A : E \to F$ is *continuous* if A maps null sequences (i.e. sequences $x_1, x_2, \ldots$ for which $\lim_{n \to \infty} \|x_n\| = 0$) in E into null sequences in F. The set $L(E, F)$ consisting of all continuous linear maps $E \to F$ is a vector space over K under the operations $(A + B)(x) := Ax + Bx$, $(\lambda A)(x) := \lambda (Ax)$ $(A, B \in L(E, F), x \in E, \lambda \in K)$. The space $L(E, E)$ is usually denoted by $L(E)$, the space $L(E, K)$ (where the norm on K equals the valuation) by E'.

PROPOSITION 13.5. *Let E, F be normed spaces over K. A K-linear map $A : E \to F$ is continuous if and only if there is $M \geqslant 0$ such that $\|Ax\| \leqslant M \|x\|$ for all $x \in E$. Let*

$$\|A\| := \inf\{M : \|Ax\| \leqslant M\|x\| \text{ for all } x \in E\} \quad (A \in L(E,F))$$

Then $(L(E,F), \| \ \|)$ *is a normed space over* K. *If* F *is a* K*-Banach space then so is* $L(E,F)$. *In particular,* E' *is a* K*-Banach space.*

Proof. Left to the reader. (For the last part the proof of Proposition 13.2 might be useful.)

DEFINITION 13.6. A *K-algebra* is a K-vector space E together with a multiplication $E \times E \to E$ making E into a ring, such that for all $\lambda \in K$, $x, y \in E$ we have $\lambda(xy) = (\lambda x)y = x(\lambda y)$. If E is also a normed vector space over K it is a *normed K-algebra* if for all $x, y \in E$ we have $\|xy\| \leqslant \|x\| \, \|y\|$. A *K-Banach algebra* (*Banach algebra over K*) is a complete normed K-algebra.

The spaces $B(X \to K)$ and $L(E)$ above are Banach algebras under the multiplication defined by

$$(fg)(x) = f(x)g(x) \qquad (f, g \in B(X \to K); x \in X)$$
$$(AB)(x) = A(Bx) \qquad (x \in E; A, B \in L(E))$$

respectively.

This completes our list of basic notions. In fact we shall need one more concept (that of an orthogonal base) which logically belongs here but will be presented at the moment when we shall apply it, in Section 50. For a good background account on non-archimedean functional analysis, see van Rooij (1978).

**Exercise* 13.D. Let E be a finite dimensional normed space over K. Show that each K-linear subspace of E is closed and that every K-linear map $E \to E$ is continuous.

**Exercise* 13.E. Let q be a seminorm on a K-vector space E. Show that $D := \{x \in E : q(x) = 0\}$ is a K-linear subspace and that the formula $\|x + D\| := q(x)$ $(x \in E)$ defines a norm on the quotient space E/D.

14. Extensions of valuations

In Sections 14 and 15 we shall prove the following two fundamental theorems.

THEOREM 14.1. (Krull's existence theorem) *Let* K *be a subfield of a field* L, *let* $| \ |$ *be a non-archimedean valuation on* K. *Then there exists a (non-archimedean) valuation on* L *that extends* $| \ |$.

THEOREM 14.2. (Uniqueness theorem) *Let* $K, L, |\ |$ *be as in Theorem* 14.1. *If* K *is complete with respect to* $|\ |$ *and if* L *is algebraic over* K *then there is a unique valuation on* L *that extends* $|\ |$.

Before presenting the formal proofs (of Theorem 14.1 here, of Theorem 14.2 in the next section) we wish to make several comments.

(1) In Sections 16 and 17 we shall use these theorems to construct the p-adic analogue of the field of complex numbers. (See the introduction to Section 11.)

(2) Implicitely, Krull's theorem furnishes many examples of non-archimedean valued fields. In fact, let L be a field of zero characteristic. It then contains $\mathbb{Q}$ and, by Krull's theorem, we can extend any p-adic valuation on $\mathbb{Q}$ to a valuation on L. Thus, for example, on the field of the complex numbers there exist infinitely many inequivalent non-archimedean valuations! In this context we also refer to the remark following Corollary 17.2. Next, suppose that L is a field with characteristic p. Then L contains the field $\mathbb{F}_p$ of p elements. If L is algebraic over $\mathbb{F}_p$ then each element of L lies in a finite subfield of L so that by Exercise 8.B (ii) the only possible valuation on L is the trivial one. But if L is not algebraic over $\mathbb{F}_p$ it contains a field isomorphic to $\mathbb{F}_p(X)$. We know from Remark 3 following Example 2.1 that $\mathbb{F}_p(X)$ admits a (non-trivial) valuation. Again, Krull's theorem shows the existence of a (non-trivial) valuation on L. We have obtained the following corollary of Krull's theorem.

COROLLARY 14.3. *Let* L *be a field that is not an algebraic extension of a finite field. Then there exists a (non-trivial) non-archimedean valuation on* L.

(3) In connection with Theorem 14.2 it should be mentioned that if K is not complete or if L is not algebraic over K the extension of the valuation is, in general, not unique.

(4) It may be interesting to point out that the answer to the extension problem for archimedean valuations is radically different. Indeed, we have the following well-known theorem of Gel'fand and Mazur (usually formulated for Banach algebras).

THEOREM. *Let* $\mathbb{C}$ *be a subfield of some field* L. *If* $\mathbb{C} \neq L$ *the absolute value function on* $\mathbb{C}$ *cannot be extended to a valuation on* L.

For a proof see any book on Banach algebras (e.g. C.E. Rickart, *General Theory of Banach Algebras*, Van Nostrand, Princeton, 1960). Actually, a refinement of the techniques used to prove Gel'fand-Mazur's theorem yields the following.

THEOREM. *Let L be a complete archimedean valued field. Then as a topological field L is isomorphic to either $\mathbb{R}$ or $\mathbb{C}$.*

(5) Several proofs of Krull's theorem are known, all being relatively hard. The proof we shall present here is inspired by our analytic approach and differs from the usual ones. The latter make use of algebraic theory of field extensions, Galois groups, etc. (see, for example Koblitz (1977); in Section 15 we shall illustrate the algebraic approach in the form of exercises). Although our proof is less constructive than the 'algebraic' ones, it is straightforward and it uses, at least to an analyst, quite natural tools. The heart of the proof can be summarized by the phrase 'smoothing an arbitrary norm on L'.

After all this talk we shall start the

Proof of Krull's theorem. A simple application of Zorn's lemma shows that it suffices to consider the case $L = K(z)$. In other words, we may assume that there is an element $z \in L$ such that the smallest field containing $\{z\}$ and K is equal to L.

First suppose that z is not algebraic over K. Then L is isomorphic to the field $K(X)$ of rational functions over K. We extend $|\ |$ to a valuation $\|\ \|$ on L as follows. For $f = a_0 + a_1 X + \ldots + a_n X^n \in K[X]$ set

$$\|f\| := \max\ \{|a_j| : 0 \leqslant j \leqslant n\}$$

Then $\|\ \|$ extends $|\ |$. Let $f, g \in K[X]$. It is easily checked that

(i) $\|f\| \geqslant 0$, $\|f\| = 0$ if and only if $f = 0$

(ii) $\|f + g\| \leqslant \max(\|f\|, \|g\|)$

We now prove the product law

(iii) $\|fg\| = \|f\|\ \|g\|$

That $\|fg\| \leqslant \|f\|\ \|g\|$ follows from a simple estimation of the coefficients of fg. To prove the opposite inequality, let $f = \sum_{j=0}^{n} a_j X^j$, $g = \sum_{j=0}^{m} b_j X^j$ be nonzero polynomials and let $s = \min\ \{j : |a_j| = \|f\|\}$, $t = \min\ \{j : |b_j| = \|g\|\}$. The coefficient c_{s+t} of X^{s+t} of fg is a finite sum of elements of the type $a_k b_l$ where $k + l = s + t$. If not $k = s$ and $l = t$ then either $k < s$ or $l < t$ so that $|a_k b_l| < \|f\|\ \|g\|$. It follows that $c_{s+t} = a_s b_t + r$ where $|r| < \|f\|\ \|g\|$. By a consequence of the strong triangle inequality (Exercise 8.C (i)) we have $|c_{s+t}| = |a_s b_t| = \|f\|\ \|g\|$. Hence $\|fg\| \geqslant \|f\|\ \|g\|$ and the product law is proved. It is not hard to prove that the formula

$$\|fg^{-1}\| := \|f\|\ \|g\|^{-1} \qquad (f, g \in K[X], g \neq 0)$$

defines a valuation on L.

In the rest of this proof we consider the remaining case $L = K(z)$ where z is algebraic over K. Then the dimension of L as a vector space over K

is finite; let $e_1, e_2, \ldots, e_n$ be a base of L. We write an element $x \in L$ as a K-linear combination $\xi_1 e_1 + \xi_2 e_2 + \ldots + \xi_n e_n$ and define

$$\|x\|_1 := \max \ \{|\xi_1|, |\xi_2|, \ldots, |\xi_n|\}$$

For $x, y \in L$ and $\lambda \in K$ we have

(i) $\|x\|_1 \geqslant 0$, $\|x\|_1 = 0$ if and only if $x = 0$
(ii) $\|\lambda x\|_1 = |\lambda| \ \|x\|_1$
(iii) $\|x + y\|_1 \leqslant \max(\|x\|_1, \|y\|_1)$

Thus, $\| \ \|_1$ has the properties of a norm, although we have to keep in mind that K may not be complete. Also, $\| \ \|_1$ already has two of the three properties required for a valuation, but it is uncertain whether the product law holds for $\| \ \|_1$. So let us find out what can be said about $\|xy\|_1$, where $x, y \in L$. If $x = \sum_{i=1}^{n} \xi_i e_i$, $y = \sum_{j=1}^{n} \eta_j e_j$ where the coefficients ξ_i and η_j are in K then

$$\|xy\|_1 = \left\| \sum_{i,j} \ \xi_i \eta_j e_i e_j \right\|_1 \leqslant \max_{i,j} |\xi_i| \ |\eta_j| \ \|e_i e_j\|_1 \leqslant C \ \|x\|_1 \ \|y\|_1$$

where $C := \max_{i,j} \|e_i e_j\|_1$. We define $\| \ \|_2 : L \to \mathbb{R}$ by the formula

$$\|x\|_2 := C \ \|x\|_1$$

Then (i), (ii), (iii) remain valid for $\| \ \|_2$ in place of $\| \ \|_1$ and we have in addition

$$\text{(iv) } \|xy\|_2 \leqslant \|x\|_2 \ \|y\|_2 \qquad (x, y \in L)$$

Still, $\| \ \|_2$ may not be a valuation on L. We introduce one more modification. Define $\nu : L \to \mathbb{R}$ by

$$\nu(x) := \overline{\lim_{n \to \infty}} \ \sqrt[n]{\|x^n\|_2} \qquad (x \in L)$$

(whose archimedean pendant, known as the 'spectral norm', plays a central role in the theory of complex Banach algebras). Since, by (iv), $\|x^n\|_2 \leqslant \|x\|_2^n$ the definition of ν makes sense. We claim that ν has the following seven properties. For all $x, y \in L$ and $\lambda \in K$

(1) $\nu(x) = \lim_{n \to \infty} \sqrt[n]{\|x^n\|_2} = \inf_{n \in \mathbb{N}} \sqrt[n]{\|x^n\|_2}$
(2) $0 \leqslant \nu(x) \leqslant \|x\|_2$
(3) $\nu(\lambda x) = |\lambda| \, \nu(x)$
(4) $\nu(xy) \leqslant \nu(x) \, \nu(y)$
(5) $\nu(1) = 1$
(6) $\nu(x^k) = \nu(x)^k$ for $k = 1, 2, \ldots$
(7) $\nu(1 + x) \leqslant \max(1, \nu(x))$

(Yes, it is true that also $\nu(x + y) \leqslant \max(\nu(x), \nu(y))$ for $x, y \in L$ but we do not need it here.) Property (6), often referred to as 'power multiplicativity',

indicates that we get closer with our search for a valuation on L. We prove the properties (1) and (7), leaving the easy proofs of (2)-(6) to the reader.
Proof of (1). Let $a := \inf_{n \in \mathbb{N}} \sqrt[n]{\|x^n\|_2}$. Let $\epsilon > 0$ and choose $n \in \mathbb{N}$ such that $\|x^n\|_2 < (a+\epsilon)^n$. Let $m \in \mathbb{N}$. Then $m = qn + r$ where $q, r \in \{0, 1, 2, \ldots\}$ and $0 \leqslant r < n$. We have

$$\|x^m\|_2 \leqslant \|x^n\|_2^q \, \|x\|_2^r \leqslant (a+\epsilon)^{nq} \, \|x\|_2^r = (a+\epsilon)^m \, ((a+\epsilon)^{-1} \, \|x\|_2)^r$$

from which it follows that $\overline{\lim}_{m \to \infty} \sqrt[m]{\|x^m\|_2} \leqslant a + \epsilon$, which proves (1).
Proof of (7). We have

$$\|(1+x)^n\|_2 = \left\| \sum_{k=0}^{n} \binom{n}{k} x^k \right\|_2 \leqslant \max_{0 \leqslant k \leqslant n} \|x^k\|_2$$

If $k = 0$ then $\|x^k\|_2 = \|1\|_2$. If $1 \leqslant k \leqslant \sqrt{n}$ then $\|x^k\|_2 \leqslant \|x\|_2^k \leqslant 1$ or $\|x\|_2^{\sqrt{n}}$ depending on whether $\|x\|_2$ is $\leqslant 1$ or > 1. In any case we have $\|x^k\|_2 \leqslant \max(1, \|x\|_2^{\sqrt{n}})$. Finally, if $\sqrt{n} < k \leqslant n$ and $\|x^k\|_2 \geqslant 1$ we have $\|x^k\|_2 \leqslant \sqrt[k]{\|x^k\|_2^n} \leqslant \sup_{s > \sqrt{n}} \sqrt[s]{\|x^s\|_2^n}$. Taking cases together we obtain

$$\|(1+x)^n\|_2 \leqslant \max\left(\|1\|_2, 1, \|x\|_2^{\sqrt{n}}, \sup_{s > \sqrt{n}} \sqrt[s]{\|x^s\|_2^n}\right)$$

from which it follows easily that $\nu(1+x) \leqslant \max(1, \nu(x))$.

Now we come to the point. Let S be the set of all functions $\nu : L \to \mathbb{R}$ having the properties (2)-(7) of above. All the trouble we went through so far results in the simple statement

$$S \text{ is not empty}$$

For $\nu_1, \nu_2 \in S$ we write $\nu_1 \leqslant \nu_2$ if $\nu_1(x) \leqslant \nu_2(x)$ for all $x \in L$. Then $\leqslant$ is a partial ordering on S. For a linearly ordered subset T of S it is easy to see by inspection that $x \mapsto \inf\{\nu(x) : \nu \in T\}$ is again an element of S. So we may apply Zorn's lemma and conclude that

$$S \text{ has minimal elements}$$

Now let τ be such a minimal element. We prove that τ is a valuation on L that extends $|\ |$. First observe that we can use the properties (3), (4) and (5) to show that τ is an extension of $|\ |$ and that if $x \in L$, $x \neq 0$ then $\tau(x) > 0$ $(1 = \tau(1) = \tau(xx^{-1}) \leqslant \tau(x)\, \tau(x^{-1}))$. These follow from the fact that τ is an element of S and have nothing to do with the minimality of τ. To prove the product law for τ, let $a \in L$, $a \neq 0$. From

$$\tau(x) \geqslant \tau(ax)\, \tau(a)^{-1} \geqslant \tau(a^2 x)\, \tau(a)^{-2} \geqslant \ldots \qquad (x \in L)$$

we infer that $\rho(x) := \lim_{n \to \infty} \tau(a^n x)\, \tau(a)^{-n}$ exists for all $x \in L$ and that $\rho \leqslant \tau$. Obviously ρ satisfies the properties (2)-(7) so that $\rho \in S$. By the minimality of τ we must have $\rho = \tau$. In other words, $\tau(x) = \tau(ax)\,\tau(a)^{-1}$ for all $x \in L$. Since a was arbitrary we arrive at

$$\tau(xy) = \tau(x)\,\tau(y) \qquad (x, y \in L)$$

The proof of the strong triangle inequality is now easy. Let $x, y \in L, x \neq 0$. Then by property (7) and the product law $\tau(x+y) = \tau(x(1+x^{-1}y)) = \tau(x)\ \tau(1+x^{-1}y) \leqslant \tau(x)\ \max(1,\ \tau(x^{-1}y)) = \max(\tau(x),\ \tau(x)\ \tau(x^{-1}y)) = \max(\tau(x), \tau(y))$.

15. Uniqueness of the extended valuation

Now we shall prove Theorem 14.2, as promised. In the following theorem we shall even be able to be more quantitative about the extended valuation.

THEOREM 15.1. *Let K be complete and let L be an algebraic field extension of K. Then there is a unique valuation $|\ |$ on L that extends the valuation on K. In fact, if $\|\ \|$ is an arbitrary norm on the K-vector space L then*

$$|x| = \lim_{n\to\infty} \sqrt[n]{\|x^n\|} \qquad (x \in L).$$

Proof. We may assume that the dimension of L, as a vector space over K, is finite. To obtain the first part of the theorem it suffices, by Krull's theorem, to prove that if $|\ |'$ and $|\ |''$ are valuations on L that extend the valuation on K then $|\ |' = |\ |''$. By Theorem 13.3 the norms $|\ |'$ and $|\ |''$ are equivalent so there are positive constants N, M such that

$$N\,|y|'' \leqslant |y|' \leqslant M\,|y|'' \qquad (y \in L)$$

After substituting x^n ($x \in L, n \in \mathbb{N}$) for y and taking nth roots we get

$$\sqrt[n]{N}\,|x|'' \leqslant |x|' \leqslant \sqrt[n]{M}\,|x|'' \qquad (x \in L, n \in \mathbb{N})$$

which, after taking limits, becomes

$$|x|' = |x|'' \qquad (x \in L).$$

To prove the second part, let $\|\ \|$ be an arbitrary norm on L. The following statements (i), (ii), (iii) follow easily from the proof of Krull's Theorem (Theorem 14.1).

(i) There is a positive constant C such that $\|xy\| \leqslant C\,\|x\|\ \|y\|$ for all $x, y \in L$.

(ii) Let $\|x\|_2 := C\,\|x\|$ ($x \in L$). Then the function $\nu : L \to \mathbb{R}$ defined by $\nu(x) := \lim_{n\to\infty} \sqrt[n]{\|x^n\|_2} = \lim_{n\to\infty} \sqrt[n]{\|x^n\|}$ exists and is an element of S.

(iii) Let τ be a minimal element of S. Then $\tau \leqslant \nu$, and τ is a valuation on L that extends the valuation on K.

By the first part of the proof there is a unique valuation $| \ |$ on L that extends the valuation on K. Hence $\tau = | \ |$. For all $x \in L$ we have

$$|x| = \tau(x) \leqslant \nu(x) \leqslant \|x\|_2$$

Now $| \ |$ and $\| \ \|_2$ are norms on L, so they are equivalent. There is a positive constant C_1 such that

$$|x| \leqslant \nu(x) \leqslant C_1 \ |x| \qquad (x \in L)$$

For all $n \in \mathbb{N}$ we have $|x^n| \leqslant \nu(x^n) \leqslant C_1 \ |x^n|$. Using the power multiplicativity of ν and taking nth roots we obtain

$$|x| \leqslant \nu(x) \leqslant \lim_{n \to \infty} \sqrt[n]{C_1} \ |x| = |x|$$

It follows that $\nu = | \ |$ (in particular, ν is a norm) and the theorem is proved.

Remark. For an explicit computation of the extended valuation the formula $|x| = \lim_{n \to \infty} \sqrt[n]{\|x^n\|}$ may not be suitable at all times. Alternative formulas for $| \ |$ will be considered in Exercises 15.E, 15.F and 15.G.

Exercise 15.A. Let $| \ |$ be the absolute value function on $\mathbb{C}$. For a complex number $z = x + iy$ $(x, y \in \mathbb{R})$ define $\|z\| := |x| + |y|$. Do we have $|z| = \lim_{n \to \infty} \sqrt[n]{\|z^n\|}$?

Exercise 15.B. Let p be a prime number. Show that $X^2 - p$ is irreducible in $\mathbb{Q}_p[X]$. Then $\mathbb{Q}_p(\sqrt{p}) := \mathbb{Q}_p[X]/(X^2 - p)$ is a field. Let $\pi : \mathbb{Q}_p[X] \to \mathbb{Q}_p(\sqrt{p})$ be the canonical map. Then $\mathbb{Q}_p(\sqrt{p}) = \{a + b\ \pi(X) : a, b \in \mathbb{Q}_p\}$. The map $a \mapsto a + 0 \cdot \pi(X)$ sends $\mathbb{Q}_p$ into a subfield of $\mathbb{Q}_p(\sqrt{p})$. According to Krull's theorem there is a unique valuation $| \ |$ on $\mathbb{Q}_p(\sqrt{p})$ that extends the p-adic valuation $| \ |_p$. Show that

$$|a + b\ \pi(X)| = \max(|a|_p, p^{-\frac{1}{2}} \ |b|_p) \qquad (a, b \in \mathbb{Q}_p)$$

Determine the residue class field and the value group of $\mathbb{Q}_p(\sqrt{p})$.

Exercise 15.C. Show that $X^2 + 1$ is irreducible in $\mathbb{Q}_3[X]$. As in the previous exercise, let $\pi : \mathbb{Q}_3[X] \to \mathbb{Q}_3[X]/(X^2 + 1) =: \mathbb{Q}_3(\sqrt{-1})$ be the canonical map. Then $\mathbb{Q}_3(\sqrt{-1}) = \{a + bi : a, b \in \mathbb{Q}_3\}$, where $i := \pi(X)$. Let $| \ |$ be the valuation on $\mathbb{Q}_3(\sqrt{-1})$ that extends the 3-adic valuation. Find a formula for $|a + bi|$ $(a, b \in \mathbb{Q}_p)$ in terms of $|a|_3$ and $|b|_3$. (Hint. Show that $|a + bi| = |a - bi|$ and that $|b|_3 = 1$ implies $|1 + b^2|_3 = 1$.) Determine the residue class field and the value group of $\mathbb{Q}_3(\sqrt{-1})$.

Exercise 15.D. Carry out the procedure of the previous exercise, but where 3 is replaced by a prime p that is congruent to 3 modulo 4. (Hint for the proof of the irreducibility of $X^2 + 1$. If $X^2 + 1$ were reducible then $X^2 + 1$

$= 0$ would have solutions in $\mathbb{F}_p$. For such a solution $a \in \mathbb{F}_p$ we have $a^4 = 1$ $(a, a^2, a^3 \neq 1)$. But also $a^{p-1} = 1$. Deduce that $p \equiv 1 \pmod 4$.) Finally, discuss the case $p = 2$.

In the next three exercises K is a complete non-archimedean valued field, L is a finite algebraic extension of K and $|\ |$ is the unique valuation on L extending the valuation on K.

Exercise 15.E. Let $a \in L$. The multiplication map $M_a : L \to L$ defined by

$$M_a(x) = ax \qquad (x \in L)$$

is a K-linear map $L \to L$, whose determinant is denoted $\det M_a$. Let n be the dimension of L over K. Show that

$$|a| = \sqrt[n]{|\det M_a|}$$

using the following steps.

(i) Denote the K-vector space consisting of all K-linear maps $L \to L$ by End (L). Then each $A \in \text{End}(L)$ is continuous (Exercise 13.D) and

$$|A| := \sup\left\{\frac{|Ax|}{|x|} : x \in L, x \neq 0\right\} \qquad (A \in \text{End}(L))$$

defines a norm on End(L). Let $e_1, \ldots, e_n$ be a base of L over K and let (a_{ij}) be the matrix of $A \in \text{End}(L)$ with respect to this base. Then

$$\|A\| := \max\ \{|a_{ij}| : 1 \leq i, j \leq n\} \qquad (A \in \text{End}(L))$$

defines another norm on End(L). Now show successively the following.

(ii) $|\det A| \leq \|A\|^n$ $(A \in \text{End}(L))$.

(iii) $\|\ \|$ and $|\ |$ are equivalent norms.

(iv) There is a positive constant c such that $|\det A| \leq c|A|^n$ $(A \in \text{End}(L))$.

(v) $|\det M_a| \leq |M_a|^n = |a|^n$ $(a \in L)$.

(vi) $|\det M_a| = |a|^n$ $(a \in L)$.

**Exercise* 15.F. Use the previous exercise to show the following. For $a \in L$ let

$$X^m + a_{m-1}X^{m-1} + \ldots + a_0$$

be the minimum polynomial in $K[X]$ of a. Then

$$|a| = \sqrt[m]{|a_0|}$$

(Hint. Choose $1, a, \ldots, a^{m-1}$ as a base of $K(a)$ over K.)

Exercise 15.G. A *K-automorphism of L* is an automorphism σ of L such that $\sigma(x) = x$ for all $x \in K$. Show that a K-automorphism σ is K-linear and that $|\sigma(x)| = |x|$ $(x \in L)$. By considering minimum polynomials of elements of L show that there are only finitely many K-automorphisms of L, say σ_1, $\ldots, \sigma_m$. Then we have obviously

$$(*) \qquad |x| = \sqrt[m]{\left|\prod_{j=1}^{m} \sigma_j(x)\right|} \qquad (x \in L)$$

Now suppose that if $x \in L$, $\sigma_j(x) = x$ for all j then $x \in K$. Conclude that

$$\prod_{j=1}^{m} \sigma_j(x) \in K$$

so that in this case formula (*) might be used to compute the extension of the valuation.

Exercise 15.H. Now that you have found three new formulas for the extended valuation (Exercises 15.E, 15.F, 15.G), reconsider Exercises 15.B, 15.C and 15.D.

16. The valuation on the algebraic closure

In this section we consider some properties of the valuation on the algebraic closure (see Appendix B) of a complete (non-archimedean valued) field K. For a field L we denote its algebraic closure by L^a. We have the following direct consequence of Theorem 15.1.

THEOREM 16.1. *Let K be complete. Then the valuation on K can uniquely be extended to a valuation on K^a.*

We shall compare the value groups and residue class fields of K and K^a. Let k be the residue class field of K and let k^a be the residue class field of K^a. The ambiguity of the notation k^a is removed by the following theorem.

THEOREM 16.2. *k^a is the algebraic closure of k. For the respective value groups $|(K^a)^\times|$, $|K^\times|$ we have $|(K^a)^\times| = \{r \in (0, \infty) : r^n \in |K^\times|$ for some $n \in \mathbb{N}\}$.*

Proof. To show that k^a is algebraic over k, let $x \in k^a$. There is $y \in K^a$, $|y| \leqslant 1$ such that $\bar{y} = x$ (Definition 11.2). So there are $a_0, a_1, \ldots, a_n \in K$ such that $a_0 + a_1 y + \ldots + a_n y^n = 0$ and not all a_j vanish. By multiplying with a suitable constant we can arrange that $\max |a_i| = 1$. Then $\bar{a}_0 + \bar{a}_1 X + \ldots + \bar{a}_n X^n \in k[X]$ is nonzero and has x as a root.

To show that k^a is algebraically closed, let $f = a_0 + a_1 X + \ldots + a_{n-1} X^{n-1} + X^n$ be a monic polynomial in $k^a[X]$; we prove that f has a root in k^a. There are $a_0, \ldots, a_n \in K^a$ such that $\bar{a}_j = a_j$ ($j = 0, \ldots, n$). The polynomial $F := a_0 + a_1 X + \ldots + a_n X^n \in K^a[X]$ has a root θ in K^a. If $|\theta|$ were > 1 then, since $|a_n| = 1$, we would have $|a_0 + a_1 \theta + \ldots + a_n \theta^n| =$

$\max_j |a_j \theta^j| = |\theta|^n \neq 0$. Hence $|\theta| \leq 1$ and therefore $\bar{a}_0 + \bar{a}_1 \bar{\theta} + \ldots + \bar{a}_n \bar{\theta}^n = \bar{0}$, i.e. $\bar{\theta}$ is a root of f.

Let $x \in K^a$, $x \neq 0$. Thanks to Exercise 15.E or 15.F, $|x|$ is an nth root of an element of $|K^\times|$, for some n. Conversely, let $n \in \mathbb{N}$, $r \in |K^\times|$ and take $a \in K$ for which $|a| = r$. There is $x \in K^a$ for which $x^n = a$. Then $|x| = \sqrt[n]{|a|} = \sqrt[n]{r}$.

COROLLARY 16.3. *The residue class field of K^a is infinite. The valuation on K^a is dense.*

COROLLARY 16.4. *The residue class field of $\mathbb{Q}_p^a$ is the algebraic closure of the field of p elements. The value group of $\mathbb{Q}_p^a$ equals $\{p^r : r \in \mathbb{Q}\}$.*

The proof of the following theorem illustrates that we can solve certain analytic problems in K by looking at K^a.

THEOREM 16.5 *Let K be complete and let $f : K \to K$ be a nonconstant polynomial function (with coefficients in K).*
(i) *If $\lambda_1, \lambda_2, \ldots$ is a sequence in K for which $\lim_{j\to\infty} |f(\lambda_j)| = 0$ then $\lambda_1, \lambda_2, \ldots$ has a subsequence that converges to a root of f.*
(ii) *If $X \subset K$ is closed then its image $f(X)$ is closed.*
(iii) *If $X \subset K$ is compact then its inverse image $f^{-1}(X)$ is compact.*
Proof. (i) There are $\xi_1, \ldots, \xi_n \in K^a$ such that $f(x) = a(x-\xi_1)(x-\xi_2)\ldots(x-\xi_n)$ $(x \in K)$ for some $a \neq 0$. It follows easily from $\lim_{j\to\infty} |f(\lambda_j)| = 0$ that there is a subsequence of $\lambda_1, \lambda_2, \ldots$ that converges to some ξ_j. As K is complete, $\xi_j \in K$.
(ii) Let $a_1, a_2, \ldots$ be in $f(X)$ and $a := \lim_{n\to\infty} a_n$. Then apply (i) to $f - a$.
(iii) Let $a_1, a_2, \ldots$ be a sequence in K for which $f(a_j) \in X$ $(j \in \mathbb{N})$. By compactness we may assume that $\lim_{j\to\infty} f(a_j)$ exists. Application of (i) to $f - \lim_{j\to\infty} f(a_j)$ yields a convergent subsequence of $a_1, a_2, \ldots$ It follows that $f^{-1}(X)$ is compact.

We return to the subject matter of this section. It is true that for a complete K we have found an algebraically closed valued field $K^a \supset K$ but is it (metrically) complete? We show that if $K = \mathbb{Q}_p$ this is not the case. Our proof is based on category arguments. For a more direct proof, see Koblitz (1977).

THEOREM 16.6. *The algebraic closure $\mathbb{Q}_p^a$ of $\mathbb{Q}_p$ is not metrically complete.*
Proof. For $n \in \mathbb{N}$ let

$$H_n := \{x \in \mathbb{Q}_p^a : \dim \mathbb{Q}_p(x) \leq n\}$$

where $\mathbb{Q}_p(x)$ denotes the smallest subfield of $\mathbb{Q}_p^a$ containing $\mathbb{Q}_p$ and $\{x\}$ and where $\dim \mathbb{Q}_p(x)$ is the dimension of $\mathbb{Q}_p(x)$ as a vector space over $\mathbb{Q}_p$.

Clearly we have $H_1 \subset H_2 \subset \ldots$ and $\bigcup_n H_n = \mathbb{Q}_p^a$. The proof runs in several steps.

(i) H_n is closed for $n \in \mathbb{N}$. In fact, let $x_1, x_2, \ldots \in H_n$, $\lim_{i\to\infty} x_i = x$. For each i there are $a_0^i, a_1^i, \ldots, a_{n-1}^i \in \mathbb{Z}_p$ such that $\max_j |a_j^i|_p = 1$ and

$$a_0^i + a_1^i x_i + \ldots + a_{n-1}^i x_i^{n-1} = 0$$

Since $\mathbb{Z}_p$ is compact we may assume (by taking a suitable subsequence) that for $j = 0, 1, \ldots, n-1$

$$a_j := \lim_{i\to\infty} a_j^i$$

exists. Then $\max\{|a_j|_p : 0 \leqslant j \leqslant n-1\} = 1$, so that $a_0, a_1, \ldots, a_{n-1}$ do not all vanish. Further we have

$$a_0 + a_1 x + \ldots + a_{n-1} x^{n-1} = 0$$

It follows that $x \in H_n$.

(ii) $H_n \neq \mathbb{Q}_p^a$ for each $n \in \mathbb{N}$. We show that for each $m \in \mathbb{N}$ there is $x \in \mathbb{Q}_p^a$ for which $\mathbb{Q}_p(x)$ has dimension m over $\mathbb{Q}_p$. In fact, choose x such that $x^m = p$. For a $\mathbb{Q}_p$-linear combination $a_0 + a_1 x + \ldots + a_{m-1} x^{m-1}$ of $1, x, \ldots, x^{m-1}$ we have (using the fact that $|x| = \sqrt[m]{p^{-1}}$)

$$\text{if } a_i \neq 0, a_j \neq 0, i \neq j \text{ then } |a_i x^i| \neq |a_j x^j|$$

It follows that $|a_0 + a_1 x + \ldots + a_{m-1} x^{m-1}| = \max\{|a_j|_p |x^j| : 0 \leqslant j \leqslant m-1\}$ (see Exercise 8.C(i)). In particular, the elements $1, x, \ldots, x^{m-1}$ are linearly independent over $\mathbb{Q}_p$.

(iii) $H_n + H_m \subset H_{nm}$ $(n, m \in \mathbb{N})$. Let $x \in H_n, y \in H_m$. Let $\tilde{m}$ be the dimension of $\mathbb{Q}_p(x, y)$ over $\mathbb{Q}_p(x)$. Then by elementary algebra we have $\tilde{m} \leqslant \dim \mathbb{Q}_p(y)$. Thus $\dim \mathbb{Q}_p(x+y) \leqslant \dim \mathbb{Q}_p(x, y) = \tilde{m} \dim \mathbb{Q}_p(x) \leqslant mn$, i.e. $x + y \in H_{mn}$.

(iv) $\mathbb{Q}_p^a$ is not complete. By the Baire category theorem (see Appendix A.1), if $\mathbb{Q}_p^a$ were complete then some H_n would contain a disc of the form

$$\{x \in \mathbb{Q}_p^a : |x - b| \leqslant \epsilon\}$$

for some $\epsilon > 0$, some $b \in \mathbb{Q}_p^a$. If $s \in \mathbb{Q}_p^a$, $|s| \leqslant \epsilon$, we have $s + b \in H_n$ so that $s \in H_n + H_n \subset H_{n^2}$. This implies that $\mathbb{Q}_p^a \subset H_{n^2}$. But this is in conflict to what we proved in (ii).

COROLLARY 16.7. *$\mathbb{Q}_p^a$ is an infinite dimensional vector space over $\mathbb{Q}_p$. $\mathbb{Q}_p$ and $\mathbb{R}$ are not isomorphic as fields.*

Proof. The first part follows from Theorem 16.6 and Theorem 13.3. For the second part observe that $\mathbb{R}^a$ is finite dimensional over $\mathbb{R}$.

In the following section we shall do something about the incompleteness of $\mathbb{Q}_p^a$.

17. Completion of the algebraic closure. $\mathbb{C}_p$

The algebraic closure of K may not be complete, its completion is. But is the latter algebraically closed? Fortunately, the answer is 'yes'.

THEOREM 17.1. *The completion of the algebraic closure of K is itself algebraically closed.*

Proof. We shall prove the theorem in the following form. Let K be complete and let L be a dense algebraically closed subfield of K. Then each monic polynomial

$f = a_0 + a_1 X + \ldots + a_{n-1} X^{n-1} + X^n \qquad (a_0, \ldots a_{n-1} \in K, n \in \mathbb{N})$

has a root in K. In fact, for each $j \in \{0, 1, \ldots, n\}$, choose $a_{ij} \in L$ such that $|a_{ij} - a_j| < 1/i$, $|a_{ij}| = |a_j|$ $(i = 1, 2, \ldots)$. Then

$$f_i := a_{i0} + a_{i1} X + \ldots + a_{in} X^n \in L[X]$$

has a root, λ_i, in L. We have

$|\lambda_i^n| = |-(a_{i0} + a_{i1}\lambda_i + \ldots a_{i,n-1}\lambda_i^{n-1})| \leqslant \max\{|a_j|\,|\lambda_i|^j : 0 \leqslant j \leqslant n-1\}$

so that

$$|\lambda_i| \leqslant \max(\sqrt[n]{|a_0|}, \sqrt[n-1]{|a_1|}, \ldots, |a_{n-1}|) =: c$$

From $|f(\lambda_i)| = |f(\lambda_i) - f_i(\lambda_i)| = \left|\sum_{j=0}^{n} (a_j - a_{ij})\lambda_i^j\right| \leqslant (1/i)\max(1, |\lambda_i|, |\lambda_i|^2, \ldots, |\lambda_i|^n) \leqslant (1/i)\max(1, c^n)$ it follows that $\lim_{i\to\infty} |f(\lambda_i)| = 0$. Now Theorem 16.5 tells us that f has a root in K.

We denote the completion of the algebraic closure of $\mathbb{Q}_p$ by $\mathbb{C}_p$ (in the literature one also encounters the symbol Ω_p), its valuation by $|\ |_p$. For reference we collect some facts about $\mathbb{C}_p$.

COROLLARY 17.2. *The completion $\mathbb{C}_p$ of the algebraic closure of $\mathbb{Q}_p$ has the following properties.*

(i) *$\mathbb{C}_p$ is algebraically closed.*

(ii) *$\mathbb{C}_p$ is infinite dimensional as a $\mathbb{Q}_p$-vector space.*

(iii) *$\mathbb{C}_p$ is not locally compact.*

(iv) *$\mathbb{C}_p$ is separable.*

(v) *The residue class field of $\mathbb{C}_p$ is the algebraic closure of the field of p elements.*

(vi) *The value group of $\mathbb{C}_p$ is $\{p^r : r \in \mathbb{Q}\}$.*

Proof. For (iv), see Exercise 17.B. The other statements are direct consequences of the previous theory.

Remark. Although $\mathbb{C}_p$ and $\mathbb{C}$ differ very much as *valued* fields one can show that $\mathbb{C}_p$ and $\mathbb{C}$ are isomorphic as fields. (See van Rooij (1978) p. 83.) This fact may become more striking if we put it as follows. *On the complex number field there exist infinitely many mutually inequivalent non-archimedean valuations each of which makes $\mathbb{C}$ into a complete valued field.*

Exercise 17.A. The valued field $\mathbb{C}$ of the complex numbers is algebraically closed and locally compact. Show that there is no non-archimedean valued field that shares these two properties with $\mathbb{C}$.

**Exercise* 17. B. ($\mathbb{C}_p$ is separable, see also Section 19) Show that the algebraic closure of $\mathbb{Q}$ in $\mathbb{C}_p$ is dense in $\mathbb{C}_p$ and conclude that $\mathbb{C}_p$ is separable. More generally, show that the completion of the algebraic closure of a separable field is again separable.

Exercise 17.C. (On distances between roots of unity in K) Let K be algebraically closed. An element $x \in K$ is a *root of unity* if there is $n \in \mathbb{N}$ such that $x^n = 1$. Show that the roots of unity form a multiplicative subgroup of the unit sphere of K. Let k be the residue class field of K. Prove the following.
(i) If K and k have equal characteristics and if x, y $(x \neq y)$ are roots of unity then $|x - y| = 1$.
(ii) If $\mathbb{Q}_p \subset K$ (see Remark 2 at the end of Section 11) and if $x^n = 1$ $(x \neq 1)$ for some $n \in \mathbb{N}$, not divisible by p, then $|x - 1| = 1$. Compute $|x - 1|$ if $x^p = 1, x \neq 1$. (Hint. Either use a polynomial identity $(1 - X)(1 + X + X^2 + \ldots + X^{m-1}) = 1 - X^m$ or expand $(1 + u)^m$ where $1 + u$ is an mth root of 1.)

PART 2: ULTRAMETRICS

18. Ultrametric spaces

In the rest of this chapter we will have a closer look at the metric and topology of K by studying 'ultrametric spaces'. Recall that a *metric space* is a set X together with a map $d: X \times X \to \mathbb{R}$ such that for all $x, y, z \in X$ (i) $d(x, y) \geqslant 0$, $d(x, y) = 0$ if and only if $x = y$, (ii) $d(x, y) = d(y, x)$, (iii) $d(x, z) \leqslant d(x, y) + d(y, z)$.

DEFINITION 18.1. A metric space (X, d) is an *ultrametric space* if the metric d satisfies the *strong triangle inequality*

$$d(x,z) \leqslant \max(d(x,y), d(y,z)) \qquad (x,y,z \in X)$$

In the sequel ultrametric spaces shall arise naturally in the form of subsets of K (with respect to the metric $(x,y) \mapsto |x-y|$) or subsets of certain normed spaces over K (with respect to $(x,y) \mapsto \|x-y\|$). In Appendix A.10 it is proved that every ultrametric space can isometrically be embedded into a suitable (non-archimedean valued field) K. (Recall that a map σ between two metric spaces X and Y is an *isometry* if for all elements $x,y \in X$ we have $d(\sigma(x), \sigma(y)) = d(x,y)$.) So, throughout the reader may have subsets of K in mind when abstract ultrametric spaces are being considered.

THROUGHOUT THE REST OF THIS CHAPTER, X IS AN ULTRAMETRIC SPACE WITH METRIC d.

PROPOSITION 18.2. (The isosceles triangle principle) *Let $x,y,z \in X$. If $d(x,y) \neq d(y,z)$ then $d(x,z) = \max\{d(x,y), d(y,z)\}$. In other words, the largest and the second largest of the numbers $d(x,y)$, $d(y,z)$, $d(x,z)$ are equal.*

Proof. Left to the reader. (See also Exercise 8.C (i)).

For the following definition compare the notion of a 'disc' in K, introduced in Section 11.

DEFINITION 18.3. Let $a \in X$, $r \in (0,\infty)$. *The 'open' ball of radius r with centre a* is the set

$$B_a(r^-) := \{x \in X : d(a,x) < r\}$$

The *'closed' ball of radius r with centre a* is

$$B_a(r) := \{x \in X : d(a,x) \leqslant r\}$$

A *ball* in X is a set of the form $B_a(r)$ or $B_a(r^-)$ for some $a \in X$, $r \in (0,\infty)$. The *diameter* of a nonempty subset A of X is

$$d(A) := \sup\{d(x,y) : x,y \in A\}$$

A is *bounded* if $d(A) < \infty$.
The *distance* between two nonempty subsets A and B of X is

$$d(A,B) := \inf\{d(x,y) : x \in A, y \in B\}$$

The *distance* between an element $x \in X$ and a nonempty subset A of X is

$$d(x,A) := d(\{x\}, A)$$

Although it is quite natural to call $B_a(r^-)$ ($B_a(r)$) an open (closed) ball of radius r with centre a we have to warn the reader that the terms open, closed, radius, centre, used in these expressions all have a certain ambiguity, as is demonstrated by the following.

PROPOSITION 18.4. *Each ball in X is both open and closed. Each point of a ball may serve as a centre. A ball may have infinitely many radii.*

Proof. Let $a \in X$ and $r \in (0, \infty)$. Owing to the strong triangle inequality the formula $d(x, y) < r$ defines an equivalence relation on X whose equivalence classes are open. Hence $B_a(r^-)$, being the complement of a union of classes, is closed. Of course, $B_a(r^-)$ is open and $B_a(r)$ is closed. To prove that $B_a(r)$ is open let $b \in B_a(r)$; we show that $B_b(r) \subset B_a(r)$. In fact, if $x \in B_b(r)$ then by the strong triangle inequality $d(x, a) \leqslant \max(d(x, b), d(b, a)) \leqslant r$, so $x \in B_a(r)$, i.e. $B_b(r) \subset B_a(r)$. We even have by symmetry ($b \in B_a(r)$ implies $a \in B_b(r)$) that each point of $B_a(r)$ is a centre of $B_a(r)$. Similarly, one proves that each point of $B_a(r^-)$ is a centre of $B_a(r^-)$. Finally, if r is strictly between 1 and p we have $B_0(1) = B_0(r) = B_0(r^-)$ in $\mathbb{Q}_p$ showing that the unit ball of $\mathbb{Q}_p$ has infinitely many radii.

Despite the surrealistic features of Proposition 18.4, in ultrametric analysis one keeps using the expression 'open' ('closed') ball (of radius r with centre a), but with the quotation marks, as in Definition 18.3.

FROM NOW ON WE USE THE WORD CLOPEN AS AN ABBREVIATION FOR CLOSED AND OPEN.

PROPOSITION 18.5. *Let B_1, B_2 be balls in X. Then either B_1 and B_2 are ordered by inclusion* (i.e. *$B_1 \subset B_2$ or $B_2 \subset B_1$*) *or B_1 and B_2 are disjoint. In the latter case we have for all $x \in B_1$, $y \in B_2$*

$$d(x, y) = d(B_1, B_2)$$

Proof. If none of the statements $B_1 \subset B_2$, $B_2 \subset B_1$, $B_1 \cap B_2$ is empty were true we could find elements $a \in B_1 \cap B_2$, $x \in B_1 \backslash B_2$, $y \in B_2 \backslash B_1$. Then a would be a centre of B_1 and B_2 and

$$d(y, a) > d(x, a) \qquad (\text{since } x \in B_1 \text{ and } y \notin B_1)$$
$$d(x, a) > d(y, a) \qquad (\text{since } y \in B_2 \text{ and } x \notin B_2)$$

which is a contradiction. To prove the second part, let $x, x' \in B_1$, $y \in B_2$. Then $d(x, x') < d(x, y)$ and $d(x, x') < d(x', y)$. By the isosceles triangle principle for the 'triangle' $\{x, x', y\}$ we have $d(x, y) = d(x', y)$. By symmetry, $d(x, y) = d(x, y')$ for all $y' \in B_2$. It follows that the function $(x, y) \mapsto d(x, y)$ ($x \in B_1$, $y \in B_2$) is constant and we are done.

For later use we prove the following decomposition theorem.

THEOREM 18.6. *Let U be a nonempty open subset of X. Then there is a partition of U into balls. More specifically, given $r_1 > r_2 > \ldots > 0$, U can be covered by disjoint balls of the form $B_a(r_n)$ ($a \in X$, $n \in \mathbb{N}$).*

Proof. We only have to prove the second statement. For each $a \in U$ we choose

$$B_a := \begin{cases} B_a(r_1) \text{ if } B_a(r_1) \subset U \\ B_a(r_n) \text{ if } B_a(r_n) \subset U, B_a(r_{n-1}) \not\subset U \end{cases}$$

Then $\{B_a : a \in U\}$ is a covering of U with balls of the prescribed form. Suppose $B_a \cap B_b$ is not empty for some $a, b \in U$. By Proposition 18.5 we may suppose $B_a \subset B_b$. But B_a is defined as the largest among the 'closed' balls with centre a and radius belonging to $\{r_1, r_2, \ldots\}$, of which B_b is one. Hence $B_b \subset B_a$, i.e. $B_a = B_b$. It follows that the collection $\{B_a : a \in U\}$ is disjoint.

**Exercise* 18.A. (i) Prove that the boundary of a ball in X is empty. For $a \in X$, $r \in (0, \infty)$ the ball $B_a(r)$ is the closure of $B_a(r^-)$ if and only if $B_a(r) = B_a(r^-)$. Give an example of a ball $B_a(r)$ ($a \in \mathbb{C}_p$, $r \in (0, \infty)$) for which $B_a(r) = B_a(r^-)$.
(ii) Show that the topology of X is zerodimensional. Deduce that X is totally disconnected. (See Appendix B.)

Exercise 18.B. (On the radii of a ball) Let B be a ball in X containing at least two points and $B \neq X$. Prove that the collection of all radii of B is of the form $[s, t]$ $(0 < s \leqslant t < \infty)$. Describe this collection in a similar way for the two remaining cases $B = \{a\}$ $(a \in X)$ and $B = X$ (if X is bounded).

**Exercise* 18.C. (Diameter versus radius) (i) Let Y be a nonempty subset of X with diameter $d(Y)$ (Definition 18.3). Show that for any $a \in Y$

$$d(Y) = \sup \{d(x, a) : x \in Y\}$$

(ii) Show that for any ball B in X

$$d(B) = \inf \{r : r \text{ is a radius of } B\}$$

where 'inf' may be replaced by 'min' if B has at least two points.
(iii) Show that if K has a dense valuation then each ball B in K has precisely one radius and that this radius equals the diameter of B.

**Exercise* 18.D. Let $Y \subset X$. Then Y, with the metric inherited from d, is an ultrametric space. Obviously, if $S \subset Y$ is a ball in the metric space Y then there is a ball B in X such that $B \cap Y = S$. Prove also the following more striking fact. If B is a ball in X and if $B \cap Y \neq \emptyset$ then $B \cap Y$ is a ball in Y.

In the exercises below we use the following notions. A function $f : X \to K$ is *continuous* if for each $a \in X$ and $\epsilon > 0$ there exists $\delta > 0$ such that $x \in X$, $d(x, a) < \delta$ implies $|f(x) - f(a)| < \epsilon$; f is *uniformly continuous* if for each

$\epsilon > 0$ there is $\delta > 0$ such that $x, y \in X$, $d(x, y) < \delta$ implies $|f(x) - f(y)| < \epsilon$. For a set $Y \subset X$ and $\epsilon > 0$, its *ϵ-neighbourhood* is the set $Y^\epsilon := \{x \in X$: there is $y \in Y$ such that $d(x, y) < \epsilon\}$.

**Exercise* 18.E. Show that a subset Y of X is clopen if and only if the (*K-valued*) *characteristic function* ξ_Y *of* Y

$$\xi_Y(x) := \begin{cases} 1 \text{ if } x \in Y \\ 0 \text{ if } x \in X \setminus Y \end{cases}$$

is continuous. Show that for each $Y \subset X$, $\epsilon > 0$ the set Y^ϵ is clopen.

Exercise 18.F. Let $Y \subset X$, $\emptyset \neq Y \neq X$. Show that the conditions (α)-(δ) are equivalent.
(α) ξ_Y is uniformly continuous.
(β) There is $\epsilon > 0$ such that $Y^\epsilon = Y$.
(γ) For some $\epsilon > 0$, Y is the union of a collection of 'open' balls of radius ϵ.
(δ) The distance between Y and its complement is positive.

Exercise 18.G. In this exercise, let us call a set $Y \subset X$ *uniformly open* if $Y^\epsilon = Y$ for some positive $\epsilon > 0$. Prove the following.
(i) The empty set and X are uniformly open. If $U_1, \ldots, U_n$ ($n \in \mathbb{N}$) are uniformly open then so are $\bigcup_j U_j$ and $\bigcap_j U_j$. If U is uniformly open then so is $X \setminus U$.
(ii) Balls are uniformly open. For each $Y \subset X$, $\epsilon > 0$ the set Y^ϵ is uniformly open.
(iii) A uniformly open set is clopen. Not every clopen subset of $\mathbb{Q}_p$ is uniformly open. If X is compact then every clopen set is uniformly open.
(iv) Each closed set is a countable intersection of uniformly open sets, each open set is a countable union of uniformly open sets. A subset U of X is clopen if and only if there are uniformly open sets $V_1 \subset V_2 \subset \ldots$ and $W_1 \supset W_2 \supset \ldots$ such that $U = \bigcup_j V_j = \bigcap_j W_j$.

Exercise 18.H. Let K be a complete (non-archimedean valued) field, and let B be a ball in K with diameter $d(B)$. Prove that $d(B, K \setminus B) = d(B)$ if the valuation is dense and that $d(B, K \setminus B) = |\pi|^{-1} d(B)$ if the valuation is discrete (where $|\pi|$ is the largest value of $|K^\times|$ that is strictly smaller than 1). More generally, show that if $Y \subset X$ is uniformly open in the sense of the preceding exercise and $\emptyset \neq Y \neq X$ then $d(Y, X \setminus Y) = \sup \{\epsilon : Y^\epsilon = Y\}$.

19. Compactness and separability

We collect a few properties of compact and separable ultrametric spaces. For facts on general metric spaces that shall be used here we refer to Appendix B. Recall that X denotes an ultrametric space and that K is a non-archimedean valued field.

DEFINITION 19.1. The metric on X is *discrete* if $x_1, x_2, \ldots \in X, y_1, y_2, \ldots \in X, d(x_1, y_1) > d(x_2, y_2) > \ldots$ implies $\lim_{n \to \infty} d(x_n, y_n) = 0$.

**Exercise* 19.A. (i) Show that the valuation on K is discrete if and only if the induced metric is discrete.
(ii) The metric on X is *dense* if for each ball B in X the set $\{d(x, y) : x, y \in B\}$ is dense in $[0, d(B)]$. Show that the valuation on K is dense if and only if the induced metric is dense.
(iii) Show that, contrary to what is true for valuations, an ultrametric may be neither discrete nor dense.

PROPOSITION 19.2. *Let X be compact. Then the following are true.*
(i) *X is complete and separable.*
(ii) *Every open covering of X has a finite refinement consisting of disjoint balls.*
(iii) *A partition of X by clopen sets is finite.*
(iv) *If $Y \subset X$ is clopen, $\emptyset \neq Y \neq X$, then $d(Y, X \setminus Y)$ is positive.*
(v) *The metric on X is discrete.*
Proof. We prove (v) leaving the proof of (i)-(iv) to the reader. If (v) were not true then by compactness we could find convergent sequences $x_1, x_2, \ldots$ and $y_1, y_2, \ldots$ such that $d(x_1, y_1) > d(x_2, y_2) > \ldots$ and $\epsilon := \lim_{n \to \infty} d(x_n, y_n) > 0$. Let $x := \lim_{n \to \infty} x_n$, $y := \lim_{n \to \infty} y_n$. Then $d(x, y) = \epsilon$ whereas $d(x, x_n) < \epsilon$, $d(y, y_n) < \epsilon$ for large n. By the isosceles triangle principle we have $\epsilon = d(x, y) = d(x_n, y) = d(x_n, y_n)$, a contradiction since the sequence $n \mapsto d(x_n, y_n)$ is strictly decreasing.

THEOREM 19.3. *An ultrametric space is separable if and only if the collection of its balls is countable.*
Proof. If the balls in X form a countable set then one obtains a countable dense subset of X by choosing a point in each ball. Conversely, assume that X is separable; let A be a countable dense subset of X. For $a \in X$, let S_a be the collection of balls containing a. Then $\bigcup_{a \in A} S_a$ is the set of all balls of X; we prove that each S_a is countable. If $B \in S_a$ then either $B = B_a(r^-)$ or $B = B_a(r)$ for some $r \in M_a$ where

$$M_a := \{r \in (0, \infty) : \text{there is } x \in X \text{ such that } d(a, x) = r\}$$

so it suffices to show that M_a is countable. For each $r \in M_a$ there is $a_r \in X$ such that $d(a, a_r) = r$. Define $B_r := \{x \in X : d(x, a_r) < r\}$, and consider the map $r \mapsto B_r$. If $r, s \in M_a$ and $r \neq s$ then B_r and B_s are disjoint (if $r < s$ and $y \in B_r \cap B_s$ then $d(y, a_r) < r$, $d(y, a_s) < s$, so $d(a_r, a_s) \leqslant \max(d(a_r, y), d(y, a_s)) < s$ but also $d(a_r, a_s) = \max(d(a_r, a), d(a, a_s)) = s$). It follows that M_a is mapped injectively into the set $\{B_r : r \in M_a\}$. But the latter set is a disjoint collection. As each B_r must contain an element of A the set $\{B_r : r \in M_a\}$ and also M_a itself is countable.

$\mathbb{Z}_p$ (more generally, balls in a locally compact field) are our standard examples of compact ultrametric spaces. $\mathbb{C}_p$ is a separable, not locally compact, space.

Exercise 19.B. (Separability versus value group and residue class field) Show the following.
(i) Separability of K does not imply discreteness of the value group of K. Conversely, a discretely valued field need not be separable.
(ii) If K is separable then its residue class field and its value group are countable. (Surprisingly the converse is false, see Exercise 20.B.)
(iii) If the residue class field of a discretely valued field is finite then K is separable.

20. Spherical completeness

An ultrametric space is complete if and only if each nested sequence of balls $B_1 \supset B_2 \supset \ldots$ for which $\lim_{n\to\infty} d(B_n) = 0$ has a nonempty intersection. Surprisingly, if we drop the condition $\lim_{n\to\infty} d(B_n) = 0$ the picture changes.

DEFINITION 20.1. An ultrametric space is *spherically complete* if each nested sequence of balls has a nonempty intersection.

We shall see in a moment that there exist complete yet non-spherically complete spaces. But first observe that spherical completeness implies ordinary completeness and that compact ultrametric spaces are spherically complete (balls are compact). We even have the following.

PROPOSITION 20.2. *If the metric on a complete ultrametric space X is discrete then X is spherically complete.*
Proof. Let $B_1 \supset B_2 \supset \ldots$ be a nested sequence of balls. We may suppose that $B_n \neq B_{n+1}$ for all n. By the discreteness of the metric (Definition 19.1) we have $\lim_{n\to\infty} d(B_n) = 0$, and by completeness the intersection of the B_n is not empty.

COROLLARY 20.3. *The following spaces are spherically complete.*
(i) *Discretely valued complete fields, in particular locally compact fields.*
(ii) *Finite dimensional normed spaces over a discretely valued complete field.*
(iii) $B(X \to K)$ (Proposition 13.2) *and all its closed subspaces if K is complete and has discrete valuation.*
Proof. (i) is clear. To prove (ii), let $\| \ \|$ be a norm on a finite dimensional space. If the induced metric were not discrete we could find $x_1, x_2, \ldots$ such that $0 < \|x_1\| < \|x_2\| < \ldots < 1$. It is easily seen that the K-linear span of $\{x_1, x_2, \ldots\}$ must be infinite dimensional. For (iii) observe that $f \in B(X \to K)$, $f \neq 0$ implies that $\|f\|_\infty$ is in the value group of K and apply Proposition 20.2.

To find examples of non-spherically complete but complete spaces we should, with an eye on Proposition 20.2, look at spaces with a dense metric (see Exercise 19.A (ii)).

PROPOSITION 20.4. *Each complete ultrametric space with a dense metric contains a subspace that is complete but not spherically complete.*
Proof. There are balls $B_1 \supset B_2 \supset \ldots$ such that $d(B_1) > d(B_2) > \ldots$ and $\inf_n d(B_n) > 0$. Now make a 'hole' in the space X by removing the (clopen) set $\bigcap_n B_n$. The resulting set $Y := X \backslash \bigcap_n B_n$ is complete, the sets $B_1 \cap Y$, $B_2 \cap Y, \ldots$ form a nested sequence of balls in Y with an empty intersection. It follows that Y is not spherically complete.

The following theorem is more interesting since it shows that $\mathbb{C}_p$ is not spherically complete.

THEOREM 20.5. *A separable ultrametric space whose metric is dense is not spherically complete.*
Proof. Let $\{a_1, a_2, \ldots\}$ be a countable dense subset of our space X. Choose real numbers $r_0, r_1, \ldots$ such that

$$r_0 > r_1 > \ldots \quad > \tfrac{1}{2} r_0$$
$$r_0 = d(a, b) \quad \text{for some } a, b \in X$$

The equivalence relation defined by the formula $d(x, y) \leqslant r_1$ divides X into at least two balls; one of them, B_1 say, does not contain a_1. Observe that $d(B_1) = r_1$. The metric on B_1 is also dense so a similar procedure yields a ball $B_2 \subset B_1$ of diameter r_2 that does not contain a_2, and so on. We obtain a sequence of balls $B_1, B_2, \ldots$ such that

$$B_1 \supset B_2 \supset \ldots$$
$$d(B_n) = r_n \text{ for each } n \in \mathbb{N}$$
$$a_n \notin B_n \quad \text{for each } n \in \mathbb{N}$$

If $\bigcap_n B_n$ were nonempty it would contain a ball B with a positive diameter. On the other hand, $a_n \notin B$ for each n. But $\{a_1 a_2, \ldots\}$ is dense, a contradiction.

COROLLARY 20.6. *$\mathbb{C}_p$ is not spherically complete. More generally, if K is complete and separable then the completion of its algebraic closure is not spherically complete.*

Proof. By Exercise 17.B the completion of the algebraic closure is separable. Now use Corollary 16.3 and apply the above theorem.

Spherical completeness is an important concept, especially in functional analysis. See Appendix A.8 for the role played by it in the ultrametric version of the Hahn-Banach theorem. The next section connects 'spherical completeness' to 'best approximations'. In this book we shall deal with spherical completeness only indirectly.

Without proof we mention the fact that for a non-archimedean valued field K spherical completeness is equivalent to maximal completeness. (K is *maximally complete* if there are no valued fields properly containing K, whose value group is $|K^{\times}|$ and whose residue class field is (naturally isomorphic to) the one of K.) For a proof, see van Rooij (1978), p. 151.

**Exercise* 20.A. (Other definitions of spherical completeness) (i) Show that X is spherically complete if and only if each collection of balls, for which each two elements have a nonempty intersection, has a nonempty intersection.
(ii) A sequence $a_1, a_2, \ldots$ in X is called a *pseudo-Cauchy sequence* if there exists an n such that either $a_n = a_{n+1} = \ldots$ or $d(a_n, a_{n+1}) > d(a_{n+1}, a_{n+2}) > \ldots$ Such a pseudo-Cauchy sequence is *pseudoconvergent* if there exists an element $a \in X$ such that $d(a, a_n) \leqslant d(a_n, a_{n+1})$ for large n. Show that X is spherically complete if and only if each pseudo-Cauchy sequence is pseudoconvergent.

Exercise 20.B. In Appendix A.9 it is shown that for every field k and every subgroup Γ of $(0, \infty)$ there is a spherically complete field K whose residue class field is isomorphic to k and whose value group equals Γ. Borrow this fact to prove the existence of a non-separable K whose value group and residue class field are both countable.

Exercise 20.C. For an ultrametric space X, construct a spherically complete space Y containing X. (Hint. Create a new point in every 'hole' of X.) Deduce that closed subspaces of spherically complete spaces need not be spherically complete.

Exercise 20.D. (Geometric character of spherical completeness) Let $[b]$ denote the entire part of $b \in \mathbb{R}$. Show that the formula

$$\rho(x, y) := \begin{cases} \exp[\log d(x, y)] & \text{if } x \neq y \\ 0 & \text{if } x = y \end{cases}$$

defines a discrete metric ρ on X for which $\rho \leqslant d \leqslant e\,\rho$. Conclude that spherical completeness is a property depending on the metric rather than the topology (or even the uniformity; two metrics induce *the same uniformity* if they induce the same collection of Cauchy sequences).

21. Best approximation

Let Y be a subset of (an ultrametric space) X. Let $a \in X$, $b \in Y$. Then b is *a best approximation of a in Y* (and a is said to have *a best approximation in Y*) if

$$d(a, b) = d(a, Y) \quad (= \inf\{d(a, y) : y \in Y\})$$

We shall use this concept (whose 'archimedean' pendant is a well-known and useful tool in analysis) in the sequel several times. An example is the approximation of a continuous function on $\mathbb{Z}_p$ by polynomials of degree $\leqslant n$, see Section 51. Contrary to the 'archimedean' case (for example the best approximation of a point of a Hilbert space in a closed convex set), in our case a best approximation is, with trivial exceptions, never unique.

PROPOSITION 21.1. *Let Y be a nonempty subset of X. Suppose that Y has no isolated points. If an element $a \in X \backslash Y$ has a best approximation in Y then it has infinitely many.*

Proof. Let b be a best approximation of a in Y. Then so is each point of $B_b(d(a, Y)) \cap Y$.

Existence of best approximation is related to spherical completeness.

THEOREM 21.2. *Let Y be a nonempty subset of X. If Y is spherically complete as a metric space then each point of X has a best approximation in Y.*

Proof. Let $a \in X$. For each $n \in \mathbb{N}$ set $B_n := \{y \in Y : d(y, a) \leqslant d(a, Y) + n^{-1}\}$. Then $B_1 \supset B_2 \supset \ldots$ and each B_n is a ball in Y (observe that B_n is not empty). Every element of $\bigcap_n B_n$ is a best approximation of a in Y.

Exercise 21.A. (The 'converse' of Theorem 21.2) Let Y be a non-spherically complete ultrametric space. Define an ultrametric space $X = Y \cup \{a\}$ ($a \notin Y$) such that a has no best approximation in Y.

Exercise 21.B. Find a nonempty clopen subset A of $\mathbb{C}_p$ such that 0 has no best approximation in A.

2

Calculus

From now on we shall use the following conventions. *Unless stated explicitly otherwise* we have

K IS A COMPLETE NON-ARCHIMEDEAN NON-TRIVIALLY VALUED FIELD WITH RESIDUE CLASS FIELD k

THE CHARACTERISTIC OF A FIELD L IS $\mathrm{char}(L)$

p IS A PRIME NUMBER

$x = \Sigma_n a_n p^n$ FOR A p-ADIC NUMBER x DENOTES THE STANDARD EXPANSION OF x (see Section 5)

IF $\mathrm{char}(K) = 0$, $\mathrm{char}(k) = p$ THEN THE VALUATION IS CHOSEN SUCH THAT $\mathbb{Q}_p$ IS A VALUED SUBFIELD OF K. THE VALUATION ON $\mathbb{C}_p$ IS DENOTED $| \ |_p$

LET K HAVE A DISCRETE VALUATION. THEN $\pi \in K$ IS AN ELEMENT SUCH THAT $|\pi| = \max |K^\times| \cap (0, 1)$

PART 1: ELEMENTARY CALCULUS

In Chapter 2 we shall develop the first principles of ultrametric calculus. Our main interest lies in calculus over $\mathbb{Q}_p$ and $\mathbb{C}_p$.

22. The classical concepts of calculus

This section is not very exciting. We shall list notions and statements that are directly borrowed from the classical analysis over $\mathbb{R}$ and $\mathbb{C}$, some of which have been used already implicitly in Chapter 1. Starting with the next section the story of ultrametric calculus is going to diverge from the 'classical' one and shall become more interesting. No proofs are given; in case of any doubt the reader may supply one as an exercise.

Recall that our assumption that K is complete means that every Cauchy sequence in K converges. A sequence $a_1, a_2, \ldots$ in K *converges* to an element

$a \in K$ if $\lim_{n\to\infty} |a - a_n| = 0$. Its *limit* a is uniquely determined and we use the expression $\lim_{n\to\infty} a_n = a$ as a synonym for $\lim_{n\to\infty} |a - a_n| = 0$. A non-convergent sequence is *divergent*. *A double sequence* in K is a map $\mathbb{N} \times \mathbb{N} \to K$. For such a double sequence $(n, m) \mapsto a_{nm}$ we confidently leave it to the reader to define expressions such as

$$\lim_{n,\, m \to \infty} a_{nm}, \quad \lim_{n+m\to\infty} a_{nm}$$

Let $a_1, a_2, \ldots$ and $b_1, b_2, \ldots$ be sequences in K. If $\lim_{n\to\infty} a_n = a$ and $\lim_{n\to\infty} b_n = b$ for some $a, b \in K$ then $\lim_{n\to\infty} (a_n + b_n) = a + b$, $\lim_{n\to\infty} a_n b_n = ab$, $\lim_{n\to\infty} |a_n| = |a|$. If in addition $b_n \neq 0$ for all n and $b \neq 0$ then $\lim_{n\to\infty} a_n b_n^{-1} = ab^{-1}$. A convergent sequence is bounded.

PROPOSITION 22.1. (On locally compact fields, see also Corollary 12.2) *The following statements are equivalent.*
(α) *K is locally compact.*
(β) *Each bounded sequence in K has a convergent subsequence.*
(γ) *Each infinite bounded subset of K has an accumulation point in K.*
(δ) *Each closed and bounded subset of K is compact.*

Let $a_1, a_2, \ldots$ be a sequence in K. By 'the series $\Sigma\, a_n$' we mean the sequence of the partial sums given by $n \mapsto s_n := \Sigma_{j=1}^{n} a_j$. The series $\Sigma\, a_n$ is *convergent with sum* $s = \Sigma_{n=1}^{\infty} a_n$ if $s = \lim_{n\to\infty} s_n$. A non-convergent series is *divergent*.

Let $X \subset K$, $f : X \to K$. If a is an accumulation point of X and $b \in K$ we write

$$\lim_{x\to a} f(x) = b$$

if for each $\epsilon > 0$ there is $\delta > 0$ such that $0 < |x - a| < \delta$, $x \in X$ implies $|f(x) - b| < \epsilon$ or, equivalently, if $a_1, a_2, \ldots \in X$, $a_n \neq a$ for large n, $\lim_{n\to\infty} a_n = a$ implies $\lim_{n\to\infty} f(a_n) = b$. It follows that we have rules for limits of functions similar to the ones given above for sequences. The reader shall have no trouble in defining $\lim_{|x|\to\infty} f(x)$, $\lim_{x\to a} |f(x)|$ and, for a function f of two variables, expressions of the type $\lim_{(x,\, y)\to(a,\, b)} f(x, y)$, etc.

Let $a \in X \subset K$ and $f : X \to K$. f is *continuous at a* if one of the following equivalent conditions (α)-(δ) is satisfied.
(α) For each neighbourhood U of $f(a)$ the set $f^{-1}(U)$ is a neighbourhood of a in X.
(β) For each $\epsilon > 0$ there is a $\delta > 0$ such that $|x - a| < \delta$, $x \in X$ implies $|f(x) - f(a)| < \epsilon$.
(γ) If $a_1, a_2, \ldots \in X$, $\lim_{n\to\infty} a_n = a$ then $\lim_{n\to\infty} f(a_n) = f(a)$.
(δ) Either a is an isolated point of X or $\lim_{x\to a} f(x) = f(a)$.

f is *continuous* if f is continuous at a for each $a \in X$. f is *uniformly continuous* if for each $\epsilon > 0$ there exists $\delta > 0$ such that $|x-y| < \delta$, $x, y \in X$ implies $|f(x) - f(y)| < \epsilon$. Similarly one defines (uniform) continuity for functions of several variables.

If $f, g : X \to K$ are (uniformly) continuous then so are $f + g$ and λf for $\lambda \in K$. It follows that $C(X \to K)$, the set of all continuous functions $X \to K$ and $UC(X \to K)$, the set of all uniformly continuous functions $X \to K$, are K-vector spaces. A uniformly continuous function is continuous. If $f, g : X \to K$ are continuous then so is their product fg. If in addition $f(x) \neq 0$ for all $x \in X$ then $1/f$ (i.e. $x \mapsto f(x)^{-1}$) is continuous. If $f : X \to K$ and $g : f(X) \to K$ are continuous then so is their composition $g \circ f$.

A sequence $f_1, f_2, \ldots$ of K-valued functions on $X \subset K$ *converges pointwise* to $f : X \to K$ (notation $\lim_{n\to\infty} f_n = f$) if $\lim_{n\to\infty} f_n(x) = f(x)$ for all $x \in X$. It *converges uniformly* to f (notation $\lim_{n\to\infty} f_n = f$ uniformly) if for each $\epsilon > 0$ there is $N \in \mathbb{N}$ such that for all $n \geqslant N$ and all $x \in X$ we have $|f(x) - f_n(x)| < \epsilon$. Uniform convergence of $f_1, f_2, \ldots$ to f is equivalent to '$f - f_n$ is bounded for large n and $\lim_{n\to\infty} \|f - f_n\|_\infty = 0$' (for $\| \ \|_\infty$ see Section 13). Similarly one defines *pointwise convergence* and *uniform convergence* of a series of functions.

Recall that the space $B(X \to K)$ consisting of all bounded functions $X \to K$ is a Banach space with respect to the supremum norm $\| \ \|_\infty$ (Proposition 13.2).

DEFINITION 22.2. Let $X \subset K$. $BC(X \to K)$ is the K-linear space consisting of all bounded continuous functions $X \to K$ normed by $\| \ \|_\infty$. $BUC(X \to K)$ is the K-linear space consisting of all bounded uniformly continuous functions $X \to K$ normed by $\| \ \|_\infty$.

THEOREM 22.3. *Let $X \subset K$. The spaces $C(X \to K)$ and $UC(X \to K)$ are uniformly closed, i.e. if $f_1, f_2, \ldots$ are in $C(X \to K)$ ($UC(X \to K)$) and if $\lim_{n\to\infty} f_n = f$ uniformly then $f \in C(X \to K)$ ($UC(X \to K)$). The spaces $BC(X \to K)$ and $BUC(X \to K)$ are closed subspaces of $B(X \to K)$, hence Banach spaces over K.*

THEOREM 22.4. *Let $X \subset K$ be compact. Then $BC(X \to K) = C(X \to K) = UC(X \to K) = BUC(X \to K)$. For $f \in C(X \to K)$ we have $\|f\|_\infty = \max \{ |f(x)| : x \in X \}$.*

THEOREM 22.5. *Let $X \subset K$ and let $f : X \to K$ be uniformly continuous. Let $\bar{X}$ be the closure of X in K. Then there is a unique continuous function $\bar{f} : \bar{X} \to K$ that extends f. This $\bar{f}$ is uniformly continuous and $\|\bar{f}\|_\infty = \|f\|_\infty$.*

COROLLARY 22.6. *If $X \subset K$ and $\bar{X}$ is compact then each uniformly continuous function $X \to K$ is bounded. In other words, $UC(X \to K) = BUC(X \to K)$.*

Let $X \subset K$, $a \in X$ be an accumulation point of X. A function $f : X \to K$ is *differentiable at a* if the *derivative* $f'(a)$ of f at a,

$$f'(a) := \lim_{x \to a} \frac{f(x) - f(a)}{x - a}$$

exists. f is *differentiable* (on X) if $f'(a)$ exists for each $a \in X$. (Observe that in that case X does not have isolated points.) The function f' is the *derivative* of f, f is an *antiderivative* of f'. The well-known rules for differentiation of sums, products, quotients and compositions (chain rule) carry over without any problem. Thus, the derivative of a polynomial function $x \mapsto \sum_{j=0}^{n} a_j x^j$ on K is $x \mapsto \sum_{j=1}^{n} j a_j x^{j-1}$. Rational functions are differentiable. A differentiable function is continuous.

Let $a_0, a_1, a_2, \ldots$ be a sequence in K. The '*power series* $\sum a_n x^n$' is the sequence of polynomial functions $s_0, s_1, \ldots$ given by

$$s_n(x) := \sum_{j=0}^{n} a_j x^j \qquad (x \in K, n \in \mathbb{N})$$

The *region of convergence* of a power series $\sum a_n x^n$ is the set

$$\{ x \in K : \sum a_n x^n \text{ converges} \}$$

Its *radius of convergence* is defined by

$$\rho := (\overline{\lim_{n \to \infty}} \sqrt[n]{|a_n|})^{-1}$$

where, by convention, $0^{-1} = \infty$, $\infty^{-1} = 0$. The next theorem shows that ρ behaves as expected. The differentiability can be proved directly by estimating $\sum_{n=1}^{\infty} \{ h^{-1} (a_n (x + h)^n - a_n x^n) - n a_n x^{n-1} \}$ for small $h \neq 0$.

THEOREM 22.7. *Let ρ be the radius of convergence of a power series $\sum a_n x^n$. Then $\sum a_n x^n$ converges on $\{ x \in K : |x| < \rho \}$ and diverges on $\{ x \in K : |x| > \rho \}$. For each $\tau \in (0, \infty)$, $\tau < \rho$ the convergence is uniform on $\{ x \in K : |x| \leqslant \tau \}$. The function*

$$x \mapsto \sum_{n=0}^{\infty} a_n x^n \qquad (|x| < \rho)$$

is differentiable. Its derivative is

$$x \mapsto \sum_{n=1}^{\infty} n a_n x^{n-1} \qquad (|x| < \rho)$$

Remarks

1. Contrary to the complex case it is not always true that for a power series $\sum a_n x^n$ in K there is a *unique* $r \in [0, \infty]$ such that the series converges on $\{x \in K : |x| < r\}$, diverges on $\{x \in K : |x| > r\}$. (Take a discretely valued field K.) The reason why we have selected ρ out of other possible candidates for 'radius of convergence' shall be explained in Exercise 40.A. For the behaviour of the power series on the 'circle' $\{x \in K : |x| = \rho\}$ which is rather surprising at first sight, see Exercise 23.F.

2. Other important tools in calculus such as monotonicity, the Darboux property, the Riemann integral, Rolle's theorem, the mean value theorem, the fundamental theorem of calculus, depend on either connectedness, local compactness or ordering of $\mathbb{R}$. As the non-archimedean fields we are dealing with in general have none of these properties we cannot simply carry over these parts of the theory. However, in the sequel we shall see that not everything is lost.

Exercise 22.A. (Functions with precompact image) For a complex valued function f the statements 'f is bounded' and 'the image of f has compact closure' are equivalent. In the ultrametric case they are not. In fact, for $X \subset K$ let $PC(X \to K)$ be the space of all continuous functions $X \to K$ for which the closure of $f(X)$ is compact ($f(X)$ is precompact). Prove the following.
(i) $PC(X \to K)$ is a closed K-linear subspace of $BC(X \to K)$, hence a K-Banach space.
(ii) If K is locally compact or X is compact then $PC(X \to K) = BC(X \to K)$.
(iii) If K is not locally compact (e.g. $K = \mathbb{C}_p$) and if $X = B_0(1)$ then $PC(X \to K)$ is a proper subset of $BC(X \to K)$.

**Exercise* 22.B. (On functions whose limits exist at every point) Let L be the set of all functions $f : \mathbb{Z}_p \to \mathbb{Q}_p$ for which

$$f^\circ(x) := \lim_{y \to x} f(y)$$

exists for all $x \in \mathbb{Z}_p$. Prove the following.
(i) A continuous function $f : \mathbb{Z}_p \to \mathbb{Q}_p$ is in L. If $f \in L$ then f° is continuous.
(ii) $f \mapsto f^\circ$ is a $\mathbb{Q}_p$-linear map of L onto $C(\mathbb{Z}_p \to \mathbb{Q}_p)$.
(iii) Each $f \in L$ can be written as $g + h$ where g is continuous and $h^\circ = 0$.
(iv) Let $f \in L$. Then $f^\circ = 0$ if and only if there are sequences $a_1, a_2, \ldots$ in $\mathbb{Z}_p$ and $\alpha_1, \alpha_2, \ldots$ in $\mathbb{Q}_p$ such that $\lim_{n \to \infty} \alpha_n = 0$ and

$$f(x) = \begin{cases} \alpha_n & \text{if } n \in \mathbb{N}, x = a_n \\ 0 & \text{if } x \notin \{a_1, a_2, \ldots\} \end{cases}$$

23. Sequences and series

A first deviation from 'classical' analysis turns up when we consider tests for convergence of series.

PROPOSITION 23.1. *Let $a_1, a_2, \dots$ be a sequence in K.*
(i) *If $\lim_{n\to\infty} a_n = a \in K$ and $a \neq 0$ then $|a_n| = |a|$ for large n.*
(ii) *$\Sigma\, a_n$ converges if and only if $\lim_{n\to\infty} a_n = 0$.*
Proof. Since $|a_n - a| < |a|$ for large n we have, by the isosceles triangle principle 18.2, $|a_n| = \max(|a_n - a|, |a|) = |a|$, which proves (i). To prove (ii) suppose $\lim_{n\to\infty} a_n = 0$. Then for large m, n and $m \geqslant n$

$$\left|\sum_{j=n}^{m} a_j\right| \leqslant \max(|a_n|, \dots, |a_m|) \leqslant \max\{|a_j| : j \geqslant n\}$$

It follows that $n \mapsto \Sigma_{j=1}^{n} a_j$ is a Cauchy sequence, hence convergent. The converse is obvious.

Part (ii) of Proposition 23.1 shows that in K we do not have sequences that behave like the harmonic sequence $n \mapsto n^{-1}$ in $\mathbb{R}$ ($\lim_{n\to\infty} a_n = 0$ and yet $\Sigma\, a_n$ divergent). For the properties of the series $n \mapsto n^{-1}$ in K see Exercise 23.C. A more fundamental conclusion of (ii) is that convergence of $\Sigma\, a_n$ implies unconditional convergence. That is, if σ is a permutation of $\mathbb{N}$ then $\Sigma\, a_{\sigma(n)}$ converges. The following two exercises illustrate the situation.

**Exercise* 23.A. (Ordinary convergence = unconditional convergence) Let T be a (countable) set and let $n \mapsto a_n$ be a map of T into K. Let $a, s \in K$. We write

$$\lim_{n \in T} a_n = a$$

if for each $\epsilon > 0$ there exists a finite subset T' of T such that $|a - a_n| < \epsilon$ for all $n \in T \setminus T'$. We write

$$\sum_{n \in T} a_n = s$$

if for $\epsilon > 0$ there is a finite subset T' of T such that for all finite sets T'' for which $T' \subset T'' \subset T$ we have $|s - \Sigma_{n \in T''} a_n| < \epsilon$ Now prove the following.

(i) Let $T = \mathbb{N}$. Then $\lim_{n \in \mathbb{N}} a_n = a$ if and only if $\lim_{n\to\infty} a_n = a$; $\Sigma_{n \in \mathbb{N}} a_n = s$ if and only if $\Sigma_{n=1}^{\infty} a_n = s$. If $\Sigma_{n=1}^{\infty} a_n = s$ and σ is a per-

mutation of $\mathbb{N}$ then $\sum_{n=1}^{\infty} a_{\sigma(n)} = s$.
(ii) $\sum_{n \in T} a_n$ exists if and only if $\lim_{n \in T} a_n = 0$.
(iii) Let $T = T_1 \cup T_2 \cup \ldots$ be a partition of T into nonempty sets. If $\sum_{n \in T} a_n$ exists then for each j the sum $\sum_{n \in T_j} a_n$ exists and

$$\sum_{n \in T} a_n = \sum_{j=1}^{\infty} \sum_{n \in T_j} a_n$$

From the existence of $\sum_{j=1}^{\infty} \sum_{n \in T_j} a_n$ it does not follow that $\sum_{n \in T} a_n$ exists.

**Exercise* 23.B. (Double sequences and series) Let $(m, n) \mapsto a_{mn}$ be a map of $\mathbb{N} \times \mathbb{N}$ into K. Show that the following conditions (α)-(δ) are equivalent.
(α) $\lim_{(m,n) \in \mathbb{N} \times \mathbb{N}} a_{mn} = 0$.
(β) $\lim_{m+n \to \infty} a_{mn} = 0$.
(γ) $\lim_{n \to \infty} a_{mn} = 0$ for each m, $\lim_{m \to \infty} a_{mn} = 0$ for each n, $\lim_{m,n \to \infty} a_{mn} = 0$.
(δ) $\lim_{n \to \infty} a_{mn} = 0$ for each m, $\lim_{m \to \infty} a_{mn} = 0$ uniformly in n (i.e. for each $\epsilon > 0$ there is N such that for $m \geqslant N$ and all $n \in \mathbb{N}$ we have $|a_{mn}| < \epsilon$).
Apply Exercise 23.A to show that if (α)-(δ) are satisfied then $\sum_{(m,n) \in \mathbb{N} \times \mathbb{N}} a_{mn}$ exists and

$$\sum_{(m,n) \in \mathbb{N} \times \mathbb{N}} a_{mn} = \sum_{m=1} \sum_{n=1} a_{mn} = \sum_{n=1} \sum_{m=1} a_{mn}$$

A series $\sum a_n$ in K may be called absolutely convergent if $\sum |a_n|$ converges (compare Exercise 13.B). In $\mathbb{C}$ the absolutely convergent series are precisely the unconditionally convergent series. In the ultrametric case, as we have seen in Exercise 23.A, *every* convergent series is unconditionally convergent. It might be for that reason that absolute convergence plays no significant role in ultrametric analysis. See also Exercises 23.0 and 23.P.

We close this section with a number of exercises. Some of them are meant to illustrate the curious behaviour of p-adic sequences and series, others are just translations of 'classical' facts.

Exercise 23.C. ('Harmonic' sequence) Show that the sequence 1, 1/2, 1/3, ... does not converge in $\mathbb{Q}_p$ but has convergent subsequences. In fact, prove that $\{1, 1/2, 1/3, \ldots\}$ is dense in $\{x \in \mathbb{Q}_p : |x|_p \geqslant 1\}$. Observe that 1, 1/2, 1/3, ... is not defined if $\mathrm{char}(K) = p$. What are the accumulation points of $\{1, 1/2, 1/3, \ldots\}$ if $\mathrm{char}(k) = 0$?

Exercise 23.D. (Geometric series) Let $a \in K$. Show that $1, a, a^2, \ldots$ converges if and only if $|a| < 1$ or $a = 1$. Show that $\sum_{n=0}^{\infty} a^n = 1/(1-a)$ for $|a| < 1$. Compute $\sum_{n=0}^{\infty} na^n$ for $|a| < 1$.

Exercise 23.E. (Logarithm) Let char(K) = 0. Show that the series $\sum_{n=1}^{\infty} x^n/n$ converges for $|x| < 1$, diverges for $|x| \geq 1$. (For more on this, see Sections 44 and 45.)

**Exercise* 23.F. (Convergence of a power series on the 'boundary') Let $\sum a_n x^n$ be a power series in K with radius of convergence $\rho < \infty$, $\rho \neq 0$. Show that it converges either everywhere or nowhere on the clopen set $\{x \in K: |x| = \rho\}$. Show that if the latter set is nonempty then it is *not* the boundary of $\{x \in K: |x| < \rho\}$. Finally, give an example of a power series in $\mathbb{C}_p$ for which $\{x \in \mathbb{C}_p: |x|_p = \rho\}$ is empty.

Exercise 23.G. Find nonzero p-adic numbers $a_0, a_1, \ldots$ such that the power series $\sum a_n x^n$ converges everywhere on $\mathbb{Q}_p$.

Exercise 23.H. Compute $\lim_{n\to\infty} 2^{3^n}$ in $\mathbb{Q}_3$. (Hint. To prove convergence set $a_n := 2^{3^n}$ and show that $|a_{n+1} - a_n|_3 \leq 3^{-1} |a_n - a_{n-1}|_3$.)

Exercise 23.I. Show that $\sum_{n=0}^{\infty} n \cdot n! = -1$ in every $\mathbb{Q}_p$. (Compare Exercise 7.D.)

Exercise 23.J. (van Hamme) Use the ideas of the previous exercise to show that in $\mathbb{Q}_p$ ($p \neq 2$ in the first formula)

$$\sum_{n=1}^{\infty} \frac{n^2 \cdot (n+1)!}{4^n + 1} = -1 \; ; \quad \sum_{n=1}^{\infty} n^2 \cdot (n+1)! = 2 \; ; \quad \sum_{n=1}^{\infty} n^5 \cdot (n+1)! = 26$$

Exercise 23.K. For each $n \in \mathbb{N}$, $n = a_0 + a_1 p + \ldots + a_s p^s$ $(a_s \neq 0)$ in base p, define $n_- := a_0 + a_1 p + \ldots + a_{s-1} p^{s-1}$ (observe that this definition depends on p). Show that $\lim_{n\to\infty} (n - n_-) = 0$ in $\mathbb{Q}_p$ and compute $\sum_{n=1}^{\infty} (n - n_-)$ in $\mathbb{Q}_p$.

Exercise 23. L. Let $a_1, a_2, \ldots \in K$. Prove that $\lim_{n\to\infty} a_n$ exists if and only if $\lim_{n\to\infty} (a_{n+1} - a_n) = 0$.

**Exercise* 23.M. Let $a_0, a_1, \ldots$ and $b_0, b_1, \ldots \in K$. If $\sum a_n$ and $\sum b_n$ converge then $\sum c_n$ converges and

$$\left(\sum_{n=0}^{\infty} a_n\right)\left(\sum_{n=0}^{\infty} b_n\right) = \sum_{n=0}^{\infty} c_n$$

where $c_n := \sum_{j=0}^{n} a_j b_{n-j}$. Prove this.

**Exercise* 23. N. (Infinite products) Let $b_1, b_2, \ldots \in K$. Set $P_n := \prod_{j=1}^{n} b_j$. If $\lim_{n\to\infty} P_n$ exists we denote it by $\prod_{j=1}^{\infty} b_j$.
(i) Show that $\prod_{j=1}^{\infty} b_j$ exists and is nonzero if and only if all b_j are nonzero and $\lim_{j=\infty} b_j = 1$.
(ii) Let $a_1, a_2, \ldots$ be a null sequence in K. Show that

$$\prod_{n=1}^{\infty} (1 + a_n) = \sum_{n=0}^{\infty} h_n$$

where $h_0 = 1$, $h_1 = \sum_{j=1}^{\infty} a_j$, $h_2 = \sum_{j_1 \neq j_2} a_{j_1} a_{j_2}, \ldots$ (In general, for $n \in \mathbb{N}$ we have $h_n = \sum a_{j_1} a_{j_2} \ldots a_{j_n}$ where the sum is taken over all $(j_1, j_2, \ldots j_n) \in \mathbb{N}^n$ for which $k \neq l$ implies $j_k \neq j_l$.)

Exercise 23.O. (On absolute convergence) Let $a_1, a_2, \ldots$ be a sequence in K. Show that convergence of $\sum |a_n|$ implies convergence of $\sum a_n$ but that the converse is false.

Exercise 23.P. (The tests of Cauchy and d'Alembert) You may have missed in this section the well-known criteria for convergence of $\sum a_n$ involving the behaviour of $\sqrt[n]{|a_n|}$ or $|a_{n+1}|/|a_n|$. If so, prove an ultrametric version of these criteria and form yourself an opinion about their importance.

Exercise 23.Q. (Cesàro convergence) A real sequence $a_1, a_2, \ldots$ is *Cesàro convergent* if the *Cesàro limit* $\lim_{n\to\infty} (a_1 + a_2 + \ldots + a_n)/n$ exists. It can be shown that if a real sequence converges then its Cesàro limit also exists and is equal to the ordinary limit. The example $a_n := (-1)^n$ shows that 'Cesàro convergence' properly extends the notion of ordinary convergence. Now let us turn to the ultrametric case. Show that the classical theorem 'if $\lim_{n\to\infty} a_n = a$ then $\lim_{n\to\infty} (a_1 + a_2 + \ldots + a_n)/n = a$' does not hold in $\mathbb{Q}_p$ but that instead we have the following. *Let* char$(K) = 0$. *and let* $a_1, a_2, \ldots \in K$. *Then* $\lim_{n\to\infty} (a_1 + a_2 + \ldots + a_n)/n = a$ *implies* $\lim_{n\to\infty} a_n = a$.

Exercise 23.R. (Nonexistence of Banach limits) Show that it is not possible to assign to every bounded sequence $\xi_1, \xi_2, \ldots$ in K an element LIM$(\xi_1, \xi_2, \ldots)$ of K such that
(i) LIM is a K-linear function $l^\infty \to K$ (Exercise 13.A),
(ii) LIM$(\xi_1, \xi_2, \ldots) = \lim_{n\to\infty} \xi_n$ if the latter exists,
(iii) LIM$(\xi_1, \xi_2, \ldots)$ = LIM$(0, \xi_1, \xi_2, \ldots)$ for all $(\xi_1, \xi_2, \ldots) \in l^\infty$.
(Hint. The sequence 1, 2, 3, ... is bounded.) This result deviates much from the corresponding real case where it is proved that such 'Banach limits' exist.

See, for example, S. Banach, *Théorie des opérations linéaires*, Warszawa, 1932.

24. Order-like structure in K

The fact that $\Sigma_{n=1}^{\infty} n \cdot n! = -1$ in $\mathbb{Q}_p$ (see Exercise 23.I) and our experiences with Exercise 3.A do not give us much hope that there exists some (partial) ordering on $\mathbb{Q}_p$ that behaves decently with respect to the algebraic and topological structure. (See Exercise 24.A.) However, it turns out that one can define a substitute for 'ordering' in (an arbitrary) K based upon the notion of 'betweenness' defined below. It enables us to speak about 'convex set', 'positive element of K', 'the sign of an element of K', 'increasing function'. For the sequel it is convenient to have this terminology at hand, which is the reason for introducing it here. In this section we are not aiming at a serious study of these concepts. We refer the interested reader to Part 3 of Chapter 4.

DEFINITION 24.1. Let $x, y, z \in K$. The smallest disc (or ball) that contains x and y is denoted by $[x, y]$. The element z is *between* x *and* y if $z \in [x, y]$, otherwise x *and* y *are at the same side of* z. *A* subset X of K is *convex* if $x, y \in X$ implies $[x, y] \subset X$.

These notions are obvious adaptations of well-known concepts in $\mathbb{R}$. Observe that they can be defined for general (ultra)metric spaces. In the following proposition we collect several immediate consequences of Definition 24.1. We leave the easy proofs to the reader.

PROPOSITION 24.2.
(i) *Let* $x, y, z, u, t \in K$. *Then* $[x, y] = [y, x] = B_x(|x-y|) = B_y(|x-y|)$. *If* u *is between* z *and* t *and if* z, t *both are between* x *and* y *then* u *is also between* x *and* y. *Further,* $z \in [x, y]$ *if and only if there exists a* $\lambda \in K$ *such that* $|\lambda| \leqslant 1$ (!) *and* $z = \lambda x + (1-\lambda)y$.
(ii) *Discs are convex. Also* $\emptyset$, K *and the singleton sets* $\{a\}$ $(a \in K)$ *are convex.*
(iii) *There are no convex subsets of* K *other than those listed in* (ii).
(iv) *If* $C \subset K$ *is convex then there is an* $r \in [0, \infty]$ *and an element* $a \in K$ *such that* $C = \{x \in K : |x-a| \leqslant r\}$ *or* $C = \{x \in K : |x-a| < r\}$.

The relation '$x \sim y$ if 0 is not in the smallest ball containing x and y' is an equivalence relation in $\mathbb{R}^\times$ and divides $\mathbb{R}^\times$ into the two equivalence classes ('sides of 0') $(0, \infty)$ and $(-\infty, 0)$. In $\mathbb{C}^\times$, $\sim$ fails to be an equivalence relation. However, in $K^\times$ it works.

PROPOSITION 24.3. *The relation* $\sim$ *defined by '*$x \sim y$ *if* x *and* y *are at the* *same side of 0' is an equivalence relation on $K^\times$. Its equivalence classes are the multiplicative cosets of the group $B_1(1^-)$.*

Proof. For $x, y \in K^\times$ we have $x \sim y \Leftrightarrow 0 \notin [x, y] \Leftrightarrow 0 \notin \{t \in K : |t-x| \leqslant |x-y|\} \Leftrightarrow |x| > |x-y| \Leftrightarrow |1-x^{-1}y| < 1 \Leftrightarrow x^{-1}y \in B_1(1^-)$. From Exercise 11.E (iii) it follows that $B_1(1^-)$ is a multiplicative subgroup of the unit sphere of K.

The sign of a nonzero real number can be interpreted as its image under the map $\mathbb{R}^\times \to \mathbb{R}^\times / \mathbb{R}^+ \simeq \{1, -1\}$, where $\mathbb{R}^+ = (0, \infty)$ is the set of the positive real numbers. With this in mind the following definition is quite natural.

DEFINITION 24.4. Let $\sim$ be as in Proposition 24.3. *A side of* 0 in K is an equivalence class of $\sim$. An element x of K is *positive* if $|1-x| < 1$. The group of all positive elements of K is denoted by K^+. The quotient group $\Sigma := K^\times / K^+$ is the *group of signs of* K. Let

$$\text{sgn} : K^\times \to K^\times / K^+$$

be the canonical homomorphism. For $x \in K^\times$ the element sgn x is the *sign* of x.

PROPOSITION 24.5. *Let $x \in K^\times$. The side of* 0 *to which x belongs equals* $\{y \in K^\times : y \sim x\} = \{y \in K^\times : \text{sgn}\, y = \text{sgn}\, x\} = xB_1(1^-) = \{y \in K : |y-x| < |x|\} = B_x(|x|^-)$. *It is the largest convex subset of K that contains x but not* 0. *Each maximal convex subset C of K with the property $0 \notin C$ is a side of* 0. *K^+ is the side of* 0 *that contains* 1. *Each side of* 0 *is clopen and bounded; there are infinitely many of them. The group Σ is infinite.*

Proof. Obvious.

Among the several notions of 'monotone function' that arise naturally from the above observations we select here the following translation of the concept of an increasing or decreasing function.

DEFINITION 24.6. Let $a \in \Sigma$ be a sign and $X \subset K$. A function $f : X \to K$ is *monotone of type a* if for all $x, y \in X, x \neq y$

$$\text{sgn}\,(f(x) - f(y)) = a\ \text{sgn}\ (x-y)$$

Such an f is *increasing* if a is the identity element of Σ.

Using a more down-to-earth terminology we can reformulate Definition 24.6 as follows.

PROPOSITION 24.7. *Let $X \subset K$ and $f : X \to K$. f is increasing if and only if for all $x, y \in X, x \neq y$*

$$\left| \frac{f(x)-f(y)}{x-y} - 1 \right| < 1$$

More generally, if there exists an element $s \in K^{\times}$ *such that for all* $x, y \in X$, $x \neq y$

$$\left| \frac{f(x)-f(y)}{x-y} - s \right| < |s|$$

then f *is monotone of type* α *where* $\alpha = \operatorname{sgn} s$.

We mention two immediate consequences.
(1) If we multiply a function f, monotone of type α with a nonzero element t of K, then the resulting function tf is monotone of type $\alpha\beta$ where $\beta = \operatorname{sgn} t$. For any s for which $\operatorname{sgn} s = \alpha$ the function $s^{-1} f$ is increasing.
(2) An increasing function is an isometry. If f is monotone of type $\alpha = \operatorname{sgn} s$ then $|f(x)-f(y)| = |s|\ |x-y|$ $(x, y \in X)$.

Exercise 24.A. Show that it is not possible to define a partial ordering $\geqslant$ on $\mathbb{Q}_p$ satisfying (i) $1 \geqslant 0 \geqslant -1$ (ii) if $a \geqslant 0, b \geqslant 0$ then $a + b \geqslant 0$ (iii) if $a_n \geqslant 0$ for all n, $\lim_{n \to \infty} a_n = a$ then $a \geqslant 0$.

Exercise 24.B. Show that $\{p^n(a + p\mathbb{Z}_p) : a \in \{1, \ldots p-1\}, n \in \mathbb{Z}\}$ is the collection of sides of 0 in $\mathbb{Q}_p$. (Draw a picture.)

Exercise 24.C. Show that the function $x \mapsto x$ is increasing. To examine the square function let $p \neq 2$. Show that $x \mapsto x^2$, restricted to some side S of 0 of $\mathbb{Q}_p$, is monotone of some type α depending on S. Conclude that $x \mapsto \frac{1}{2}x^2$ is increasing on $\mathbb{Q}_p^+$. However, the function $x \mapsto x^2$, restricted to any side of 0 of $\mathbb{Q}_2$, is not injective and therefore fails to be monotone. Prove this.

25. (Locally) analytic functions

We have seen (Theorem 22.7) that for a given sequence $a_0, a_1, \ldots$ in K the power series $\Sigma\, a_n x^n$ converges for $|x| < \rho$, diverges for $|x| > \rho$ where $\rho = (\overline{\lim}_{n \to \infty} \sqrt[n]{|a_n|})^{-1}$. Exercise 23.F tells us that the behaviour on the 'boundary' $\{x \in K : |x| = \rho\}$ is much simpler than in the complex case. Thus, the region of convergence of a power series is a convex set C containing 0. It is an easy matter to show that if $C \neq \{0\}$ the function $x \mapsto \Sigma_{n=0}^{\infty} a_n x^n$ is differentiable on C and that its derivative is $x \mapsto \Sigma_{n=1}^{\infty} n a_n x^{n-1}$ (this is a slight extension of the statement made in Theorem 22.7). For any $u \in K$

the region of convergence of $\Sigma\, a_n(x-u)^n$ $(:=\{x \in K:\ \Sigma\, a_n(x-u)^n$ converges $\}$) is a convex set containing u.

Now let us start with an open convex subset D of K and a function $f: D \to K$. f is *analytic* on D if there are elements $u \in D$ and $a_0, a_1, \ldots \in K$ such that

$$f(x) = \sum_{n=0}^{\infty} a_n(x-u)^n \qquad (x \in D)$$

An analytic function is infinitely many times differentiable. With f as above we have $f'(x) = \Sigma_{n=1}^{\infty} na_n(x-u)^{n-1}$ for $x \in D$. The following theorem says that, contrary to the complex case, it does not matter which $u \in D$ we choose in the definition of analyticity.

THEOREM 25.1. *Let D be an open convex subset of K and let $f: D \to K$ be analytic. Then for each $v \in D$ there exist $b_0, b_1, \ldots \in K$ such that $f(x) = \Sigma_{n=0}^{\infty} b_n(x-v)^n$ for all $x \in D$.*

Proof. f is analytic so there are $u \in D$ and $a_0, a_1, \ldots \in K$ such that $f(x) = \Sigma_{n=0}^{\infty} a_n(x-u)^n$ for all $x \in D$. Now

$$(x-u)^n = (x-v+v-u)^n = \sum_{j=0}^{n} \binom{n}{j} (x-v)^j (v-u)^{n-j}$$

Set $t_{jn} := a_n\binom{n}{j}(x-v)^j(v-u)^{n-j}$ if $j \leqslant n$, $t_{jn} := 0$ if $j > n$. Then for all j, n we have $|t_{jn}| \leqslant |a_n| \max(|x-v|, |v-u|)^n \leqslant |a_n| \max(|x-u|, |v-u|)^n$. Since $\Sigma\, a_n(x-u)^n$ and $\Sigma\, a_n(v-u)^n$ converge we have $\lim_{n\to\infty} |a_n| \cdot \max(|x-u|, |v-u|)^n = 0$. It follows that $\lim_{n\to\infty} t_{jn} = 0$ uniformly in j. Obviously $\lim_{j\to\infty} t_{jn} = 0$ for all n. By Exercise 23.B we may conclude that

$$f(x) = \sum_{n=0}^{\infty} \sum_{j=0}^{\infty} t_{jn} = \sum_{j=0}^{\infty} \sum_{n=0}^{\infty} t_{jn} = \sum_{j=0}^{\infty} b_j(x-v)^j$$

where $b_j = \Sigma_{n=j}^{\infty} a_n\binom{n}{j}(v-u)^{n-j}$. The theorem follows.

A peculiar consequence of this theorem is that the recipe of 'analytic continuation' often used in complex function theory to extend the domain of an analytic function does not work in the ultrametric case. In fact, suppose we have an analytic function

$$f(x) = \sum_{n=0}^{\infty} a_n x^n \qquad (x \in D)$$

where D is the region of convergence of the power series $\Sigma\, a_n x^n$. If we choose any $v \in D$ and develop f in a neighbourhood $U \subset D$ of v into a power series in $x - v$

$$f(x) = \sum_{n=0}^{\infty} b_n (x-v)^n \qquad (x \in U)$$

then by Theorem 25.1 the region of convergence of our new power series $\Sigma\ b_n(x-v)^n$ is again D. So, by this procedure we 'never get out of D'. This leads to the following definition.

DEFINITION 25.2. Let U be an open subset of K. A function $f : U \to K$ is *locally analytic* if for each $a \in U$ there is a (convex) neighbourhood $V \subset U$ of a such that $f | V$ is analytic.

One can obtain examples of locally analytic functions on U by decomposing U into discs (Theorem 18.6) and choosing (arbitrary) analytic functions on these discs. For later use we need also the following.

DEFINITION 25.3 A function $f : \mathbb{Z}_p \to \mathbb{C}_p$ is *locally analytic of order* $h \in \{0, 1, 2, \ldots\}$ if the restriction of f to any disc of radius p^{-h} is analytic.

Analytic functions on $\mathbb{Z}_p$ are locally analytic of order 0. Each locally analytic function on $\mathbb{Z}_p$ is locally analytic of order h for some $h \in \{0, 1, 2, \ldots\}$

Exercise 25.A. Let D be an open convex subset of K. Show that $\{f : D \to K : f \text{ analytic}\}$ is closed for sums and products. Obtain a similar conclusion for the space of all locally analytic functions defined on an open subset of K.

**Exercise* 25.B. (On the uniqueness of the power series expansions) Let f be analytic on some open convex set D containing 0, say, $f(x) = \Sigma_{n=0}^{\infty} a_n x^n$ for $x \in D$. Prove that $a_n n! = f^{(n)}(0)$ for all n. Deduce that if $\text{char}(K) = 0$ the coefficients $a_0, a_1, \ldots$ are determined by f. Now let $\text{char}(K) = p$. Although the formula $a_n n! = f^{(n)}(0)$ for $n \geq p$ is not of much help anymore prove nevertheless that also in this case the coefficients $a_0, a_1, \ldots$ are determined by f. Obtain as a by-product that, if $\text{char}(K) = p$, the pth derivative of a locally analytic function is identically 0.

Exercise 25.C. (Sequel to the previous exercise) Let $\text{char}(K) = 0$. Show that $f' = 0$, f analytic implies f is constant. Describe for the case $\text{char}(K) = p$ the set $\{f : B_0(1) \to K : f \text{ analytic}, f' = 0\}$.

**Exercise* 25.D. (On the zeros of an analytic function) Let $D \subset K$ be open and convex and let $f : D \to K$ be a nonzero analytic function. Prove that $\{x \in D : f(x) = 0\}$ has no accumulation points. (Hint. To include also the case $\text{char}(K) = p$, assume $0 \in D$, $\lim_{n \to \infty} x_n = 0$, $f(x_n) = 0$ for all n, $f(x)$

$= \Sigma_{n=0}^{\infty} a_n x^n$. Prove successively $a_0 = 0$, $a_1 = 0, \ldots$) Deduce that if $f(x) = 0$ for all x in a nonempty open subset of D then f is identically 0.

An important analytic function in analysis is the exponential function, so we shall consider the power series $\Sigma x^n/n!$ in K. Because of the factorials in the denominator we have to require that $\mathrm{char}(K) = 0$. But even then there is a difficulty as the series diverges for $x = 1$ since $1/n!$ does not tend to 0.

DEFINITION 25.4. Let $\mathrm{char}(K) = 0$. The *exponential function* is given by

$$\exp x = \sum_{n=0}^{\infty} \frac{x^n}{n!} \qquad (x \in E)$$

where E is the region of convergence of the power series $\Sigma x^n/n!$.

In order to find its radius we shall estimate $|n!|_p$ using the following lemma.

LEMMA 25.5. *Let $n \in \mathbb{N}$ be written using the base p*

$$n = a_0 + a_1 p + \ldots + a_s p^s$$

Define the sum of the digits s_n of n by

$$s_n := \sum_{j=0}^{s} a_j$$

Then

$$|n!|_p = p^{-\lambda(n)}$$

where $\lambda(n)$, the number of factors p in $n!$, equals

$$(*) \qquad \lambda(n) = \sum_{j=1}^{\infty} \left[\frac{n}{p^j}\right] = \frac{n - s_n}{p-1}$$

Proof. There are $[n/p]$ numbers in $\{1, 2, \ldots, n\}$ that are divisible by p, $[n/p^2]$ numbers in $\{1, 2, \ldots, n\}$ that are divisible by p^2, etc. It follows that $\lambda(n) = \Sigma_{j=1}^{s} [n/p^j] = \Sigma_{j=1}^{\infty} [n/p^j]$. For $j \in \{1, 2, \ldots, s\}$ we have $[n/p^j] = a_j + a_{j+1} p + \ldots + a_s p^{s-j} = p^{-j} \Sigma_{i=j}^{s} a_i p^i$ so that $\Sigma_{j=1}^{s} [n/p^j] = \Sigma_{j=1}^{s} p^{-j} \Sigma_{i=j}^{s} a_i p^i = \Sigma_{i=1}^{s} \Sigma_{j=1}^{i} p^{-j} a_i p^i = (p-1)^{-1} \Sigma_{i=0}^{s} (1 - p^{-i}) a_i p^i = (p-1)^{-1} (n - s_n)$.

THEOREM 25.6. *Let* $\mathrm{char}(K) = 0$. *For the region of convergence E of the exponential function we have* (recall that k is the residue class field of K)

$$E = \{x \in K : |x| < p^{1/(1-p)}\} \text{ if } \mathrm{char}(k) = p$$

$$E = \{x \in K : |x| < 1\} \text{ if } \mathrm{char}(k) = 0$$

Proof. Let char $(k) = p$. Then, by the assumption at the beginning of this chapter, $\mathbb{Q}_p \subset K$. From $s_n = a_0 + a_1 + \ldots + a_s \leqslant p(s+1)$ one obtains the (real) limit $\lim_{n\to\infty} s_n/n = 0$ so that

$$\lim_{n\to\infty} \frac{\lambda(n)}{n} = \frac{1}{p-1}$$

It follows that $\lim_{n\to\infty} \sqrt[n]{|n!|_p} = p^{1/(1-p)}$. According to Theorem 22.7 the radius of convergence of the power series $\Sigma\ x^n/n!$ is equal to $p^{1/(1-p)}$. We proceed to prove that for $x \in K$, $|x| = p^{1/(1-p)}$ the series diverges. In fact, for such x and for n a power of p we have $s_n = 1$ and $|x^n/n!| = p^{1/(1-p)}$, so that the sequence $n \mapsto x^n/n!$ does not tend to 0.

If $\mathrm{char}(k) = 0$ then $|n!| = 1$ for all n. Obviously, $1 \notin E$. The theorem follows.

It is a disappointing fact that the natural domain of the exponential function is strictly contained in the 'closed' unit disc! As a consequence we do not have an element in K that plays the role of $e = \exp 1$ in $\mathbb{R}$. We refer the reader who is interested in efforts to define ultrametric analogues of the fundamental real constants e, π, γ (the Euler constant) to Exercises 25.I(vi), 33.C, 45.E, (the preamble to) Exercise 37.A, Section 36, Proposition 46.6(v).

Analogues of several functions from classical calculus other then exp can be defined by means of power series as follows.

DEFINITION 25.7. Let $\mathrm{char}(K) = 0$ and let E be as in Definition 25.4.

$$\log(1+x) := \sum_{n=1}^{\infty} (-1)^{n+1} \frac{x^n}{n} \qquad (|x| < 1)$$

$$\sin x := \sum_{n=0}^{\infty} (-1)^n \frac{x^{2n+1}}{(2n+1)!} \qquad (x \in E)$$

$$\cos x := \sum_{n=0}^{\infty} (-1)^n \frac{x^{2n}}{(2n)!} \qquad (x \in E)$$

$$\sinh x := \sum_{n=0}^{\infty} \frac{x^{2n+1}}{(2n+1)!} \qquad (x \in E)$$

$$\cosh x := \sum_{n=0}^{\infty} \frac{x^{2n}}{(2n)!} \qquad (x \in E)$$

$$\arctan x := \sum_{n=0}^{\infty} (-1)^n \frac{x^{2n+1}}{2n+1} \quad (|x|<1)$$

We shall discuss properties of these and related functions in Part 3 of this chapter. At this moment we content ourselves with the results of the following exercises.

Exercise 25.E. Use Exercise 23.M to show that if $\mathrm{char}(K) = 0$ then $\exp(x + y) = (\exp x)(\exp y)$ for all $x, y \in E$.

Exercise 25.F. Show that in $\mathbb{Q}_p$ we have $E = \{x \in \mathbb{Q}_p : |x|_p < 1\}$ if $p \neq 2$ but that $E = \{x \in \mathbb{Q}_2 : |x|_2 < \frac{1}{2}\}$ in $\mathbb{Q}_2$.

Exercise 25.G. Let $n \in \mathbb{N}$, $n \geq 2$ and let $\mathrm{char}(K) = 0$. Prove that $x_1, \ldots, x_n \in E$ implies $|x_1 x_2 \ldots x_{n-1}/n!| < 1$ and $x_1 x_2 \ldots x_n/n! \in E$. In particular, if $x \in E$ then $|x^{n-1}/n!| < 1$ and $x^n/n! \in E$.

**Exercise* 25.H. Show that exp is an increasing function (Proposition 24.7) on E and that its range is contained in $1 + E$. (In fact, it is precisely $1 + E$, see Exercise 27.D.) Observe that it follows that exp is an isometry, in other words $|\exp x - \exp y| = |x - y|$ for all $x, y \in K$.

Exercise 25.I. (Elementary properties of exp, log, sin, . . .) Let $\mathrm{char}(K) = 0$. Prove the following.
(i) exp is differentiable on E, $\exp' = \exp$.
(ii) log is differentiable on K^+ with derivative $x \mapsto x^{-1}$ $(x \in K^+)$.
(iii) $\sin^2 x + \cos^2 x = 1$, $\cosh^2 x - \sinh^2 x = 1$ $(x \in E)$.
(iv) Let i be a solution of the equation $x^2 + 1 = 0$ (if in K there are no such solutions then use Krull's theorem to extend K). Then $\exp ix = \cos x + i \sin x$ for all $x \in E$.
(v) sin is an isometry of E into $1 + E$, cos is not locally injective at 0 (i.e. for every neighbourhood U of 0 the restriction of cos to U is not injective).
(vi) (Analogue of $\pi = 3.14\ldots$?) cos has no zeros, sin has no zeros except 0, there is no $t \in K$ for which $\sin x = \cos(x + t)$ $(x \in E)$.

Exercise 25.J. (No exp if $\mathrm{char}(K) = p$) Let $\mathrm{char}(K) = p$ and let f be an analytic function defined on some open convex set D containing 0. Show that each one of the following properties (i), (ii) implies $f = 0$.
(i) $f' = f$.
(ii) $f(x + y) = f(x) f(y)$ $(x, y \in D)$.

Exercise 25.K. Discuss the existence of $\lim_{n \to \infty} (n^2)!/(n!)^n$ in $\mathbb{Q}_p$ and in $\mathbb{R}$.

26. Continuity and differentiability

In this section we turn to 'ordinary' continuous and differentiable functions and discuss a few simple properties in so far as typical ultrametric features are concerned.

A basic property of continuous functions in ultrametric calculus is the fact that they can uniformly be approximated by locally constant functions.

DEFINITION 26.1. Let $X \subset K$. A function $f : X \to K$ is *locally constant* if for each $x \in X$ there is a neighbourhood U of x such that f is constant on $U \cap X$.

The (K-valued) *characteristic function* ξ_U of a relatively clopen set $U \subset X$ defined by

$$\xi_U(x) := \begin{cases} 1 \text{ if } x \in U \\ 0 \text{ if } x \in X \setminus U \end{cases}$$

is locally constant. If $f : X \to K$ is locally constant then X admits a partition into relatively clopen sets U_i, where i runs through some indexing set, such that f is constant on U_i for each i. Locally constant functions are continuous. The locally constant functions on X form a K-linear subspace of $C(X \to K)$.

THEOREM 26.2. *Let $X \subset K$ and $f \in C(X \to K)$, $\epsilon > 0$. Then there is a locally constant function $g : X \to K$ for which $|f(x) - g(x)| < \epsilon$ for all $x \in X$. The bounded locally constant functions form a dense subspace of $BC(X \to K)$.*

Proof. We only need to prove the first statement. The relation $\sim$ on X defined by

$$x \sim y \text{ if } |f(x) - f(y)| < \epsilon$$

is an equivalence relation whose equivalence classes U_i $(i \in I)$ are relatively clopen. For each $i \in I$ choose an element $a_i \in U_i$ and define $g : X \to K$ by

$$\text{if } x \in U_i \text{ then } g(x) = f(a_i) \qquad (i \in I)$$

Then g is constant on each U_i and $|g(x) - f(x)| < \epsilon$ for all $x \in X$.

Exercise 26.A. (On step functions) Let $X \subset K$. A function $f : X \to K$ is a *step function* if it is locally constant and has only finitely many values. Prove that the step functions form a dense K-linear subspace of $PC(X \to K)$. (See Exercise 22.A.)

Next, we consider differentiability. For a locally constant function g defined on a subset X of K without isolated points we have obviously that for each $a \in X$ the difference quotient $(g(x) - g(a))/(x - a)$ is 0 if x is sufficiently close to a. So g is differentiable and $g'(a) = 0$ for all $a \in X$! It follows that there are 'many' nonconstant functions whose derivative is identically 0. (Compare this to the behaviour of analytic functions, see Exercise 25.C.)

COROLLARY 26.3. *Let X be a nonempty subset of K without isolated points. Let $f \in C(X \to K)$ and $\epsilon > 0$. Then there is a function $g : X \to K$ for which $g' = 0$ and $|f(x) - g(x)| < \epsilon$ for all $x \in X$.*

The set $\{f : X \to K : f' = 0\}$ (whose elements are called 'pseudo-constants' by some authors) is a K-linear subspace of $C(X \to K)$, closed for products, and containing the locally constant functions. The following example destroys the conjecture that such 'pseudo-constants' would be locally constant or 'almost' locally constant in some sense.

EXAMPLE 26.4. *There is an injective $f : \mathbb{Z}_p \to \mathbb{Z}_p$ whose derivative is 0.*

Proof. For $x = \sum_{n=0}^{\infty} a_n p^n \in \mathbb{Z}_p$ set

$$f(x) := \sum_{n=0}^{\infty} a_n p^{2n}$$

To prove that f satisfies the requirements, let $x = \sum_{n=0}^{\infty} a_n p^n$ and $y = \sum_{n=0}^{\infty} b_n p^n$ be elements of $\mathbb{Z}_p$ and $|x - y|_p = p^{-j}$ for some $j \in \{0, 1, 2, \ldots\}$. Then $a_0 = b_0$, $a_1 = b_1, \ldots, a_{j-1} = b_{j-1}$ and $a_j \neq b_j$. It follows that $|f(x) - f(y)|_p = p^{-2j}$ and we have proved that

$$|f(x) - f(y)|_p = |x - y|_p^2 \qquad (x, y \in \mathbb{Z}_p)$$

From this formula one reads directly that f is injective and that $f' = 0$.

We can still do better (or worse). To state it in an appropriate language we shall borrow the following classical concept.

DEFINITION 26.5. Let $X \subset K$, $\alpha > 0$. A function $f : X \to K$ *satisfies a Lipschitz condition of order* α if there exists a constant $M > 0$ (*a Lipschitz constant of* f) such that

$$|f(x) - f(y)| \leq M |x - y|^{\alpha} \qquad (x, y \in X)$$

The K-linear space consisting of all functions $f : X \to K$ satisfying a Lipschitz condition of order α is denoted $\mathrm{Lip}_\alpha(X \to K)$

Our function f of Example 26.4 is in $\mathrm{Lip}_2(\mathbb{Z}_p \to \mathbb{Q}_p)$. The following exercise shows that we can make our example much more 'extreme'.

**Exercise* 26.B. Let $f : \mathbb{Z}_p \to \mathbb{Z}_p$ be defined by the formula

$$f\left(\sum_{n=0}^{\infty} a_n p^n\right) := \sum_{n=0}^{\infty} a_n p^{n!}$$

Show that f is injective, that $f' = 0$ and that $f \in \mathrm{Lip}_a(\mathbb{Z}_p \to \mathbb{Q}_p)$ for each positive a.

Remarks.

1. In real analysis, functions satisfying Lipschitz conditions of order > 1 are trivial. In fact, if f is a real valued function defined on a real interval such that $|f(x) - f(y)| \leqslant |x-y|^a$ for all x, y belonging to that interval and $a > 1$ then $f' = 0$ so that f is necessarily constant.
2. The spaces $\mathrm{Lip}_a(X \to K)$ shall figure in the sequel occasionally.
3. An important consequence of the preceding theory is that there is no such thing as an 'ultrametric mean value theorem'. Indeed, if $f' = 0$ and $f(x) - f(y) = f'(t)(x-y)$ for some $x, y, t \in K$ then $f(x) = f(y)$. We see that *for our f of Example* 26.4 *there is no triple $x, y, t \in \mathbb{Z}_p$ for which $x \neq y$ and* $f(x) - f(y) = f'(t)(x-y)$. There is not much hope that we can remove this obstruction, so instead let us face reality.

Another deviation from 'classical' analysis turns up when considering the subject of local invertibility of (continuously) differentiable functions at points where the derivative is nonzero. We have the following striking example.

EXAMPLE 26.6. *There exists a differentiable function $f: \mathbb{Z}_p \to \mathbb{Q}_p$ such that $f'(x) = 1$ for all $x \in \mathbb{Z}_p$ but for which $f(p^n) = f(p^n - p^{2n})$ for all $n \in \mathbb{N}$ so that f is injective on no neighbourhood of* 0.

Proof. For each $n \in \mathbb{N}$ let $B_n := \{x \in \mathbb{Z}_p : |x - p^n|_p < p^{-2n}\}$. Then $x \in B_n$ implies $|x|_p = p^{-n}$ so the discs $B_1, B_2, \ldots$ are pairwise disjoint. Define

$$f(x) := \begin{cases} x - p^{2n} & \text{if } n \in \mathbb{N}, x \in B_n \\ x & \text{if } x \in \mathbb{Z}_p \setminus \bigcup_n B_n \end{cases}$$

Since $p^n \in B_n$ we have $f(p^n) = p^n - p^{2n}$. On the other hand, $p^n - p^{2n}$ is in no B_m hence also $f(p^n - p^{2n}) = p^n - p^{2n}$. It remains to prove that $f' = 1$, i.e. that the function g defined by

$$g(x) := x - f(x) = \begin{cases} p^{2n} & \text{if } n \in \mathbb{N}, x \in B_n \\ 0 & \text{if } x \in \mathbb{Z}_p \setminus \bigcup_n B_n \end{cases}$$

has zero derivative. Now g is locally constant on $\mathbb{Z}_p \setminus \{0\}$ so we only have to check that $g'(0) = 0$. For $x \in \mathbb{Z}_p, x \neq 0$ we see that $|(g(x) - g(0))/x|_p = |g(x)|_p \, |x^{-1}|_p$ is either 0 (if x is in no B_m) or p^{-n} (if $x \in B_n$) and it follows that $g'(0) = \lim_{x \to 0} g(x)x^{-1} = 0$.

In the next section we shall overcome this difficulty by modifying the definition of 'continuous differentiability'.

Exercise 26.C. Let $f, g : K \to K$. Prove that if g is differentiable and $f' = 0$ then $(g \circ f)' = (f \circ g)' = 0$. Does the conclusion remain valid if we replace the differentiability condition on g by $g \in \mathrm{Lip}_1(K \to K)$?

Exercise 26.D. Find an example of a function which is in $\mathrm{Lip}_{\frac{1}{2}}(\mathbb{Z}_p \to \mathbb{Q}_p)$ but not in $\mathrm{Lip}_1(\mathbb{Z}_p \to \mathbb{Q}_p)$. Let $\alpha > \beta > 0$. Then $\mathrm{Lip}_\alpha(\mathbb{Z}_p \to \mathbb{Q}_p) \subset \mathrm{Lip}_\beta(\mathbb{Z}_p \to \mathbb{Q}_p)$, but is it a proper inclusion?

**Exercise* 26.E. Let X be a bounded subset of K. Show that the formula

$$\|f\|_1 := \|f\|_\infty \vee \sup\left\{\left|\frac{f(x)-f(y)}{x-y}\right| : x, y \in X, x \neq y\right\}$$

defines a norm on $\mathrm{Lip}_1(X \to K)$ for which the latter is a K-Banach space.

Exercise 26.F. Show that the following ultrametric version of a well known 'classical' theorem is *false*. Let $f, f_1, f_2, \ldots : \mathbb{Z}_p \to \mathbb{Q}_p$ be differentiable with continuous derivatives. If $\lim_{n\to\infty} f_n = f$ uniformly and $g := \lim_{n\to\infty} f_n'$ exists uniformly then $f' = g$. (See however Exercise 27.C.)

Exercise 26.G. Is it the domain or the range of the function that is responsible for the 'ultrametric features' in this section?

27. Continuously differentiable functions

In order to get back some form of a local invertibility theorem for 'C^1-functions' we should not simply define continuous differentiability of a function f by 'f is differentiable and f' is continuous' (Example 26.6). The following definition suits our purpose. See Exercise 27.C and Theorem 27.5.

DEFINITION 27.1. Let X be a nonempty subset of K without isolated points, let $f : X \to K$. The (first) *difference quotient* $\Phi_1 f$ of f is the function of two variables given by

$$\Phi_1 f(x, y) = \frac{f(x)-f(y)}{x-y} \qquad (x, y \in X, x \neq y)$$

defined on $X \times X \setminus \Delta$ where $\Delta := \{(x, x) : x \in X\}$. f is *continuously differentiable at a point $a \in X$ (f is C^1 at a)* if

$$\lim_{(x,y)\to(a,a)} \Phi_1 f(x, y)$$

exists. In other words, f is C^1 at a if f is differentiable at a and if for each $\epsilon > 0$ there exists a $\delta > 0$ such that if $|x-a| < \delta$, $|y-a| < \delta$, $(x, y) \in X \times X \setminus \Delta$ then $|(f(x) - f(y))/(x-y) - f'(a)| < \epsilon$. f is *continuously dif-*

ferentiable (*f is C^1, f is a C^1-function*) if f is C^1 at a for all $a \in X$. The set of all C^1-functions $X \to K$ is denoted $C^1(X \to K)$. For $f : X \to K$ set

$$\|f\|_1 := \|f\|_\infty \vee \|\Phi_1 f\|_\infty$$

and let $BC^1(X \to K) := \{f \in C^1(X \to K) : \|f\|_1 < \infty\}$.

Remarks.

1. The crucial point of the above definition is of course that in taking the limit of the difference quotient we let x and y tend to a independently. A C^1-function has a continuous derivative. However the converse is not true. In fact, let f be as in Example 26.6. Then $\lim_{n\to\infty} (f(p^n) - f(p^n - p^{2n}))\, p^{-2n} = 0 \neq 1 = f'(0)$. Thus in general $C^1(X \to K)$ is strictly contained in $\{f : X \to K : f \text{ is differentiable and } f' \text{ is continuous}\}$.
2. Observe that for a real valued function f defined on some subinterval I of $\mathbb{R}$ the continuity of f' already guarantees the existence of $\lim_{(x,y)\to(a,a)} (f(x) - f(y))/(x-y)$ (where the limit is taken over all $x, y \in I$ for which $x \neq y$) since by the mean value theorem $(f(x) - f(y))/(x-y) = f'(t)$ for some t between x and y.
3. In the literature also the terms 'strictly differentiable' and sometimes 'uniformly differentiable' are used to indicate what we have called 'continuously differentiable'. We prefer the term 'continuously differentiable' because of Remark 2 above and Proposition 27.2 (β) below, and we shall reserve the expression 'uniform differentiability' for a stronger property (see Exercise 28.E).
4. $C^1(X \to K)$ is a K-vector space closed for products. The function $\| \ \|_1$ is easily seen to be a norm on $BC^1(X \to K)$ (see also Exercise 27.C).

The following restatements of Definition 27.1 shall be used frequently in the sequel.

PROPOSITION 27.2. *Let X be a nonempty subset of K without isolated points, let $f : X \to K$. The following statements are equivalent.*

(α) *f is continuously differentiable.*

(β) *The function $\Phi_1 f$ of Definition* 27.1 *can* (*uniquely*) *be extended to a continuous function $\overline{\Phi}_1 f$ on $X \times X$.*

(γ) *There is a continuous function $R : X \times X \to K$ such that*

$$f(x) = f(y) + (x-y)R(x,y) \qquad (x, y \in X)$$

Proof. Left to the reader.

Exercise 27.A. (Analytic functions are C^1) Let f be a K-valued analytic function defined on an open convex set $D \subset K$. Show that $f \in C^1(D \to K)$.

(For a stronger result see Corollary 29.11.)

Exercise 27.B. (Lipschitz versus C^1) Let $X \subset K$ have no isolated points and let $f: X \to K$. Prove that if f satisfies a Lipschitz condition of order > 1 then f is a C^1-function. Give an example of an $f \in \mathrm{Lip}_{\frac{1}{2}}(\mathbb{Z}_p \to \mathbb{Q}_p)$ that is nowhere differentiable. (Consider your solution of Exercise 26.D.)

**Exercise* 27.C. ('Solution' of the problem of Exercise 26.F) Let $X \subset K$ have no isolated points and let $f_1, f_2, \ldots$ be a sequence of C^1-functions on X. Suppose that $f := \lim_{n\to\infty} f_n$ exists and that $\lim_{n\to\infty} \Phi_1 f_n$ exists uniformly on $X \times X \backslash \Delta$. Prove that f is a C^1-function and that $\lim_{n\to\infty} \overline{\Phi}_1 f_n = \overline{\Phi}_1 f$ uniformly on $X \times X$. Deduce that $BC^1(X \to K)$ is a K-Banach space with respect to the norm $\| \; \|_1$. (Comment. This result is a first indication that our new definition of a C^1-function is better than the traditional one.)

We shall now look into the local invertibility of C^1-functions. It is easy to see that dramatic examples such as Example 26.6 do not occur in $C^1(X \to K)$. In fact we have the following result.

PROPOSITION 27.3. (Local injectivity of C^1-functions) *Let X be a non-empty subset of K without isolated points, let $f: X \to K$ be C^1 at some point $a \in X$. If $f'(a) \neq 0$ there is a neighbourhood U of a such that*

$$|f(x) - f(y)| = |f'(a)|\,|x - y| \qquad (x, y \in X \cap U)$$

In other words, $f/f'(a)$ is an isometry on a (relative) neighbourhood of a. In particular, f is injective on a neighbourhood of a.

Proof. The statement follows from the simple observation that if $x, y \in X$ $(x \neq y)$ are sufficiently close to a then

$$\left| \frac{f(x) - f(y)}{x - y} - f'(a) \right| < |f'(a)|$$

Remark. Observe that we in fact have proved that $f' | X \cap U$ is monotone of type $\operatorname{sgn} f'(a)$ in the sense of Definition 24.6.

How about the local image of f? Useful statements on this for arbitrary X cannot be expected, so let us assume that X is an open subset of K. If f is as above does it map small neighbourhoods of a onto (full) neighbourhoods of $f(a)$ in K? It should be noticed that the mere fact that f is locally a scalar times an isometry is not enough to prove it. (See also Section 75 for bad behaviour of isometries.) To prove the following key lemma we shall need a property of f slightly stronger then local monotony.

LEMMA 27.4. (Newton approximation) *Let f be a K-valued function defined* *on a disc $B := B_a(r)$ in K. Suppose there is an $s \in K$ such that*

$$\sup \left\{ \left| \frac{f(x)-f(y)}{x-y} - s \right| : x, y \in B, x \neq y \right\} < |s|$$

Then $s^{-1} f$ is an isometry and f maps discs onto discs. More precisely, for each $b \in B$ and $r_1 \in (0, r]$ f maps the disc $B_b(r_1)$ onto the disc $B_{f(b)}(|s|\, r_1)$.

Proof. It follows directly that $|f(x)-f(y)| = |s|\,|x-y|$ for all $x, y \in B$ so that $f(B) \subset B_{f(a)}(|s|\,r)$. We shall prove that f maps B onto $B_{f(a)}(|s|\,r)$. (Application of this result to subdiscs yields the rest of theorem.) Choose $c \in B_{f(a)}(|s|\,r)$; we show that $x \mapsto f(x) - c$ has a zero in B using Newton's method. Define $g(x) := x - s^{-1}(f(x)-c)$ $(x \in B)$. It is easily seen that g maps B into B. If $x, y \in B$ then $|g(x) - g(y)| = |x-y-s^{-1}(f(x)-f(y))| = |s^{-1}(x-y)|\,|(f(x)-f(y))/(x-y) - s| \leqslant \tau |x-y|$ for some $\tau \in (0, 1)$. By the contraction theorem (Appendix A.1) we may conclude that g has a fixed point $z \in B$. It follows that $f(z) = c$.

Remark. It is clear from the proof of the contraction theorem that a solution of $f(x) = c$ can be obtained by iteration as follows. Choose an arbitrary $x_1 \in B$ and set $x_{n+1} = x_n - s^{-1}(f(x_n) - c)$ for $n \in \mathbb{N}$. Then $f(\lim_{n\to\infty} x_n) = c$.

As a corollary we obtain the following.

THEOREM 27.5. (Local invertibility theorem for C^1-functions) *Let f be a K-valued function defined on some neighbourhood of $a \in K$. If f is C^1 at a and $f'(a) \neq 0$ then for sufficiently small $r \in (0, \infty)$ the disc $B_a(r)$ is mapped by f onto $B_{f(a)}(|f'(a)|\,r)$. The local inverse g of f*

$$g : B_{f(a)}(|f'(a)|\,r) \to B_a(r)$$

is C^1 at $f(a)$ and $g'(f(a)) = f'(a)^{-1}$.

Proof. If r is small enough then (with the notation $\Phi_1 f$ as in Definition 27.1)

$$\sup \{ |\Phi_1 f(x, y) - f'(a)| : x, y \in B_a(r), x \neq y \} \leqslant \tfrac{1}{2}\,|f'(a)|$$

and the previous lemma tells us that f maps $B_a(r)$ onto $B_{f(a)}(|f'(a)|\,r)$. The function g is a scalar multiple of an isometry and hence continuous. To prove that g is C^1 at $f(a)$, let $z, t \in B_{f(a)}(|f'(a)|\,r)$, $z \neq t$. Then

$$\Phi_1 g(z, t) = \Phi_1 f(g(z), g(t))^{-1}$$

so that $g'(f(a)) = \lim_{(z,t)\to(f(a), f(a))} \Phi_1 g(z, t) = \lim_{(u,v)\to(a,a)} (\Phi_1 f(u, v))^{-1} = f'(a)^{-1}$ (where the limits are restricted to $z \neq t$, $u \neq v$). The statement follows.

Another corollary of Lemma 27.4 states roughly that if an analytic function has an approximate zero then it has a zero.

THEOREM 27.6. (Hensel's lemma) *Let f be an analytic function on $B_0(1)$ given by*

$$f(x) = \sum_{n=0}^{\infty} a_n x^n \qquad (x \in B_0(1))$$

Suppose that $|a_n| \leqslant 1$ for all n, and that there is an element $a \in B_0(1)$ for which $|f(a)| < 1$ and $|f'(a)| = 1$. Then there is $b \in B_0(1)$ such that $|b-a| \leqslant |f(a)|$ and $f(b) = 0$.

Proof. We may assume $r := |f(a)| > 0$. We shall apply Lemma 27.4 to $f|B_a(r)$ with $s = f'(a)$. First we develop f into a power series in $x - a$ (Theorem 25.1)

$$f(x) = b_0 + b_1(x-a) + b_2(x-a)^2 + \ldots \qquad (|x| \leqslant 1)$$

Observe that $b_0 = f(a)$ and $b_1 = f'(a)$ and that $|b_n| \leqslant 1$ for all n. If $x, y \in B_a(r)$, $x \neq y$ then

$$|\Phi_1 f(x,y) - f'(a)| = \left| b_2 \frac{(x-a)^2 - (y-a)^2}{x-y} + b_3 \frac{(x-a)^3 - (y-a)^3}{x-y} + \ldots \right|$$

$$\leqslant \max_{n \geqslant 2} |x-y|^{-1} \, |(x-a)^n - (y-a)^n|$$

$$\leqslant \max_{n \geqslant 2} \left\{ \left| \frac{u^n - v^n}{u - v} \right| : |u| \leqslant r, \ |v| \leqslant r, \ u \neq v \right\}$$

$$\leqslant \max_{n \geqslant 2} r^{n-1} = r = |f(a)| < |f'(a)|$$

It follows that $f|B_a(r)$ satisfies the conditions of Lemma 27.4 and we may conclude that f maps $B_a(r)$ onto $B_{f(a)}(r)$. But $0 \in B_{f(a)}(r)$ so there is an element $b \in B_a(r)$ for which $f(b) = 0$ and we are done.

Remarks.

1. Hensel's lemma is often formulated for polynomials in a more algebraic way. See Exercise 27.J.
2. For further applications of Lemma 27.4, see Exercises 27.D-27.H.
3. Other tools to obtain information on the zeros of analytic functions on $\mathbb{C}_p$ are the 'Newton polygons', see Amice (1975) or Koblitz (1977).
4. In Part 1 of Chapter 4 we shall study differentiable functions that are not necessarily C^1.

**Exercise* 27.D. Let exp, E be as in Definition 25.4. Show that exp maps every 'closed' subdisc D of E containing 0 isometrically onto $1 + D$. Deduce that exp maps E onto $1 + E$.

Exercise 27.E. (On $\sqrt{-1}$ in $\mathbb{Q}_p$) In Exercise 15.D you were asked to prove that the equation $x^2 + 1 = 0$ has no solutions in $\mathbb{Q}_p$ for $p \equiv 3 \pmod 4$ and for $p = 2$. In this exercise we consider the remaining case $p \equiv 1 \pmod 4$. Show that there is an element in the residue class field $\mathbb{F}_p$ of $\mathbb{Q}_p$ whose square is -1. Now use Hensel's lemma to show that *if* $p \equiv 1 \pmod 4$ *then the equation* $x^2 + 1 = 0$ *has two solutions in* $\mathbb{Q}_p$.

Exercise 27.F. (On the 'circle' $x^2 + y^2 = 1$ in $\mathbb{Q}_p$) Use the previous exercise to show that the 'circle' $\{ (x, y) \in \mathbb{Q}_p^2 : x^2 + y^2 = 1 \}$ is a compact subset of $\mathbb{Q}_p^2$ if and only if $p = 2$ or $p \equiv 3 \pmod 4$.

**Exercise* 27.G. (The $(p-1)$th roots of unity in $\mathbb{Q}_p$) Show that for $x, y \in \mathbb{Z}_p$ with $|x - y|_p < 1$ we have $|x^p - y^p|_p \leqslant p^{-1}\, |x - y|_p$ and use the contraction theorem to show that in each additive coset of $p\mathbb{Z}_p$ in $\mathbb{Z}_p$ there is a unique p-adic number a for which $a^p = a$. Deduce that the equation $x^p = x$ has precisely p roots in $\mathbb{Q}_p$ and that they are equidistant. Finally, conclude that for each $b \in \mathbb{Z}_p$ the sequence $b, b^p, b^{p^2}, \ldots$ converges to the only element a for which $a^p = a$ and $|b - a|_p < 1$. (Compare Exercise 23.H.)

Exercise 27.H. Arrive at the same result as in the previous exercise by applying Hensel's lemma to the function $x \mapsto x^{p-1} - 1$.

**Exercise* 27.I. (The Teichmüller representation) Exercise 27.G leads to an alternative way to represent p-adic numbers (the *Teichmüller representation*) which is – in a sense – more canonical than the standard p-adic expansion. In fact, the solution set of $x^p = x$ in $\mathbb{Q}_p$ is $\{0, \theta, \theta^2, \ldots, \theta^{p-1}\}$ where θ is a primitive $(p-1)$th root of unity. Show that each $x \in \mathbb{Q}_p$ can uniquely be written as $\sum_{n=-\infty}^{\infty} b_n p^n$ where $b_n \in \{0, \theta, \theta^2, \ldots, \theta^{p-1}\}$ and $b_{-n} = 0$ for large n and that, conversely, each such series represents a p-adic number.

Exercise 27.J. (Algebraic form of Hensel's lemma) Let k and $\bar{x}$ be as in Definition 11.2. Prove the following theorem. Let $P := a_0 + a_1 X + \ldots + a_n X^n \in K[X]$, $\max \{ |a_j| : 0 \leqslant j \leqslant n \} = 1$. Let $\bar{P} := \bar{a}_0 + \bar{a}_1 X + \ldots + \bar{a}_n X^n \in k[X]$ have a simple root s in k (that is, $\bar{P}$ is divisible by $X - s$, not by $(X - s)^2$). Then P has a root $b \in B_0(1)$ for which $\bar{b} = s$. (See Amice (1975) for a more general version of this theorem.)

28. Twice continuously differentiable functions

In the next section we shall define C^n-functions for arbitrary $n \in \mathbb{N}$. As an intermediate we shall consider C^2-functions.

In order to define C^2-functions in the spirit of the previous section we consider the first difference quotient of a function $f: X \to K$ $(X \subset K)$

$$\Phi_1 f(x, y) = \frac{f(x) - f(y)}{x - y} \qquad (x, y \in X, x \neq y)$$

and again form difference quotients in one of the variables (which one is immaterial) obtaining the *second order difference quotient* $\Phi_2 f$ given by

$$\Phi_2 f(x, y, z) := \frac{\Phi_1 f(x, y) - \Phi_1 f(x, z)}{y - z} =$$

$$(y - z)^{-1} \left(\frac{f(x) - f(y)}{x - y} - \frac{f(x) - f(z)}{x - z} \right)$$

Observe that $\Phi_2 f$ is a symmetric function of its three variables x, y, z and is defined on $\{ (x, y, z) \in X^3 : x \neq y, y \neq z, x \neq z \}$. The latter set is dense in X^3 if X has no isolated points.

DEFINITION 28.1. Let X be a nonempty subset of K without isolated points. Let $a \in X$. A function $f : X \to K$ is C^2 *(twice continuously differentiable) at a* if

$$\lim_{(x, y, z) \to (a, a, a)} \Phi_2 f(x, y, z)$$

exists. (Here, of course, the limit is taken only with respect to those (x, y, z) that are in the domain of $\Phi_2 f$.) f is a *C^2-function (twice continuously differentiable function)* if f is C^2 at every point of X. The set of all C^2-functions $X \to K$ is denoted $C^2(X \to K)$. For $f : X \to K$ set

$$\|f\|_2 = \|f\|_\infty \vee \|\Phi_1 f\|_\infty \vee \|\Phi_2 f\|_\infty$$

and let $BC^2(X \to K) := \{ f \in C^2(X \to K) : \|f\|_2 < \infty \}$

If there is any justice in the world the following should be true.

PROPOSITION 28.2. *Let X be a nonempty subset of K without isolated points, let $f : X \to K$.*
(i) If f is C^2 at some point $a \in X$ then f is C^1 at a.
(ii) $C^2(X \to K)$ and $BC^2(X \to K)$ are K-linear subspaces of $C^1(X \to K)$ and $BC^1(X \to K)$ respectively. $\| \ \|_2$ is a norm on $BC^2(X \to K)$.
(iii) $C^2(X \to K)$ and $BC^2(X \to K)$ are closed for products.
(iv) $f \in C^2(X \to K)$ if and only if $\Phi_2 f$ can (uniquely) be extended to a continuous function $\bar{\Phi}_2 f$ on X^3.
Proof. If f is C^2 at a then $\Phi_2 f$ is bounded on some neighbourhood of (a, a, a), i.e. $|\Phi_2 f|$ is bounded by some $M > 0$ on a set of the form $\{ (x, y, z) \in X^3 : |x - a| < \delta, |y - a| < \delta, |z - a| < \delta, x \neq y, y \neq z, x \neq z \}$. This is the information we need to prove that f is C^1 at a. In fact, if x, y, z, t are sufficiently close to a and pairwise distinct then

$$|\Phi_1 f(x,y) - \Phi_1 f(z,t)| \leqslant |\Phi_1 f(x,y) - \Phi_1 f(y,z)| \vee |\Phi_1 f(y,z) - \Phi_1 f(z,t)|$$
$$\leqslant |x-z|\,|\Phi_2 f(x,y,z)| \vee |y-t|\,|\Phi_2 f(y,z,t)|$$
$$\leqslant M \max(|x-z|, |y-t|)$$

and from this it follows easily that f is C^1 at a. This proves (i). (ii) is simple and (iii) follows from the formula

$$\Phi_2 fg(x,y,z) = f(x)\,\Phi_2 g(x,y,z) + \Phi_1 f(x,y)\,\Phi_1 g(y,z) + \Phi_2 f(x,y,z)\,g(z)$$

(where $f, g : X \to K$ and $x, y, z \in X$ are pairwise distinct).

Finally, to prove (iv) (only one half is interesting) let $f \in C^2(X \to K)$. By (i), f is also in C^1. From that it follows that we can extend $\Phi_2 f$ to a function $\tilde{\Phi}_2 f$ defined on $X^3 \setminus \{(x,x,x) : x \in X\}$, for example by letting x tend to y in

$$\Phi_2 f(x,y,z) = \frac{\Phi_1 f(x,y) - \Phi_1 f(x,z)}{y-z}$$

obtaining

$$\tilde{\Phi}_2 f(y,y,z) = \frac{f'(y) - \Phi_1 f(y,z)}{y-z}$$

etc. It is easily seen that $\tilde{\Phi}_2 f$ is continuous on its domain and that the fact that $f \in C^2(X \to K)$ enables us to further extend $\tilde{\Phi}_2 f$ to a continuous function $\overline{\Phi}_2 f$ on X^3. (See Theorem 29.9 for a less sketchy proof.)

The proof of the above proposition, although quite elementary, shows that we have to do some work in order to arrive at results whose 'archimedean' counterparts are trivial. We shall meet the same state of things in the next section.

Where is the second derivative for a C^2-function f? We had $\overline{\Phi}_1 f(x,x) = f'(x)$, so what is $\overline{\Phi}_2 f(x,x,x)$? By the definition of $\overline{\Phi}_2 f$ we have for $x, z \in X, x \neq z$

$$\overline{\Phi}_2 f(x,x,z) = (x-z)^{-1}\,(f'(x) - \Phi_1 f(x,z))$$
$$\overline{\Phi}_2 f(z,x,z) = (x-z)^{-1}\,(\Phi_1 f(x,z) - f'(z))$$

Addition yields

$$\overline{\Phi}_2 f(x,x,z) + \overline{\Phi}_2 f(z,x,z) = (x-z)^{-1}\,(f'(x) - f'(z))$$

It follows that

$$(*) \qquad 2\overline{\Phi}_2 f(x,x,x) = f''(x)$$

Apparently $f \in C^2(X \to K)$ implies that f is twice differentiable and that f'' is continuous. The converse is, of course, not true (as we have seen there are functions with zero derivative that are not even C^1-functions, see also Exercise 28.B.).

If $\operatorname{char}(K) \neq 2$ it follows from (*) that

$$\overline{\Phi}_2 f(x, x, x) = \tfrac{1}{2} f''(x)$$

but if $\mathrm{char}(K) = 2$ we can draw another interesting conclusion, namely

$$f \in C^2(X \to K) \text{ implies } f'' = 0$$

Observe that in this case $\overline{\Phi}_2 f(x, x, x)$ may be nonzero! For example, let $f(x) = x^2$ for all $x \in K$. Then $\overline{\Phi}_2 f(x, x, x)$ is identically 1, which is the coefficient of x^2 in the 'Taylor expansion' of f. In general, $\overline{\Phi}_2 f(x, x, x)$ shall give us more information than f'' and it is for that reason that we prefer to work with $\overline{\Phi}_2 f(x, x, x)$ rather than f''. (In the next section we shall encounter similar features in arbitrary prime characteristic.)

Exercise 28.A. Show that locally analytic functions are C^2-functions. (Consider also the last sentence of Exercise 25.B.)

Exercise 28.B. (i) Let $\mathrm{char}(K) \neq 2$ and let $f : K \to K$ satisfy (see Example 26.4)

$$|f(x) - f(y)| = |x - y|^2 \qquad (x, y \in K)$$

Show that f is infinitely many times differentiable and that f is a C^1- but not a C^2-function.
(ii) Give an example showing that the clause '$\mathrm{char}(K) \neq 2$' in (i) is necessary.

Exercise 28.C. Let $X \subset K$ have no isolated points and let $f : X \to K$. Show that boundedness of $\Phi_2 f$ implies $f \in C^1(X \to K)$.

Exercise 28.D. Prove that $BC^2(X \to K)$ (Definition 28.1) is a K-Banach space. (An easy way to do it is by interpreting $BC^2(X \to K)$ as a closed subspace of $B(X \to K) \times B(X^2 \to K) \times B(X^3 \to K)$ via the map

$$f \mapsto (f, \overline{\Phi}_1 f, \overline{\Phi}_2 f) \qquad (f \in BC^2(X \to K))$$

Here B is as in Section 13; if E, F are Banach spaces then $E \times F$ is a Banach space with respect to the norm $(x, y) \mapsto \max(\|x\|, \|y\|)$.)

Exercise 28.E. (On uniform differentiability) In this exercise we consider two notions that deserve the name 'uniform differentiability'. Both are stronger than 'continuous differentiability' which we have defined in Section 27. For simplicity, let $f : B_0(1) \to K$. f is *uniformly differentiable* (u.d.) if f is differentiable and

$$\lim_{(x, y) \to (a, a)} \Phi_1 f(x, y) = f'(a) \qquad \text{uniformly in } a \in B_0(1)$$

f is *strongly uniformly differentiable* (s.u.d.) if $\Phi_1 f$ can be extended to a uniformly continuous function $\overline{\Phi}_1 f$ on $B_0(1) \times B_0(1)$.

Prove the following assertions (i)-(vi).

(i) A s.u.d. function is u.d.; a u.d. function is C^1.
(ii) If K is locally compact then the above two notions of uniform differentiability coincide with continuous differentiability.
(iii) If K is not locally compact there is a u.d. function that is not s.u.d; there is a C^1-function that is not u.d.
(iv) If f is u.d. then f' is uniformly continuous.
(v) If $\Phi_2 f$ is bounded then f is s.u.d.
(vi) If f is analytic then f is s.u.d.
Now let $f : B_0(1) \to K$ be bounded and uniformly differentiable. Show that f is strongly uniformly differentiable. (Hint. First prove that f' is bounded, next that $\Phi_1 f$ is bounded and finally that f is s.u.d.) Does your proof remain valid if $B_0(1)$ is replaced by an arbitrary nonempty subset of K without isolated points?

We conclude this section with a discussion concerning the Taylor formula for C^2-functions. Proposition 27.2 (γ) may be regarded as a Taylor formula for C^1-functions which leads to the following proposition.

PROPOSITION 28.3. (Taylor formula for C^2-functions) *Let $X \subset K$ have no isolated points, let $f \in C^2(X \to K)$. Then there is a continuous function $R_2 : X \times X \to K$ such that*

$$f(x) = f(y) + (x-y)f'(y) + (x-y)^2 R_2(x,y) \qquad (x, y \in X)$$

Proof. Choose $R_2(x,y) = \overline{\Phi}_2 f(x,y,y)$.

A little harder to prove is the following converse of Proposition 28.3.

PROPOSITION 28.4. *Let $X \subset K$ have no isolated points and let $f : X \to K$. Suppose that there are continuous functions $\lambda : X \to K$ and $\Lambda : X \times X \to K$ such that*

$$f(x) = f(y) + (x-y)\lambda(y) + (x-y)^2 \Lambda(x,y) \qquad (x, y \in X)$$

Then $f \in C^2(X \to K)$, $\lambda = f'$, $\Lambda(x,y) = \overline{\Phi}_2 f(x,y,y)$ $(x, y \in X)$.

Proof. First observe that $f \in C^1(X \to K)$ and that $\lambda = f'$. Now let $x, y, z \in X$ $(x \neq y, y \neq z, x \neq z)$. We may suppose that, say, $|y-z|$ is the largest among the numbers $|x-y|$, $|x-z|$, $|y-z|$. We have

$$\Phi_1 f(x,y) = f'(x) + (y-x)\Lambda(y,x)$$
$$\Phi_1 f(x,z) = f'(x) + (z-x)\Lambda(z,x)$$

so that

$$\begin{aligned}\Phi_2 f(x,y,z) &= (y-z)^{-1}(\Phi_1 f(x,y) - \Phi_1 f(x,z)) \\ &= \mu_1(x,y,z)\,\Lambda(y,x) + \mu_2(x,y,z)\,\Lambda(z,x)\end{aligned}$$

where $|\mu_1| \leqslant 1$, $|\mu_2| \leqslant 1$ and $\mu_1 + \mu_2 = 1$. It follows that if each one of the triple x, y, z is close to some $a \in X$ then $\Lambda(y, x)$ and $\Lambda(z, x)$ are close to $\Lambda(a, a)$, hence so is $\Phi_2 f(x, y, z)$, i.e. f is C^2 at a. That $\overline{\Phi}_2 f(x, y, y) = \Lambda(x, y)$ is now easy to prove.

For the problem of finding a similar converse for C^3-functions we refer to Section 83. In Section 84 we shall touch upon the notions of C^1- and C^2-functions of two variables.

Exercise 28.F. Find an example of an injective function $f : \mathbb{Z}_p \to \mathbb{Z}_p$ for which

$$|f(x) - f(y)|_p = |x - y|_p^3 \qquad (x, y \in \mathbb{Z}_p)$$

Show that such an f is a C^2-function. (See Exercise 28.B.)

Exercise 28.G. The derivative of a C^2-function is a C^1-function. Take the trouble to verify this.

29. C^n-functions

We now define C^n-functions.

IN THIS SECTION X IS A NONEMPTY SUBSET OF K WITHOUT ISOLATED POINTS

DEFINITION 29.1. For $n \in \mathbb{N}$ set

$$\nabla^n X := \{(x_1, x_2, \ldots, x_n) \in X^n : \text{if } i \neq j \text{ then } x_i \neq x_j\}$$

The *nth (order) difference quotient* $\Phi_n f : \nabla^{n+1} X \to K$ of a function $f : X \to K$ is inductively given by $\Phi_0 f := f$ and, for $n \in \mathbb{N}$, $(x_1, x_2, \ldots, x_{n+1}) \in \nabla^{n+1} X$ by

$$\Phi_n f(x_1, x_2, \ldots, x_{n+1}) = \frac{\Phi_{n-1} f(x_1, x_3, \ldots, x_{n+1}) - \Phi_{n-1} f(x_2, x_3, \ldots, x_{n+1})}{x_1 - x_2}$$

f is a *C^n-function* (f *is* C^n) if $\Phi_n f$ can be extended to a continuous function $\overline{\Phi}_n f : X^{n+1} \to K$. We then set

$$D_n f(a) := \overline{\Phi}_n f(a, a, \ldots, a) \qquad (a \in X)$$

The set of all C^n-functions $X \to K$ is denoted $C^n(X \to K)$. Let $C^\infty(X \to K) := \bigcap_{n=1}^\infty C^n(X \to K)$. The elements of $C^\infty(X \to K)$ are *C^∞-functions*.

Remarks.

1. For $n = 1, 2$ the definitions of $\Phi_n f$, $C^n(X \to K)$, $\overline{\Phi}_n f$ tie in with the contents of Sections 27 and 28.

2. Since X has no isolated points the set $\nabla^{n+1} X$ is dense in X^{n+1} so that for a C^n-function f the extension of $\Phi_n f$ is unique.

In this introductory chapter we shall not treat the theory of C^n-functions systematically (for this, see the second part of Chapter 4) but restrict ourselves to checking a few properties in order to justify our definition. Thus in this section we shall prove that $C^n(X \to K)$ is a K-vector space closed for products, that

$$C(X \to K) \supset C^1(X \to K) \supset \ldots \qquad \supset C^\infty(X \to K)$$

and that (locally) analytic functions are C^∞. Further, we shall see that being in C^n is a local property, that $f \in C^n(X \to K)$ implies f is n times differentiable and $n!\, D_n f = f^{(n)}$. Also we shall establish a Taylor formula for C^n-functions.

All these facts are not surprising, their proofs are elementary but somewhat laborious. However, in this context we should not think too little of 'obvious' looking statements. For example, the reader may prove that the following 'propositions' are *false*.

$$f' \text{ is } C^n \Rightarrow f \text{ is } C^{n+1}$$

A C^n-function has a C^{n+1}-antiderivative.

We first develop some machinery involving $\Phi_n f$.

LEMMA 29.2. (Computational rules for $\Phi_n f$) *Let $f, g : X \to K$, let $\lambda, \mu \in K$ and $n \in \mathbb{N}$. Then we have the following.*

(i) *If $(x, y, z, x_1, \ldots, x_{n-1}) \in \nabla^{n+2}(X)$ then*

$$(x-y)\,\Phi_n f(x, y, x_1, \ldots, x_{n-1}) + (y-z)\,\Phi_n f(y, z, x_1, \ldots, x_{n-1}) = (x-z)\,\Phi_n f(x, z, x_1, \ldots, x_{n-1})$$

(ii) *$\Phi_n f$ is a symmetric function of its $n+1$ variables.*

(iii) *If $(x_1, \ldots, x_n, a_1, \ldots, a_n) \in \nabla^{2n} X$ then*

$$\Phi_{n-1} f(x_1, \ldots, x_n) - \Phi_{n-1} f(a_1, \ldots, a_n) = \sum_{j=1}^{n} (x_j - a_j)\,\Phi_n f(a_1, \ldots, a_j, x_j, \ldots, x_n)$$

(iv) $\Phi_n(\lambda f + \mu g) = \lambda \Phi_n f + \mu \Phi_n g$.

(v) *If $(x_1, \ldots, x_{n+1}) \in \nabla^{n+1} X$ then*

$$\Phi_n(fg)(x_1, \ldots, x_{n+1}) = \sum_{j=0}^{n} \Phi_j f(x_1, \ldots, x_{j+1})\,\Phi_{n-j} g(x_{j+1}, \ldots, x_{n+1})$$

(vi) *If $f(x) \neq 0$ for all $x \in X$, $g = 1/f$, $(x_1, \ldots, x_{n+1}) \in \nabla^{n+1} X$ then*

$$\Phi_n g(x_1,\ldots,x_n) = -f(x_1)^{-1} \sum_{j=1}^{n} \Phi_j f(x_1,\ldots,x_{j+1})\Phi_{n-j} g(x_{j+1},\ldots,x_{n+1})$$

(vii) *f is a polynomial function of degree $\leqslant n$ if and only if $\Phi_{n+1}f = 0$.*

Proof. We shall not take pains in carrying out the proof in full detail.

(i) is a direct consequence of the definition of $\Phi_n f$. The proof of (ii) runs by induction. The hypothesis that $\Phi_{n-1}f$ is symmetric implies already the invariance of $\Phi_n f(x_1,\ldots,x_{n+1})$ under permutations of $x_3,\ldots,x_{n+1}$ and of x_1, x_2. Thus it suffices to show that $\Phi_n f(x_1, x_2, x_3, \ldots) = \Phi_n(x_3, x_2, x_1, \ldots)$ and this is a consequence of (i). Property (iii) follows from the definition of $\Phi_{n+1}f$ and the formula $\Phi_n f(x_1,\ldots,x_{n+1}) - \Phi_n f(a_1,\ldots, a_{n+1}) = (\Phi_n f(x_1,\ldots,x_{n+1}) - (\Phi_n f(a_1,x_2,\ldots,x_{n+1})) + (\Phi_n f(a_1,x_2, \ldots,x_{n+1}) - \Phi_n f(a_1,a_2,x_3,\ldots,x_{n+1})) + \ldots + (\Phi_n f(a_1,\ldots,a_n,x_{n+1}) - \Phi_n f(a_1,\ldots,a_{n+1}))$. (iv) is clear, and (v) follows by induction on n. Rule (vi) follows from (v) and the fact that $\Phi_n(fg) = 0$ for $n \geqslant 1$. Finally, consider (vii). Let x^n denote the function $x \mapsto x^n$. From (v) it follows that $\Phi_n x^n = 1$, hence $\Phi_{n+1}f = 0$ for any polynomial function of degree $\leqslant n$. Conversely, if $\Phi_{n+1}f = 0$ then $\Phi_n f$ is a constant c, so that $\Phi_n(f - c x^n) = 0$. By the induction hypothesis, $f - c x^n$ is a polynomial function of degree $\leqslant n-1$.

Exercise 29.A. (A symmetric formula for $\Phi_n f$) Let $n \in \mathbb{N}$, $f : X \to K$. Show that

$$\Phi_n f(x_1,\ldots,x_{n+1}) = \sum_{i=1}^{n+1} \left\{ \prod_{j \neq i} (x_i - x_j)^{-1} \right\} f(x_i)$$
$$((x_1,\ldots,x_{n+1}) \in \nabla^{n+1} X)$$

Exercise 29.B. Let $0 \notin X$ and $f(x) := x^{-1}$ $(x \in X)$. Prove that for $n \in \mathbb{N}$

$$\Phi_n f(x_1,\ldots,x_{n+1}) = (-1)^n \prod_{j=1}^{n+1} x_j^{-1} \quad ((x_1,\ldots,x_{n+1}) \in \nabla^{n+1} X)$$

COROLLARY 29.3. *$C^{n-1}(X \to K) \supset C^n(X \to K)$ for each n. If in the rules (i)-(vi) of Lemma 29.2 we replace $\Phi.$ by $\bar{\Phi}.$ and $\nabla^{\cdot} X$ by $X^{\cdot}$ everywhere and assume that $f, g \in C^n(X \to K)$ then the resulting rules are again true. $C^n(X \to K)$ is a linear space, closed for products. If $f \in C^n(X \to K)$ and $f(x) \neq 0$ for all $x \in X$ then $1/f \in C^n(X \to K)$. Polynomial functions are C^∞.*

Proof. The right-hand side of Lemma 29.2 (iii) can, in the case $f \in C^n(X \to K)$, be extended to a continuous function of $(x_1,\ldots,x_n)$. It follows that $\Phi_{n-1}f$ can be extended to a continuous function of $(x_1,\ldots,x_n) \in X^n$, i.e. $f \in C^{n-1}(X \to K)$. A similar proof works for 'f is C^n, $f(x) \neq 0$ for all

$x \Rightarrow 1/f$ is C^{n}' (induction and rule (vi)). The rest follows by simple continuity arguments.

**Exercise* 29.C. (A Banach space of C^n-functions) Let $BC^n(X \to K)$ be the space of all C^n-functions $f : X \to K$ for which

$$\|f\|_n := \max \{\|\Phi_j f\|_\infty : 0 \leqslant j \leqslant n\}$$

is finite. Prove that $(BC^n(X \to K), \|\ \|_n)$ is a K-Banach space (see Exercise 28.D).

THEOREM 29.4. (Taylor formula for C^n-functions) *Let $f \in C^n(X \to K)$. Then for all $x, y \in X$*

$$f(x) = f(y) + \sum_{j=1}^{n-1} (x-y)^j D_j f(y) + (x-y)^n \bar{\Phi}_n f(x, y, y, \ldots, y)$$

$$= f(y) + \sum_{j=1}^{n} (x-y)^j D_j f(y) + (x-y)^n (\bar{\Phi}_n f(x, y, y, \ldots, y) - D_n f(y))$$

Proof. The Taylor formula is true for $n = 1, 2$ (Propositions 27.2. and 28.3). Suppose it is true for $n-1$, and let $f \in C^n(X \to K)$. Then, by Corollary 29.3, $f \in C^{n-1}(X \to K)$ so by the induction hypothesis

$$f(x) = f(y) + \sum_{j=0}^{n-2} (x-y)^j D_j f(y) + (x-y)^{n-1} \bar{\Phi}_{n-1} f(x, y, y, \ldots, y)$$

$$(x, y \in X)$$

The required formula for f now follows from

$$\bar{\Phi}_{n-1} f(x, y, y, \ldots, y) = D_{n-1} f(y) + (x-y) \bar{\Phi}_n f(x, y, y, \ldots, y) \quad (x, y \in X)$$

THEOREM 29.5. *Let $f \in C^n(X \to K)$. Then f is n times differentiable and $j!\, D_j f = f^{(j)}$ for $1 \leqslant j \leqslant n$.*

Proof. We first show that $D_{n-1} f$ is C^1 and $(D_{n-1} f)' = n D_n f$. By Corollary 29.3 $f \in C^{n-1}(X \to K)$ so that for $x, y \in X$, $x \neq y$

$$\Phi_1 D_{n-1} f(x, y) = (x-y)^{-1} (D_{n-1} f(x) - D_{n-1} f(y)) =$$
$$(x-y)^{-1} (\bar{\Phi}_{n-1} f(x, x, \ldots, x) - \bar{\Phi}_{n-1} f(y, y, \ldots, y))$$

By the (extended) rule (iii) of Lemma 29.2 the right-hand side equals

$$\bar{\Phi}_n f(x, y, \ldots, y) + \bar{\Phi}_n f(x, x, y, \ldots, y) + \ldots + \bar{\Phi}_n f(x, x, \ldots, x, y)$$

which is a continuous function of $(x, y) \in X^2$. It follows that $D_{n-1} f$ is C^1. After letting y tend to x we arrive at $(D_{n-1} f)'(x) = n\, \bar{\Phi}_n f(x, x, \ldots, x) = n D_n f(x)$. Since f is also a C^j-function for $1 \leqslant j \leqslant n$ we have by the same token that $(D_{j-1} f)' = j D_j f$ so

$$j!D_jf = ((j-1)!D_{j-1}f)' = ((j-2)!D_{j-2}f)'' = \ldots = (D_1f)^{(j-1)} = f^{(j)}$$

The theorem follows.

Remark. Theorem 29.5 does not state that $f' \in C^{n-1}(X \to K)$. For a proof of this, see Section 78.

COROLLARY 29.6. *Let* $\operatorname{char}(K) = p$ *and* $f \in C^p(X \to K)$. *Then* $f^{(p)} = 0$.

We now turn to the question as to whether being in C^n is a local property.

LEMMA 29.7. *Let* $f \in C^{n-1}(X \to K)$ *for some* $n \in \mathbb{N}$. *Then* $\Phi_n f$ *can be extended to a continuous function* $\tilde{\Phi}_n f$ *on* $X^{n+1} \setminus \Delta$ *where the diagonal* Δ *is defined by* $\Delta := \{(x, x, \ldots, x) \in X^{n+1} : x \in X\}$.

Proof. For $1 \leqslant i, j \leqslant n+1, i \neq j$ set

$$U_{ij} := \{(x_1, x_2, \ldots, x_{n+1}) \in X^{n+1} : x_i \neq x_j\}$$

Then each U_{ij} is open in X^{n+1} and their union is $X^{n+1} \setminus \Delta$. Define $h_{ij} : U_{ij} \to K$ by

$$h_{ij}(x_1, x_2, \ldots, x_{n+1}) := (x_i - x_j)^{-1} (\overline{\Phi}_{n-1} f(x_1, \ldots, x_{j-1}, x_{j+1}, \ldots, x_{n+1}) - \overline{\Phi}_{n-1} f(x_1, \ldots, x_{i-1}, x_{i+1}, \ldots, x_{n+1}))$$

Each h_{ij} is a continuous extension of $\Phi_n f$. We glue these functions together by defining

$$\tilde{\Phi}_n f(x_1, \ldots, x_{n+1}) := h_{ij}(x_1, \ldots, x_{n+1}) \quad ((x_1, \ldots, x_{n+1}) \in U_{ij})$$

One checks easily that $\tilde{\Phi}_n f$ is a well-defined continuous extension of $\Phi_n f$ on $X^{n+1} \setminus \Delta$.

DEFINITION 29.8. Let $n \in \mathbb{N} \cup \{0\}$. A function $f : X \to K$ is C^n *at a point* $a \in X$ if the limit

$$\lim_{v \to \mathbf{a}} \Phi_n f(v) \qquad (\mathbf{a} := (a, a, \ldots, a) \in X^{n+1})$$

(where v is restricted to $\nabla^{n+1} X$) exists.

THEOREM 29.9. *A function* $f : X \to K$ *is a* C^n*-function if and only if it is* C^n *at* a *for each* $a \in X$.

Proof. We only need to prove the 'if' part for $n \geqslant 1$. For $n = 1$ this is Proposition 27.2 so assume that the statement is true for $n - 1$. Existence of the above limit implies boundedness (say, by M) of $\Phi_n f$ on $U \cap \nabla^{n+1} X$ where U is some neighbourhood of $\mathbf{a}$ in X^{n+1}. By rule (iii) of Lemma 29.2, if $(x_1, x_2, \ldots, x_n, y_1, y_2, \ldots, y_n) \in \nabla^{2n} X$ and if for each i the distances $|x_i - a|$ and $|y_i - a|$ are sufficiently small

$$|\Phi_{n-1}f(x_1,\ldots,x_n)-\Phi_{n-1}f(y_1,\ldots,y_n)| \leqslant M \max\{|x_i-y_i| : 1\leqslant i\leqslant n\}$$

It follows that $\lim_{w\to a}\Phi_{n-1}f(w)$ exists for $a \in X^n$. By the induction hypothesis, $f\in C^{n-1}(X\to K)$. By Lemma 29.7, $\Phi_n f$ can be extended to a continuous function $\tilde{\Phi}_n f$ on $X^{n+1}\setminus \Delta$. It is not hard to prove that the function $\bar{\Phi}_n f$ defined by

$$\bar{\Phi}_n f(a_1,\ldots,a_{n+1}) = \begin{cases} \tilde{\Phi}_n f(a_1,\ldots,a_{n+1}) \text{ if } (a_1,\ldots,a_{n+1})\in X^{n+1}\setminus\Delta \\ \lim\limits_{v\to a} \Phi_n f(v) \text{ if } a=(a_1,\ldots,a_{n+1})\in\Delta \end{cases}$$

is a continuous extension of $\Phi_n f$ to X^{n+1}. Hence, $f\in C^n(X\to K)$.

COROLLARY 29.10. *Let $f : X\to K$, $n\in\mathbb{N}\cup\{0\}$. Suppose X can be covered by (relatively) open sets such that the restriction of f to each such open set is C^n. Then f is itself a C^n-function. In particular, locally constant functions are C^∞.*

COROLLARY 29.11. *Locally analytic functions are C^∞-functions.*

Proof. In view of the previous corollary it suffices to prove the statement for analytic functions f. Without loss of generality we may assume that

$$f(x) = \sum_{j=0}^{\infty} a_j x^j \qquad (|x|\leqslant r)$$

for some $r\in|K^\times|$. Denoting the functions $x\mapsto x^j\,(|x|\leqslant r)$ by $\mathbf{x}^j$ we have, because of

$$\|\Phi_n(\mathbf{x}^j)\|_\infty \leqslant \begin{cases} r^{j-n} \text{ if } 0\leqslant n\leqslant j \\ 0 \quad\quad \text{if } n>j \end{cases}$$

that $\|\Phi_n f\|_\infty = \|\sum_{j=0}^{\infty} a_j\,\Phi_n(\mathbf{x}^j)\|_\infty \leqslant \sup\{|a_j|r^{j-n} : j\geqslant n\} \leqslant r^{-n} \sup\{|a_j|\,r^j : j\geqslant 0\}<\infty$, since the power series converges on the nonempty set $\{x\in K : |x|=r\}$. We see that $\Phi_n f$ is bounded for each n. Formula (iii) of Lemma 29.2 shows that $\Phi_{n-1}f$ is uniformly continuous on $\nabla^n X$ for each $n\geqslant 1$. A multidimensional version of Theorem 22.5 leads to the desired result $f\in C^{n-1}(X\to K)$ for each $n\geqslant 1$.

**Exercise* 29.D. Show that for an analytic function

$$f(x) = \sum_{n=0}^{\infty} a_n x^n \qquad (x\in D)$$

where D is an open convex set containing 0 we have, as expected

$$a_n = D_n f(0,0,\ldots,0) \qquad (n=0,1,2,\ldots)$$

(Do not forget the case $\mathrm{char}(K)=p$.)

As an application of the preceding theory we shall 'characterize' the C^n-functions with vanishing derivative, more precisely $\{f \in C^n(X \to K) : D_1 f = D_2 f = \ldots = D_n f = 0\}$. This reduces the number of occurring variables from $n+1$ to 2.

THEOREM 29.12. *Let* $f : X \to K$, $n \in \mathbb{N}$. *The following conditions* (α), (β) *are equivalent.*

(α) $f \in C^n(X \to K)$ *and* $D_1 f = D_2 f = \ldots = D_n f = 0$.

(β)

$$\lim_{(x,y)\to(a,a)} \frac{f(x)-f(y)}{(x-y)^n} = 0 \text{ for each } a \in X$$

If $\operatorname{char}(K) = 0$ *then conditions* (α), (β) *are equivalent to*

(γ) $f \in C^n(X \to K)$ *and* $f' = 0$.

Proof. We shall prove that (β) implies $f \in C^n(X \to K)$. The other statements follow directly from the Taylor formula (Theorem 29.4) and Theorem 29.5. Thus, let $a \in X$, $\epsilon > 0$. There is $\delta > 0$ such that

$$|(x-y)^{-n}(f(x)-f(y))| < \epsilon \quad (x, y \in B_a(\delta) \cap X, x \neq y)$$

We shall prove by induction on j that for $1 \leqslant j \leqslant n$, $x_1, \ldots, x_{j+1} \in B_a(\delta)$, $(x_1, \ldots, x_{j+1}) \in \nabla^{j+1} X$

$$|\Phi_j f(x_1, \ldots, x_{j+1})| < \epsilon\, d(\{x_1, \ldots, x_{j+1}\})^{n-j}$$

(where, as in Definition 18.3, $d(\{x_1, \ldots, x_{j+1}\})$ is the diameter of $\{x_1, \ldots, x_{j+1}\}$). The statement is true for $j = 1$. For the step from $j-1$ to j let $(x_1, \ldots, x_{j+1}) \in \nabla^{j+1} X$, $x_1, \ldots, x_{j+1} \in B_a(\delta)$. By symmetry we may suppose that

$$d := d(\{x_1, \ldots, x_{j+1}\}) = |x_1 - x_2|$$

Then by the induction hypothesis

$$\begin{aligned} |\Phi_j f(x_1, \ldots, x_{j+1})| &= |x_1 - x_2|^{-1}\, |\Phi_{j-1} f(x_1, x_3, \ldots, x_{j+1}) \\ &- \Phi_{j-1} f(x_2, x_3, \ldots, x_{j+1})| < d^{-1} \epsilon \max(d(\{x_1, x_3, \ldots, x_{j+1}\}), \\ &d(\{x_2, x_3, \ldots, x_{j+1}\}))^{n-j+1} \leqslant d^{-1} \epsilon\, d^{n-j+1} = \epsilon\, d^{n-j} \end{aligned}$$

Applying this result for $j = n$ we may conclude that if $(x_1, \ldots, x_{n+1}) \in \nabla^{n+1} X$ and $x_1, \ldots, x_{n+1}$ are sufficiently close to a then $|\Phi_n f(x_1, \ldots, x_{n+1})| < \epsilon$. In other words, f is C^n at a for each $a \in X$ and, by Theorem 29.9, f is in $C^n(X \to K)$.

**Exercise* 29.E. Let $f : X \to K$. Show that the following conditions (α), (β) are equivalent.

(α) f satisfies a Lipschitz condition of order α for each $\alpha > 0$.

(β) $f \in C^\infty(X \to K)$, $D_n f = 0$ for all $n \in \mathbb{N}$.

If, in addition, $\operatorname{char}(K) = 0$ prove that (α), (β) are equivalent to
(γ) $f \in C^\infty(X \to K), f' = 0$.

Exercise 29.F. In Exercise 28.F you were asked to provide an example of a function $f : \mathbb{Z}_p \to \mathbb{Z}_p$ for which $|f(x) - f(y)|_p = |x - y|_p^3$ $(x, y \in \mathbb{Z}_p)$. Show that f is C^2 but not C^3. Utilize the underlying idea to show that the inclusions in

$$C(\mathbb{Z}_p \to \mathbb{Q}_p) \supset C^1(\mathbb{Z}_p \to \mathbb{Q}_p) \supset \dots$$

are strict.

Exercise 29.G. Find an injective $f \in C^\infty(\mathbb{Z}_p \to \mathbb{Q}_p)$ with vanishing derivative.

30. Antiderivation and integration

In real analysis antiderivation and integration (of continuous functions $f: \mathbb{R} \to \mathbb{R}$) are connected by the formula

$$(*) \qquad F(b) - F(a) = \int_a^b f(x)\mathrm{d}x$$

where F is an antiderivative of f. In the ultrametric case, however, we do not know yet whether an 'integral' exists with similar properties. Although we shall define notions deserving the name 'integral' later on (Section 55, Exercise 62.G, Appendix A.5), a simple connection between antiderivation and 'integration' such as the one given by (*) is lost. We therefore treat the problems of antiderivation and integration separately.

We first turn to antiderivation. For simplicity, let us consider functions $f : \mathbb{Z}_p \to \mathbb{Q}_p$. (For a more detailed study, see Part 3 of Chapter 3 and Part 1 of Chapter 4.) If F is an antiderivative of f and if $g' = 0$ then $F + g$ is also an antiderivative of f. Hence, *if f has an antiderivative then* it has 'many' in the sense that *the set of all antiderivatives of f is dense in* $C(\mathbb{Z}_p \to \mathbb{Q}_p)$. This is an easy consequence of Corollary 26.3.

Which functions *do* have an antiderivative? Let f be analytic. Does it have an *analytic* antiderivative? Not always, since if

$$f(x) = \sum_{n=0}^{\infty} p^n x^{p^n - 1} \qquad (x \in \mathbb{Z}_p)$$

then an analytic antiderivative of f would (up to an additive constant) have the form

$$x \mapsto \sum_{n=0}^{\infty} x^{p^n}$$

but the latter power series does not converge for $x = 1$. However, such f always have locally analytic antiderivatives, as is stated in the following exercise.

**Exercise* 30.A. Let $f : \mathbb{Z}_p \to \mathbb{Q}_p$ be locally analytic of order h (Definition 25.3). Show that f has an antiderivative that is locally analytic of order $h + 1$.

The question as to whether C^n-functions have C^{n+1}-antiderivatives and whether C^∞-functions have C^∞-antiderivatives will be treated in Part 2 of Chapter 4. In this section we consider the following more modest question. Does every continuous function $f : \mathbb{Z}_p \to \mathbb{Q}_p$ have an antiderivative? To answer it we shall approximate f by locally constant functions f_n and choose antiderivatives F_n of f_n. One then may hope that the F_n converge to an antiderivative of f. Without further restrictions, however, this procedure is too rough and will not work. We shall have to be more careful.

LEMMA 30.1. *Let $f_1, f_2, \ldots$ be bounded functions on $\mathbb{Z}_p$ such that $f := \sum_{n=1}^{\infty} f_n$ uniformly. Suppose each f_n has an antiderivative F_n such that*

$$\|F_n\|_1 = \max(\|F_n\|_\infty, \|\Phi_1 F_n\|_\infty) \leqslant \|f_n\|_\infty$$

Then $\sum F_n$ converges uniformly and $F := \sum_{n=1}^{\infty} F_n$ is an antiderivative of f.

Proof. By uniform convergence we have $\lim_{n\to\infty} \|f_n\|_\infty = 0$, so certainly $\lim_{n\to\infty} \|F_n\|_\infty = 0$ and F is a well-defined continuous function. Let $\epsilon > 0$, let N be such that $\|f_n\|_\infty < \epsilon$ for $n > N$. For these n we have

$$|\Phi_1 F_n(s,t) - f_n(t)|_p \leqslant \max(\|\Phi_1 F_n\|_\infty, \|f_n\|_\infty) < \epsilon$$

uniformly in $s, t \in \mathbb{Z}_p$, $s \neq t$. If x is sufficiently close to a then

$$|\Phi_1 F_n(x,a) - f_n(a)|_p < \epsilon \qquad \text{for } n = 1, 2, \ldots, N$$

and by the foregoing this holds for all $n \in \mathbb{N}$ so

$$\left| \frac{F(x) - F(a)}{x - a} - f(a) \right|_p \leqslant \sup_n |\Phi_1 F_n(x,a) - f_n(a)|_p \leqslant \epsilon$$

and the lemma is proved.

**Exercise* 30.B. Let, in the above lemma, all F_n be C^1-functions. Show (by modifying the proof) that F is a C^1-function.

THEOREM 30.2. (Dieudonné) *Each continuous function $\mathbb{Z}_p \to \mathbb{Q}_p$ has an antiderivative (in $C^1(\mathbb{Z}_p \to \mathbb{Q}_p)$).*

Proof. In virtue of the lemma and exercise above it suffices to prove that if $f : \mathbb{Z}_p \to \mathbb{Q}_p$ is locally constant then f has a C^1-antiderivative F for which $\|F\|_1 \leqslant \|f\|_\infty$. f has the form

$$\sum_{m=0}^{p^n-1} \lambda_m \, \xi_{m+p^n \mathbb{Z}_p}$$

for some $n \in \mathbb{N}$, $\lambda_0, \ldots, \lambda_{p^n-1} \in \mathbb{Q}_p$. Set

$$F(x) := \sum_{m=0}^{p^n-1} \lambda_m (x-m) \, \xi_{m+p^n \mathbb{Z}_p} \qquad (x \in \mathbb{Z}_p)$$

Obviously $F' = f$ and $\|F\|_\infty \leqslant \max |\lambda_m|_p = \|f\|_\infty$. It remains to prove that $|\Phi_1 F(x, y)|_p \leqslant \|f\|_\infty$ for all $x, y \in \mathbb{Z}_p$, $x \neq y$. Let $x \in m + p^n \mathbb{Z}_p$ and $y \in m' + p^n \mathbb{Z}_p$. If $m' = m$ then $F(x) - F(y) = \lambda_m (x-y)$, hence $|\Phi_1 F(x, y)|_p \leqslant \|f\|_\infty$. If $m \neq m'$ then $|x-y|_p \geqslant \max(|x-m|_p, |y-m'|_p)$ so that $|F(x) - F(y)|_p = |\lambda_m (x-m) + \lambda_{m'} (y-m')|_p \leqslant \|f\|_\infty |x-y|_p$. This finishes the proof.

Exercise 30.C. (Antiderivation map) Show that there does not exist a map $P : C(\mathbb{Z}_p \to \mathbb{Q}_p) \to C^1(\mathbb{Z}_p \to \mathbb{Q}_p)$ satisfying the following 'reasonable' requirements.
(i) Pf is an antiderivative of f for each $f \in C(\mathbb{Z}_p \to \mathbb{Q}_p)$.
(ii) P is a continuous linear map $(C(\mathbb{Z}_p \to \mathbb{Q}_p), \| \ \|_\infty) \to (C^1(\mathbb{Z}_p \to \mathbb{Q}_p), \| \ \|_1)$.
(iii) P maps x^n into $x^{n+1}/(n+1)$ $(n \in \{0, 1, \ldots\})$.

Certain noncontinuous functions $\mathbb{Z}_p \to \mathbb{Q}_p$ also have antiderivatives, which is hardly surprising. However, one might not expect that the following function has one.

EXAMPLE 30.3. *The function $f : \mathbb{Z}_p \to \mathbb{Z}_p$ given by*

$$f(x) := \begin{cases} 0 \text{ if } x \neq 0 \\ 1 \text{ if } x = 0 \end{cases}$$

has an antiderivative.

Proof. Let $F : \mathbb{Z}_p \to \mathbb{Q}_p$ be as follows

$$F\left(\sum_{j=0}^{\infty} a_j p^j\right) := \begin{cases} a_n p^n + \ldots + a_{2n} p^{2n} & \text{if } a_n \neq 0, a_j = 0 \text{ for } j < n \\ 0 & \text{if } a_j = 0 \text{ for all } j \end{cases}$$

We prove that $F' = f$. F is locally constant on $\mathbb{Z}_p \setminus \{0\}$, so that if $x \neq 0$ we have $F'(x) = 0 = f(x)$. Now let $x \in \mathbb{Z}_p$, $x = a_n p^n + \ldots$ where $a_n \neq 0$. Then $|\Phi_1 F(x, 0) - 1|_p = |x^{-1}|_p \, |F(x) - x|_p = p^n |a_{2n+1} p^{2n+1} + \ldots|_p \leqslant p^{-n-1}$. Consequently $F'(0) = 1$.

In Section 70 we shall determine the class of all derivative functions.

Next we consider integration. Here we cannot just formally take over the classical definitions.

One might try to set up a theory of integration by means of antiderivatives of continuous functions. Thus if $P : C(\mathbb{Z}_p \to \mathbb{Q}_p) \to C(\mathbb{Z}_p \to \mathbb{Q}_p)$ is some well-behaved antiderivation map (see, however, Exercise 30.C) the formula

$$\int_a^b f(x)dx \; := Pf(b) - Pf(a) \qquad (a, b \in \mathbb{Z}_p)$$

defines some 'definite integral'. This approach however has not yet been proved useful.

Another way is to use measure theory. We shall assign to every open compact subset A of $\mathbb{Q}_p$ an element $m(A)$ of $\mathbb{Q}_p$ (*not* of $\mathbb{R}$!) such that m somewhat resembles the Lebesgue measure. The following requirements (i), (ii), (iii) seem reasonable.

(i) $m(A \cup B) = m(A) + m(B)$ $(A \cap B = \emptyset)$ ('additivity')

(ii) $m(a + A) = m(A)$ $(a \in \mathbb{Q}_p)$ ('translation invariance')

(iii) $\sup \{ |m(A)|_p : A \subset \mathbb{Z}_p, A \text{ compact open} \} < \infty$ ('boundedness')

('Positivity' of a measure does not make sense, instead we assume (iii).) Once we are able to produce an m satisfying (i), (ii), (iii) we can define the integral of the characteristic function ξ_A of a compact open subset A of $\mathbb{Q}_p$ by $I(\xi_A) := m(A)$ and confidently extend I by linearity and continuity to a larger class of functions. However, it turns out that this approach is too optimistic.

PROPOSITION 30.4. (No translation invariant measure on $\mathbb{Q}_p$) *If m is a function assigning a p-adic number to every compact open subset of $\mathbb{Q}_p$ and if m satisfies* (i), (ii), (iii) *then* $m = 0$.

Proof. For each $n \in \mathbb{N}$ the p^n additive cosets of $p^n \mathbb{Z}_p$ in $\mathbb{Z}_p$ form a partition of $\mathbb{Z}_p$. So by (i) and (ii) we have $m(\mathbb{Z}_p) = p^n m(p^n \mathbb{Z}_p)$. By (iii) $\sup \{ |m(p^n \mathbb{Z}_p)|_p : n \in \mathbb{N} \} < \infty$ so $m(\mathbb{Z}_p) = \lim_{n \to \infty} p^n m(p^n \mathbb{Z}_p) = 0$. Then also $m(p^n \mathbb{Z}_p) = p^{-n} m(\mathbb{Z}_p) = 0$ for each $n \in \mathbb{Z}$. By translation invariance $m(B) = 0$ for each ball B in $\mathbb{Q}_p$. Any compact open subset A of $\mathbb{Q}_p$ is a finite disjoint union of balls (Proposition 19.2 if you wish) so $m(A) = 0$.

Exercise 30.D. (No translation invariant integral on $C(\mathbb{Z}_p \to \mathbb{Q}_p)$) For $f : \mathbb{Z}_p \to \mathbb{Q}_p$, $s \in \mathbb{Z}_p$ set $f_s(x) := f(x + s)$ $(x \in \mathbb{Z}_p)$. Use Proposition 30.4 to derive the following result. *If $\int$ is a $\mathbb{Q}_p$-valued function on $C(\mathbb{Z}_p \to \mathbb{Q}_p)$ satisfying*

(i) $\int (\lambda f + \mu g) = \lambda \int f + \mu \int g$ $(\lambda, \mu \in \mathbb{Q}_p, f, g \in C(\mathbb{Z}_p \to \mathbb{Q}_p))$

(ii) $\int f_s = \int f$ $(s \in \mathbb{Z}_p, f \in C(\mathbb{Z}_p \to \mathbb{Q}_p))$

(iii) *there is a constant c such that*

$$|\textstyle\int f|_p \leqslant c\|f\|_\infty \qquad (f \in C(\mathbb{Z}_p \to \mathbb{Q}_p))$$

then $\int f = 0$ *for all* $f \in C(\mathbb{Z}_p \to \mathbb{Q}_p)$. (Comment. The same conclusion holds even if the continuity condition (iii) is dropped, see Exercise 34.B.)

The fact that the Lebesgue integral or measure does not have an immediate analogue in $\mathbb{Q}_p$ is a serious obstacle! How can we save part of the theory and shall we have to pay for it? One may think of several proposals.
(1) Weakening of the boundedness condition (iii). It is easy to see that the conditions (i), (ii) for a measure m determine m already up to a (multiplicative) constant. So, by requiring also that $m(\mathbb{Z}_p) = 1$ our m is completely determined. Then the 'measure' of $p^n \mathbb{Z}_p$ is equal to p^{-n}. More generally, $m(B_a(p^{-n})) = p^{-n}$ for each $a \in \mathbb{Q}_p$ and $n \in \mathbb{Z}$. This 'measure' behaves quite strangely since $p^n \mathbb{Z}_p$ 'shrinks to $\{0\}$' for $n \to \infty$ whereas $|m(p^n \mathbb{Z}_p)|_p \to \infty$. Yet, m is bounded in the following sense. There is a constant $c > 0$ such that

$$|m(A)|_p \leqslant c\,\|\xi_A\|_1 \qquad (A \text{ compact open}, A \subset \mathbb{Q}_p)$$

**Exercise* 30.E. Prove the above statement and show that in fact we may choose $c = p$. (Hint. Consider a partition of A into maximal convex subsets.)

This fact is the starting point of the integration theory of Chapter 3. We shall see there that the 'Riemann sums'

$$\sum_{j=0}^{p^n-1} f(j)\, m(j + p^n \mathbb{Z}_p) = p^{-n} \sum_{j=0}^{p^n-1} f(j)$$

approach a limit for $n \to \infty$ if f is continuously differentiable but, in general, diverge if f is just continuous.
(2) Dropping of the condition of translation invariance (ii). This leads to a reasonable theory of integration on (locally) compact spaces. As we do not need it in the main theory the basic ideas are presented in Appendix A.5.
(3) Changing of the range of m. Let us require that m takes its values in $\mathbb{Q}_q$ where $q \neq p$ rather than $\mathbb{Q}_p$. It is easy to see that there exists a unique additive, translation invariant, bounded $\mathbb{Q}_q$-valued function m, defined on the collection of all compact open subsets of $\mathbb{Q}_p$, for which $m(\mathbb{Z}_p) = 1$. This leads to a non-trivial translation invariant integral on $C(\mathbb{Z}_p \to \mathbb{Q}_q)$. See Exercise 62.G. (However, many people feel that functions $\mathbb{Z}_p \to \mathbb{Q}_p$ are more interesting than functions $\mathbb{Z}_p \to \mathbb{Q}_q$, which is understandable.) For a detailed account of this approach and the Fourier theory emerging from it we refer to van Rooij (1978).

Remark. For an integral in the style of Cauchy's line integral for complex analytic functions see Koblitz (1980) p. 129.

PART 2 : INTERPOLATION

31. The idea of interpolation

The set $\{1, 2, 3, \ldots\}$ is dense in $\mathbb{Z}_p$. If we are given a sequence $a_1, a_2, \ldots$ where the a_n are elements of some K then there exists at most one continuous function $f : \mathbb{Z}_p \to K$ such that $f(n) = a_n$ for all $n \in \mathbb{N}$. Of course, a similar conclusion holds for two-sided sequences $\ldots, a_{-1}, a_0, a_1, \ldots$ and 'sequences' such as $a_0, a_1, \ldots$ and $a_5, a_6, a_7, \ldots$ In general, let $A \subset \mathbb{Z}$ be p-adically dense in $\mathbb{Z}$ and let $n \mapsto a_n (n \in A)$ be a 'sequence' in K. We say that it *can* (p-adically) *be interpolated* if there exists a continuous function f : $\mathbb{Z}_p \to K$ such that $f(n) = a_n$ for all $n \in A$. From an abstract point of view the situation is quite simple. In order that $n \mapsto a_n$ can be interpolated it is necessary and sufficient that $n \mapsto a_n$ is uniformly continuous.

**Exercise* 31.A. In this exercise let $\mathbb{N}$ be equipped with the p-adic metric. Show that the restriction map $C(\mathbb{Z}_p \to K) \to BUC(\mathbb{N} \to K)$ (see Corollary 22.6) is an isomorphism of K-Banach spaces, i.e. a K-linear surjective isometry.

In practice the approach is often not an abstract one. The reason behind the introduction of the term 'interpolation' lies in the fact that several natural concrete rational sequences admit interpolation yielding interesting p-adic (continuous) functions, such as the p-adic gamma and zeta functions, see Sections 35, 61.

**Exercise* 31.B. (i) Prove that for each $j \in \mathbb{N}$ the (p-adic) sequences

$$1^j, 2^j, 3^j, \ldots$$
$$[1/p^j], [2/p^j], [3/p^j], \ldots$$

can be interpolated.
(ii) The p-adic sequence $n \mapsto (-1)^n$ can be interpolated if and only if $p = 2$.
(iii) Let n_- be as in Exercise 23.K. Show that a p-adic sequence $a_1, a_2, \ldots$ can be interpolated if and only if $\lim_{n \to \infty} (a_n - a_{n_-}) = 0$.

A sequence $a_0, a_1, \ldots$ in K can be interpolated if and only if for each $\epsilon > 0$ there is an N such that $|n-m|_p \leqslant p^{-N}$ implies $|a_n - a_m| < \epsilon$. To check the latter it is not necessary to consider *all* n, m for which $|n-m|_p \leqslant p^{-N}$ but it suffices to know that $|a_n - a_m| < \epsilon$ only for pairs n, m for which $n = m + p^N$. In fact, let $s, t \in \mathbb{N} \cup \{0\}$ and $|s-t|_p \leqslant p^{-N}$. Suppose $s > t$. Then $s-t$ is divisible by p^N so that $s = t + bp^N$ for some $b \in \mathbb{N}$. We have

$$a_s - a_t = \sum_{j=1}^{b} (a_{t+jp^N} - a_{t+(j-1)p^N})$$

The value of each of the summands is less than ϵ. By the strong triangle inequality $|a_s - a_t| < \epsilon$. In a less precise formulation we can state it as follows. $a_0, a_1, \ldots$ can be interpolated if and only if

'if n, m differ by a large power of p then $a_n - a_m$ is small'

If $a_n \in \mathbb{Z}$ for all n and $K = \mathbb{Q}_p$ we may reformulate this condition by

'if n, m differ by a large power of p then $a_n - a_m$ is divisible by a large power of p'

Part (i) of the following exercise gives an exact formulation.

**Exercise* 31.C. Let $a_0, a_1, \ldots \in K$. Prove the following.
(i) $a_0, a_1, \ldots$ can be interpolated if and only if $\lim_{j \to \infty} \sup_{n \in \mathbb{N} \cup \{0\}} |a_{n+p^j} - a_n| = 0$.
(ii) If $a_0, a_1, \ldots$ can be interpolated and $\lim_{n \to \infty} a_n$ exists then $a_0, a_1, \ldots$ is constant.
(iii) Suppose that for each $x \in \mathbb{Z}_p$

$$h(x) := \lim_{n \to x} a_n$$

exists. Then h is continuous and $n \mapsto h(n) - a_n$ is a null sequence. Conversely, if $h : \mathbb{Z}_p \to K$ is continuous and if $b_0, b_1, \ldots$ is a null sequence in K then $\lim_{n \to x} b_n = 0$ for every $x \in \mathbb{Z}_p$ so that

$$\lim_{n \to x} (h(n) + b_n)$$

exists. (Compare Exercise 22.B.)

Remark. The term 'interpolation' also appears in ultrametric literature in the following closely related context. If $a_0, a_1, \ldots$ is a (p-adic) sequence one can form for each n its *interpolation polynomial* P_n (the unique polynomial function P_n of degree $\leqslant n$ for which $P_n(0) = a_0, \ldots, P_n(n) = a_n$). We shall see in Exercise 51.D that the *interpolation series* $P_0 + \sum_{n=0}^{\infty}$

$(P_{n+1} - P_n)$ converges to a continuous function f for which $f(n) = a_n$ if and only if $a_0, a_1, \ldots$ can be interpolated.

32. *p*-adic exponents

Our aim in this section is to find out for which $a \in K$ the sequence $1, a, a^2, \ldots$ can be interpolated yielding a continuous 'exponential' function $x \mapsto a^x$ defined on $\mathbb{Z}_p$ with values in K. Our main interest lies in the case where $K \supset \mathbb{Q}_p$. First we do some estimating on powers. Recall that the residue class field of K is k and that K^+, the set of the positive elements of K, equals $B_1(1^-)$.

LEMMA 32.1. *Let* $\operatorname{char}(k) = p$, *let* $0 < \epsilon < 1$. *Then* $|y - 1| \leqslant \epsilon$ *implies* $|y^p - 1| \leqslant \tau |y - 1|$, *where* $\tau = \max(\epsilon, p^{-1})$.

Proof. Set $y = 1 + a$; then $|a| \leqslant \epsilon$ and $y^p - 1 = \binom{p}{1}a + \binom{p}{2}a^2 + \ldots + \binom{p}{p} \cdot a^p = (y-1)\left(\binom{p}{1} + \binom{p}{2}a + \ldots + \binom{p}{p}a^{p-1}\right)$. Now $|\binom{p}{j}a^{j-1}| \leqslant p^{-1}$ for $j = 1, \ldots, p-1$ and $|a^{p-1}| \leqslant \epsilon^{p-1} \leqslant \epsilon$. So we get $|y^p - 1| \leqslant |y - 1| \max(p^{-1}, \epsilon)$.

THEOREM 32.2. *Let* $\operatorname{char}(k) = p$ *and* $a \in K$. *Then the following are equivalent.*

(α) *a is positive.*

(β) $\lim_{n \to \infty} a^{p^n} = 1$.

Proof. Suppose (α). For each n we have $|a^{p^n} - 1| \leqslant |a - 1| < 1$ so that we may apply the lemma for $\epsilon = |1 - a|$ and $y = a^{p^{n-1}}$. We get

$$|a^{p^n} - 1| \leqslant \tau |a^{p^{n-1}} - 1|$$

Inductively we arrive at

$$|a^{p^n} - 1| \leqslant \tau^n \, |a - 1| \qquad (n \in \mathbb{N})$$

and (β) follows. Conversely, suppose (β). Obviously $|a| = 1$ and if a were not in K^+ we would have $|a - 1| = 1$. Then $1 = |(a-1)^p| = |a^p - \binom{p}{1}a^{p-1} + \ldots + (-1)^p|$. Now $|\binom{p}{j}a^{p-j}| < 1$ for $j = 1, \ldots, p-1$ so that $1 = |a^p + (-1)^p| =$ (even if $p = 2$) $= |a^p - 1|$. By induction $|a^{p^n} - 1| = 1$ for all n, a contradiction.

COROLLARY 32.3. *Let* $\operatorname{char}(k) = p$. *If* $a^p \in K^+$ *then* $a \in K^+$.

It follows easily from Theorem 32.2 that $1, a, a^2, \ldots$ can be interpolated if $a \in K^+$. Indeed, $|a^{j+p^n} - a^j| = |a^{p^n} - 1|$ tends to 0 uniformly in j. Now use Exercise 31.C(i).

THEOREM 32.4. (p-adic powers of elements of K) *Let* $\mathrm{char}(k) = p$ *and let* $a \in K$. *Then* $1, a, a^2, \ldots$ *can be interpolated if and only if* a *is positive. Set*

$$a^x := \lim_{n \to x} a^n \qquad (x \in \mathbb{Z}_p, a \in K^+)$$

Then for all $x, y \in \mathbb{Z}_p, a \in K^+$

$$a^x \in K^+$$
$$a^{x+y} = a^x a^y$$
$$a^{-x} = (a^x)^{-1}$$
$$|a^x - a^y| \leqslant \tau^{\mathrm{ord}_p(x-y)} |a - 1|$$

where $\tau = \max(|a-1|, p^{-1})$.

Proof. Only the last statement may need some explanation. First observe that $|a^n - 1| \leqslant |a - 1|$ for all $n \in \mathbb{N}$. If n is divisible by p, say $n = pm$, then by Lemma 32.1 $|a^n - 1| = |(a^m)^p - 1| \leqslant \tau |a^m - 1|$. Repeated application leads to $|a^n - 1| \leqslant \tau^{\mathrm{ord}_p(n)} |a - 1|$ (where ord_p is as in Definition 4.3). The formula $|a^n - a^m| = |a^{n-m} - 1|$ and continuity yield the desired estimate for $|a^x - a^y|$.

Exercise 32.A. Let $\mathrm{char}(k) = 0$. Show that the sequence $1, a, a^2, \ldots$ $(a \in K)$ can p-adically be interpolated if and only if $a^{p^n} = 1$ for some $n \in \mathbb{N}$.

Exercise 32.B. Let $\mathrm{char}(k) = p$. Are the following formulas true?

$$(ab)^x = a^x b^x \qquad (a, b \in K^+, x \in \mathbb{Z}_p)$$
$$(a^x)^y = a^{xy} \qquad (a \in K^+, x, y \in \mathbb{Z}_p)$$

Exercise 32.C. Show that for each $a \in T_p := \{x \in \mathbb{Z}_p : |x|_p = 1\}$ the sequence $1, a^{p-1}, a^{2(p-1)}, \ldots$ can be interpolated. Let $f : \mathbb{Z}_p \to \mathbb{Q}_p$ be a continuous function for which $f(n) = a^{n(p-1)}$ for all $n \in \mathbb{N} \cup \{0\}$. Is $f(1/(p-1)) = a$?

Exercise 32.D. Show that for $p \neq 2$

$$\exp(px) = (\exp p)^x \qquad (x \in \mathbb{Z}_p)$$

**Exercise* 32.E. (a^x if $K \supset \mathbb{Q}_p$) Let $K \supset \mathbb{Q}_p$ and let $a \in 1 + E$ (recall that $E = \{x \in K : |x| < p^{1/(1-p)}\}$ is the region of convergence of exp). Show that the estimate of $|a^x - a^y|$ in Theorem 32.4. can be sharpened to

$$|a^x - a^y| = |x - y|_p \, |a - 1| \qquad (x, y \in \mathbb{Z}_p)$$

and conclude that there are no roots of unity in $1 + E$ except 1. The set $K^+ \backslash 1 + E$ may very well contain roots of unity, see Exercise 33.A. If $b \in K^+$ is not a root of unity then $x \mapsto b^x$ is a homeomorphism of $\mathbb{Z}_p$ into K^+. Prove this.

Exercise 32.F. (a^x if char$(K) = p$) Let char$(K) = p$, $a \in K^+$, $a \neq 1$. Show that for $n \in \{0, 1, 2, \ldots\}$ and $x, y \in \mathbb{Z}_p$

$$\text{if } |x - y|_p = p^{-n} \text{ then } |a^x - a^y| = |a - 1|^{p^n}$$

Conclude that $x \mapsto a^x$ is a homeomorphism of $\mathbb{Z}_p$ into K^+ and that therefore K^+ does not contain roots of unity except 1.

**Exercise* 32.G. (Discreteness of the group of roots of unity) Use the results of the two previous exercises and complete the proof of the following theorem. *Let K be a non-archimedean valued complete field. Then there is a neighbourhood U of 1 such that U does not contain roots of unity except* 1. In fact, if K and k have equal characteristics we may choose $U := B_1(1^-)$. Otherwise we can take $U := 1 + E$. See also Exercise 33.A.

Exercise 32.H. (The derivative of $x \mapsto a^x$) Let $a \in \mathbb{C}_p^+$. Prove that for $n \in \mathbb{N}$

$$\frac{a^n - 1}{n} = (a-1) + \frac{1}{2}\binom{n-1}{1}(a-1)^2 + \frac{1}{3}\binom{n-1}{2}(a-1)^3 + \ldots + \frac{1}{n}\binom{n-1}{n-1}(a-1)^n$$

(Hint. $a^n = (a - 1 + 1)^n$). Use this result to show that (see Definition 25.7)

$$\lim_{x \to 0} \frac{a^x - 1}{x} = \sum_{n=1}^{\infty} (-1)^{n-1} \frac{(a-1)^n}{n} = \log a$$

Obtain the formula $\log a + \log b = \log ab$ $(a, b \in \mathbb{C}_p^+)$. Find the derivative of $x \mapsto a^x$ $(x \in \mathbb{Z}_p, a \in \mathbb{C}_p^+)$.

Exercise 32.I. Let $a \in \mathbb{C}_p^+$. Prove the formula

$$a^x = \sum_{j=0}^{\infty} \binom{x}{j} (a-1)^j \qquad (x \in \mathbb{Z}_p)$$

(where $\binom{x}{j} := x(x-1)\ldots(x-j+1)/j!$) and conclude that a^x is an analytic function of a (for the local analyticity of a^x as a function of x see Section 47). Find the derivative of $a \mapsto a^x$.

**Exercise* 32.J. Obtain the following inequality. If $x, y \in \mathbb{C}_p$, $|x|_p = |y|_p = 1$, $|x - y|_p < p^{-1}$ then

$$(*) \qquad |x^n - y^n|_p \leq |n|_p |x - y|_p \qquad (n \in \mathbb{N})$$

In particular, (*) holds for all $x, y \in a + p\mathbb{Z}_p$ $(a \in \{1, \ldots, p-1\})$.

33. Roots of unity in $\mathbb{C}_p$. The Teichmüller character

In Exercise 27.G we have seen that the equation $x^{p-1} = 1$ has $p - 1$ roots in $\mathbb{Q}_p$ and that for each $a \in \mathbb{Q}_p$, $|a|_p = 1$ the sequence $a, a^p, a^{p^2}, \ldots$ converges to such a root. We shall carry out something similar in $\mathbb{C}_p$ using the theory of the previous section.

Let Γ be the group of all roots of unity of $\mathbb{C}_p$. It has two interesting subgroups Γ_u (u = unramified), Γ_r (r = ramified) given by

$$\Gamma_u := \{\theta \in \mathbb{C}_p : \theta^n = 1 \text{ for some } n \in \mathbb{N} \text{ not divisible by } p\}$$
$$\Gamma_r := \{\theta \in \mathbb{C}_p : \theta^{p^m} = 1 \text{ for some } m \in \mathbb{N}\}$$

By elementary group theory we have $\Gamma = \Gamma_u \cdot \Gamma_r$, $\Gamma_u \cap \Gamma_r = \{1\}$.

LEMMA 33.1. $\Gamma_u \cap \mathbb{C}_p^+ = \{1\}$, $\Gamma_r \cap (1+E) = \{1\}$, $\Gamma_r \subset \mathbb{C}_p^+$.

Proof. The last two statements follow from Corollary 32.3 and Exercise 32.E. To prove the first one, let $\theta \in \Gamma_u \cap \mathbb{C}_p^+$. Then $\theta = 1 + u$ where $|u|_p < 1$ and $(1+u)^n = 1$ for some n not divisible by p. Now $1 = (1+u)^n = 1 + \binom{n}{1}u + \binom{n}{2}u^2 + \ldots + u^n$ so that $|u|_p\, |\binom{n}{1} + \binom{n}{2}u + \ldots + u^{n-1}|_p = 0$. But $|\binom{n}{1}|_p = 1$ and $|\binom{n}{2}u + \ldots + u^{n-1}|_p < 1$. By the isosceles triangle principle $|\binom{n}{1} + \binom{n}{2}u + \ldots + u^{n-1}|_p = 1$. Hence, $u = 0$, i.e. $\theta = 1$.

It follows from Lemma 33.1 that $\theta_1, \theta_2 \in \Gamma_u$, $\theta_1 \neq \theta_2$ implies $|\theta_1 - \theta_2|_p = 1$. In the following exercise we compute distances between elements of Γ_r. (See also Exercise 17.C.)

Exercise 33.A. (On distances between p^nth roots of unity in $\mathbb{C}_p$) If $\theta \in \mathbb{C}_p$ is a primitive p^nth root of unity for some $n \in \mathbb{N}$ then

$$|\theta - 1|_p = p^{1/(p^{n-1} - p^n)}$$

Prove this. (Hint. Use the polynomial identity $1 + X + \ldots + X^{p^n - 1} = \prod_{j=1}^{p^n - 1} (X - \theta^j)$.)

LEMMA 33.2. *Each coset of* $\mathbb{C}_p^+$ *in* $\{x \in \mathbb{C}_p : |x|_p = 1\}$ *contains precisely one element of* Γ_u.

Proof. Let $x \in \mathbb{C}_p$, $|x|_p = 1$. Then $\bar{x}$ (see Definition 11.2) is an element of the residue class field of $\mathbb{C}_p$, hence algebraic over $\mathbb{F}_p$ (the field of p elements). This implies the existence of a number q (a power of p) such that $\bar{x}^q = \bar{x}$, hence $|x^q - x|_p < 1$. Application of Hensel's lemma (Theorem 27.6) to the function $z \mapsto z^q - z$ with $a := x$ leads to the existence of a θ such that $|\theta - x|_p < 1$ and $\theta^q = \theta$, i.e. $\theta^{q-1} = 1$. Now $q - 1$ is not divisible by p, so $\theta \in \Gamma_u$. The uniqueness follows from Lemma 33.1.

For a given $x \in \mathbb{C}_p$, $|x|_p = 1$, how can we compute the unique $\theta \in \Gamma_u$ for which $|x - \theta|_p < 1$? We have $\bar{x}^q = \bar{x}$ where q is some power of p. Then $|x^{q-1} - 1|_p < 1$. By Theorem 32.2 $\lim_{n \to \infty} (x^{q-1})^{p^n} = 1$ so certainly $\lim_{n \to \infty} (x^{q-1})^{q^n} = 1$. In other words, $\lim_{n \to \infty} |x^{q^{n+1}} - x^{q^n}|_p = 0$.

We see that the sequence $x, x^q, x^{q^2}, \ldots$ converges. Set

$$\theta := \lim_{n \to \infty} x^{q^n}$$

Then clearly $\theta^q = \theta$, $|\theta - x|_p < 1$ and we have found our θ. To avoid the presence of q (which depends on x) in this story we define:

DEFINITION 33.3. The *Teichmüller character* $\omega_p : \{x \in \mathbb{C}_p : |x|_p = 1\} \to \Gamma_u$ is given by

$$\omega_p(x) = \lim_{n \to \infty} x^{p^{n!}}$$

THEOREM 33.4. (Properties of the Teichmüller character)
(i) *ω_p is a homomorphism of the unit sphere of $\mathbb{C}_p$ onto Γ_u.*
(ii) *ω_p maps an $x \in \mathbb{C}_p$ ($|x|_p = 1$) into the unique element θ of Γ_u for which $|\theta - x|_p < 1$.*
(iii) $\{\omega_p(1), \omega_p(2), \ldots, \omega_p(p-1)\} = \{x \in \mathbb{Q}_p : x^{p-1} = 1\}$.
(iv) *If L is a closed subfield of $\mathbb{C}_p$, $\mathbb{Q}_p \subset L \subset \mathbb{C}_p$ then ω_p maps $\{x \in L : |x|_p = 1\}$ into L.*
(v) *If $x \in \mathbb{Q}_p$, $|x|_p = 1$ then $\omega_p(x) = \lim_{n \to \infty} x^{p^n}$.*
Proof. Obvious.

**Exercise* 33.B. Show that, if $p \neq 2$, the roots of unity of $\mathbb{Q}_p$ are precisely the $(p-1)$th roots of unity and that $1, -1$ are the only roots of unity of $\mathbb{Q}_2$. Use this to prove the following statement. *If p, q are distinct primes then $\mathbb{Q}_p$ and $\mathbb{Q}_q$ are not isomorphic as fields.* (Use the equation $x^2 + 2 = 0$ to distinguish $\mathbb{Q}_2$ and $\mathbb{Q}_3$.)

Exercise 33.C. Let $s \in \mathbb{Z}$, $|s|_p = |s+1|_p = 1$. Show that $\lim_{n \to \infty} (1 + 1/(s + p^n))^{s+p^n} = (1 + 1/s)^s \, \omega_p(1 + 1/s)$ in $\mathbb{Q}_p$.

**Exercise* 33.D. (Decompositions in $\mathbb{C}_p^\times$) (i) Prove that an element $x \in \mathbb{C}_p$ for which $|x|_p = 1$ can uniquely be written as yz where $y \in \Gamma_u$, $z \in \mathbb{C}_p^+$ (choose $y = \omega_p(x)$, $z = x\,\omega_p(x^{-1})$).
(ii) The fact that $\Gamma_r \cap (1 + E) = \{1\}$ leads to the question as to whether we can further decompose an element of $\mathbb{C}_p^+$ into a product uv where $u \in \Gamma_r$, $v \in 1 + E$. Show, by considering multiplicative cosets of $1 + E$ in $\{x \in \mathbb{C}_p : |1 - x|_p \leqslant p^{1/(1-p)}\}$, that this is not always so. (See Exercise 45.A.)
(iii) For any rational number r choose an element a_r of $\mathbb{C}_p$ for which $|a_r| = p^r$. Show that for each $x \in \mathbb{C}_p^\times$ there are unique $r \in \mathbb{Q}$, $\theta \in \Gamma_u$, $x' \in \mathbb{C}_p^+$ such that $x = a_r\,\theta\,x'$.

34. $\sum_{n=0}^{x} a_n$ for a p-adic integer x

We return to interpolation. We shall prove the following fundamental result which shall be used quite often in the subsequent sections. If a p-adic sequence $a_0, a_1, \ldots$ can be interpolated then so can the sequence of its partial sums $n \mapsto \sum_{j=0}^{n} a_j$.

IN THIS SECTION WE SUPPOSE $K \supset \mathbb{Q}_p$.

THEOREM 34.1. *Let $f \in C(\mathbb{Z}_p \to K)$. Then there is a unique function $F \in C(\mathbb{Z}_p \to K)$ such that*

$$F(x+1) - F(x) = f(x) \qquad (x \in \mathbb{Z}_p)$$
$$F(0) = 0$$

Proof. We are forced to define $F(n) := \sum_{j=0}^{n-1} f(j)$ $(n \in \mathbb{N})$ so the uniqueness of F is established. It suffices to prove that the sequence $F(0), F(1), \ldots$ can be interpolated. Consider $F(n+p^j) - F(n)$ where $n \in \{0, 1, 2, \ldots\}$, $j \in \mathbb{N}$. We have

$$F(n+p^j) - F(n) = f(n) + f(n+1) + \ldots + f(n+p^j-1)$$

The numbers $n, n+1, \ldots, n+p^j-1$ form a complete set of representatives in $\mathbb{Z}$ modulo p^j. Let $1 \leqslant s \leqslant j$ and divide $\{n, n+1, \ldots, n+p^j-1\}$ into p^s classes modulo p^s each having p^{j-s} elements. Let V be such a class and $v \in V$. then

$$\sum_{x \in V} f(x) = \left\{ \sum_{x \in V} (f(x) - f(v)) \right\} + p^{j-s} f(v)$$

The summand $f(x) - f(v)$ can be estimated by

$$|f(x) - f(v)| \leqslant \sup \{|f(u) - f(t)| : |u - t| \leqslant p^{-s}\} =: \rho_s$$

and we get

$$\left| \sum_{x \in V} f(x) \right| \leqslant \max(\rho_s, p^{s-j} \|f\|_\infty)$$

Summation over all classes V yields

$$|F(n+p^j) - F(n)| \leqslant \max(\rho_s, p^{s-j} \|f\|_\infty)$$

which can be made arbitrarily small by choosing j large and, for example, $s = [j/2]$ (that $\lim_{s \to \infty} \rho_s = 0$ follows from the uniform continuity of f) and the theorem is proved.

COROLLARY 34.2. *If a sequence in K can be interpolated then so can the sequence of its partial sums.*

Proof. Let $f \in C(\mathbb{Z}_p \to K)$ and $a_n = f(n)$ $(n = 0, 1, 2, \ldots)$. Let F be as in Theorem 34.1. Then $x \mapsto F(x+1)$ is a continuous function interpolating $n \mapsto \Sigma_{j=0}^{n} a_j$.

DEFINITION 34.3. The *indefinite sum* of a continuous function $f : \mathbb{Z}_p \to K$ is the continuous function Sf interpolating $n \mapsto \Sigma_{j=0}^{n-1} f(j)$ $(n \in \mathbb{N})$. Instead of $Sf(x)$ $(x \in \mathbb{Z}_p)$ we sometimes shall write

$$\sum_{j=0}^{x-1} f(j) \qquad \left(= \lim_{\substack{n \to x \\ n \in \mathbb{N}}} \sum_{j=0}^{n-1} f(j) \right)$$

Exercise 34.A. Find the indefinite sum of each of the following functions (i) $f_1(x) = 1$ $(x \in \mathbb{Z}_p)$, (ii) $f_2(x) = x^2$ $(x \in \mathbb{Z}_p)$, (iii) $f_3(x) = a^x$ $(x \in \mathbb{Z}_p)$ where $a \in \mathbb{C}_p^+$.

Exercise 34.B. (No translation invariant linear function on $C(\mathbb{Z}_p \to K)$) Show that if I is a K-linear function on $C(\mathbb{Z}_p \to K)$ (continuous or not) such that $I(f_s) = I(f)$ for all $s \in \mathbb{Z}_p$ then $I = 0$. (See Exercise 30.D.)

Exercise 34.C. Show that $\lim_{n \to \infty} \Sigma'^{\,p^n}_{j=1} 1/j = 0$ where Σ' restricts the summation to those j between 1 and p^n that are not divisible by p.

The last four exercises of this section concern a generalization of Corollary 34.2.

Exercise 34.D. (On convolution of sequences) For two elements $f = (f(0), f(1), \ldots)$, $g = (g(0), g(1), \ldots)$ of l^∞ (see Exercise 13.A) define their *convolution* $f * g$ by

$$(f * g)(n) = \sum_{j=0}^{n} f(j)\, g(n-j) \qquad (n \in \{0, 1, 2, \ldots\})$$

Show that $f * g \in l^\infty$, that $\|f * g\| \leqslant \|f\|\,\|g\|$ and prove the formula
$(*)$ $(f * g)(s+t) = (f_s * g)(t) + (f * g_t)(s) - f(s)g(t)$ $(s, t \in \{0, 1, 2, \ldots\})$
where $f_s = (f(s), f((s+1), \ldots)$ and $g_t = (g(t), g(t+1), \ldots)$.

Exercise 34.E. Use the previous exercise to prove the following theorem. *If $f, g \in l^\infty$ can be interpolated then so can $f * g$.* See also Exercise 52.J. (Hint. Consider the space
$V := \{(f, g) \in C(\mathbb{Z}_p \to K)^2 : (f * g)(0), (f * g)(1), \ldots \text{ can be interpolated}\}$
where $(f * g)(n) := \Sigma_{j=0}^{n} f(j)\, g(n-j)$. V is a linear subspace of $C(\mathbb{Z}_p \to K)^2$; if $(f_1, g_1), (f_2, g_2), \ldots$ are in V and $\lim_{n \to \infty} f_n = f$, $\lim_{n \to \infty} g_n = g$ both uniformly then $(f, g) \in V$. Use formula $(*)$ of Exercise 34.D to show that it suffices to prove that $(\xi_A, \xi_B) \in V$ where $A = p^m \mathbb{Z}_p$ for some m and B

is a disc containing A. Prove that $(\xi_A * \xi_B)(n) = [n/p^m]\, \xi_B(n)$ for $n \in \{0, 1, 2, \ldots\}$ and use Exercise 31.B(i).)

Exercise 34.F. By Exercise 34.E the convolution $*$ in l^∞ induces a multiplication, again denoted $*$, in $C(\mathbb{Z}_p \to K)$. For $a \in \mathbb{C}_p^+$ denote the function $x \mapsto a^x$ $(x \in \mathbb{Z}_p)$ by a^*. Prove the formula

$$(f * g)a^* = (fa^*) * (ga^*) \qquad (f, g \in C(\mathbb{Z}_p \to K), a \in \mathbb{C}_p^+)$$

and compute also $a^* * b^*$ $(a, b \in \mathbb{C}_p^+)$.

Exercise 34.G. Obtain Theorem 34.1 as a corollary of Exercise 34.E by considering $f * \xi_{\mathbb{Z}_p}$ for a continuous function $f : \mathbb{Z}_p \to K$.

35. The p-adic gamma function

We shall use the technique of interpolation to obtain a p-adic analogue of the classical gamma function Γ. In order to understand the contents of Sections 35-39 it is not necessary to have a deep knowledge about Γ; it is enough to know that Γ is a sensible extension of $n \mapsto (n-1)!$ On the other hand, we shall sometimes compare the behaviour of the p-adic and the complex gamma function, in which case we state the properties of Γ without proof.

The classical gamma function Γ is a meromorphic function on $\mathbb{C}$ with simple poles in $0, -1, -2, \ldots$ satisfying the functional relation

$$\Gamma(1) = 1, \Gamma(z+1) = z\,\Gamma(z) \qquad (z \in \mathbb{C} \backslash \{0, -1, -2, \ldots\})$$

so that $\Gamma(n+1) = n!$ for $n \in \mathbb{N}$. To find a p-adic version of Γ our first thought could be to interpret $n \mapsto n!$ p-adically and hope that it can be interpolated. However, if $x \in \mathbb{Z}_p$ and $n \in \mathbb{N}$ $(n \neq x)$ approaches x then n becomes large in the ordinary sense so that $|n!|_p$ is small. Thus we have $\lim_{n \to x} n! = 0$ p-adically and there is no continuous function $f : \mathbb{Z}_p \to \mathbb{Q}_p$ satisfying $f(n) = n!$ for $n = 1, 2, \ldots$.

To overcome this difficulty two ways have been invented. The one of Overholtzer and Morita boils down to modifying $n!$. Diamond's starting point is to stay out of $\mathbb{Z}_p$. In this chapter we take the first approach. Diamond's function and its connection with the one of Overholtzer and Morita shall be discussed in Section 60.

The idea is to remove in

$$n! = 1.2.\ldots.n$$

those factors that are divisible by p. Let us define provisionally

$$(n!)_p := \prod_{1 \leqslant j \leqslant n}' j$$

(where as in Exercise 34.C the prime in Π' indicates that we multiply only those factors that are not divisible by p. Without explicit definition we shall henceforth use expressions such as $\prod_{j=1}^{\prime\infty} a_j$, $\Sigma_{j=1}^{\prime n} a_j$, $\Sigma_{j=1}^{\prime\infty} a_j$ etc.). Then

$$|(n!)_p|_p = 1$$

In order to find out whether $n \mapsto (n!)_p$ can be interpolated we shall compare the numbers $((n+p^s)!)_p$ and $(n!)_p$ for large s. We have

$$((n+p^s)!)_p = (n!)_p \prod_{j=1}^{p^s}{}' (n+j)$$

The following proposition tells us something about the occurring product.

PROPOSITION 35.1 (Generalization of Wilson's theorem) *Let* $p \neq 2, n \in \mathbb{Z}$, $s \in \mathbb{N}$. *Then*

$$\prod_{j=0}^{p^s-1}{}' (n+j) \equiv -1 \pmod{p^s}$$

Proof. The numbers $n, n+1, \ldots, n+p^s-1$ form a complete set of representatives in $\mathbb{Z}$ modulo $p^s\mathbb{Z}$. Let G be the (multiplicative) group of units of $\mathbb{Z}/p^s\mathbb{Z}$. Then for each $n \in \mathbb{N}$ the factors occurring in the product form a complete set of representatives of G. So, if φ denotes the homomorphism $\mathbb{Z} \to \mathbb{Z}/p^s\mathbb{Z}$ we have

$$\varphi\left(\prod_{j=0}^{p^s-1}{}' (n+j)\right) = \prod_{g \in G} g$$

Now, if $g \neq g^{-1}$ the terms g and g^{-1} in the product cancel and we have

$$\prod_{g \in G} g = \prod_{g^2=1} g$$

If $g^2 = 1$ in $\mathbb{Z}/p^s\mathbb{Z}$ then $g = 1$ or $g = -1$. (In fact let $g \neq 1, -1$ and $(g-1)\cdot(g+1) = 0$. Then $g-1$ and $g+1$ are both zero divisors in $\mathbb{Z}/p^s\mathbb{Z}$, so $g \equiv 1 \pmod{p}$ and $g \equiv -1 \pmod{p}$. As $p \neq 2$ this is impossible.) We see that

$$\prod_{g \in G} g = -1$$

from which it follows that

$$\prod_{j=0}^{p^s-1}{}' (n+j) \equiv -1 \pmod{p^s}$$

Observe that for $s=1, n=0$ we obtain the familiar result $(p-1)! \equiv -1 \pmod p$, which is known as Wilson's theorem.

Back to $((n+p^s)!)_p$ and $(n!)_p$. We may conclude that

$$((n+p^s)!)_p \equiv -(n!)_p \pmod{p^s}$$

so the sign is wrong. We therefore make a second modification. Define

$$g(n) := (-1)^{n+1}(n!)_p \qquad (n \in \mathbb{N})$$

From the congruences (remember that p is odd)

$$g(n+p^s) \equiv -(-1)^{n+1}((n+p^s)!)_p \equiv (-1)^{n+1}(n!)_p \equiv g(n) \pmod{p^s}$$

it follows that g can be interpolated.

DEFINITION 35.2. The *p-adic gamma function* Γ_p is the continuous extension to $\mathbb{Z}_p$ of $n \mapsto g(n-1) = (-1)^n \prod'_{1 \leqslant j < n} j$ $(n \geqslant 2)$. (For $p=2$ see Exercise 35.A.)

PROPOSITION 35.3. *Let $p \neq 2$. Then Γ_p has the following properties.*
(i) *For all $x \in \mathbb{Z}_p$*

$$\Gamma_p(x+1) = h_p(x)\,\Gamma_p(x)$$

where

$$h_p(x) := \begin{cases} -x \text{ if } |x|_p = 1 \\ -1 \text{ if } |x|_p < 1 \end{cases}$$

(ii) *For all $x, y \in \mathbb{Z}_p$*

$$|\Gamma_p(x) - \Gamma_p(y)|_p \leqslant |x-y|_p$$

(iii) $\Gamma_p(0) = 1$, $\Gamma_p(1) = -1$, $\Gamma_p(2) = 1$. *For all $x \in \mathbb{Z}_p$ we have $|\Gamma_p(x)|_p = 1$.*

Proof. Inspection and continuity.

**Exercise* 35.A. (Save the 2-adic gamma function!) The example $1 \cdot 3 \cdot 5 \cdot 7 \equiv 1 \pmod 8$ shows that Proposition 35.1 is not correct for $p=2$. However, in this exercise we shall prove that *the sequence* $\Gamma_2(2), \Gamma_2(3), \ldots$ where

$$\Gamma_2(n) := (-1)^n \prod_{1 \leqslant j < n}{}' j$$

can be extended to a continuous function on $\mathbb{Z}_2$.
(i) As in the proof of Proposition 35.1 introduce φ and G and prove that for $n, s \in \mathbb{N}$

$$\Gamma_2(n+2^s) = \Gamma_2(n) \prod_{0 \leqslant j < 2^s}{}' (n+j)$$

$$\varphi\Big(\prod_{j<2^s}{}'(n+j)\Big) = \prod_{g^2=1} g$$

(ii) This time we hope that the product $\prod_{g^2=1} g$ is congruent to $1 \pmod{2^s}$ instead of -1. However, the statement is false for $s = 2$.
(iii) Show that for $s \geq 2$ the solutions of $g^2 = 1$ in $\mathbf{Z}/2^s\mathbf{Z}$ are precisely $1, -1, 2^{s-1}+1, 2^{s-1}-1$. Deduce that for $n, m \geq 2$

$$|\Gamma_2(n) - \Gamma_2(m)|_2 \leq |n-m|_2 \quad (|n-m|_2 \leq 2^{-3})$$

(iv) It follows from (iii) that the sequence $\Gamma_2(2), \Gamma_2(3), \ldots$ can be extended to a continuous function (again denoted Γ_2) : $\mathbf{Z}_2 \to \mathbf{Q}_2$. Show that properties (i), (iii) of Proposition 35.3 remain valid for $p = 2$ and that instead of (ii) we have

$$|\Gamma_2(x) - \Gamma_2(y)|_2 \leq |x-y|_2 \left(x, y \in \mathbf{Z}_2, |x-y|_2 \neq \tfrac{1}{4}\right)$$
$$|\Gamma_2(x) - \Gamma_2(y)|_2 \leq 2|x-y|_2 \left(x, y \in \mathbf{Z}_2, |x-y|_2 = \tfrac{1}{4}\right)$$

36. A *p*-adic Euler constant

In Section 58 we shall prove that Γ_p is locally analytic. In this little section (whose logical place is in Chapter 3) we borrow the differentiability of Γ_p. Differentiation of the functional relation $\Gamma_p(x+1) = \Gamma_p(x)\, h_p(x)$ (Proposition 35.3) yields

$$\frac{\Gamma_p'(x+1)}{\Gamma_p(x+1)} - \frac{\Gamma_p'(x)}{\Gamma_p(x)} = \frac{h_p'(x)}{h_p(x)} = \begin{cases} 1/x & \text{if } |x|_p = 1 \\ 0 & \text{if } |x|_p < 1 \end{cases}$$

There is a constant c such that

$$\frac{\Gamma_p'(x)}{\Gamma_p(x)} = c + L_p(x) \qquad (x \in \mathbf{Z}_p)$$

where L_p denotes the indefinite sum (Definition 34.3) of h_p'/h_p. After substituting $x = 1$ and using the fact that $L_p(1) = 0$ we obtain $c = \Gamma_p'(1)/\Gamma_p(1)$ so that $\Gamma_p'(x)/\Gamma_p(x) = \Gamma_p'(1)/\Gamma_p(1) + L_p(x)$ $(x \in \mathbf{Z}_p)$. In particular

$$\frac{\Gamma_p'(m)}{\Gamma_p(m)} = \frac{\Gamma_p'(1)}{\Gamma_p(1)} + \sum_{j<m}{}' \frac{1}{j} \qquad (m \in \mathbf{N})$$

This resembles the formula

$$\frac{\Gamma'(m)}{\Gamma(m)} = \frac{\Gamma'(1)}{\Gamma(1)} + \sum_{j<m} \frac{1}{j} \qquad (m \in \mathbf{N})$$

for the complex gamma function Γ. The classical Euler constant γ is defined as $\gamma = -\Gamma'(1)/\Gamma(1)$. Thus we define the *p-adic Euler constant* γ_p to be

$$\gamma_p := -\frac{\Gamma_p'(1)}{\Gamma_p(1)} = \Gamma_p'(1) = -\Gamma_p'(0)$$

I do not know of results concerning (ir)rationality of γ_p.

Exercise 36.A. Show that $|\gamma_p|_p \leq 1$ and that in $\mathbb{Q}_p$

$$\gamma_p = \lim_{n \to \infty} p^{-n} \left\{ 1 - (-1)^p \frac{p^n!}{p^{n-1}!\, p^{p^{n-1}}} \right\}$$

37. Values of Γ_p in $\frac{1}{2}$, 0, −1, −2, . . .

A formula for $\Gamma_p(-n)$ $(n \in \mathbb{N})$ is given by

PROPOSITION 37.1. $\Gamma_p(-n) = (-1)^{n+1-[n/p]} (\Gamma_p(n+1))^{-1}$ $(n \in \mathbb{N})$.
Proof. We have $1 = \Gamma_p(0) = \Gamma_p(-1)\, h_p(-1) = \ldots$ so that

$$\Gamma_p(-n)^{-1} = \prod_{j=1}^{n} h_p(-j)$$

Among $1, 2, \ldots, n$ there are $[n/p]$ numbers divisible by p. So by the definition of h_p

$$\Gamma_p(-n)^{-1} = (-1)^{[n/p]} \prod_{j=1}^{n}{}' j = (-1)^{[n/p]} (-1)^{-n-1} \Gamma_p(n+1)$$

and we are done.

PROPOSITION 37.2. *If* $p \neq 2$ *then*

$$\Gamma_p(x)\, \Gamma_p(1-x) = (-1)^{l(x)} \qquad (x \in \mathbb{Z}_p)$$

and for $p = 2$

$$\Gamma_2(x)\, \Gamma_2(1-x) = (-1)^{\sigma_1(x)+1} \qquad (x \in \mathbb{Z}_2)$$

where $l: \mathbb{Z}_p \to \{1, 2, \ldots, p\}$ *assigns to* $x \in \mathbb{Z}_p$ *its residue* $\in \{1, 2, \ldots, p\}$ *modulo* $p\mathbb{Z}_p$ *and where* σ_1 *is defined by the formula*

$$\sigma_1\left(\textstyle\sum_{j=0}^{\infty} a_j 2^j \right) = a_1$$

Proof. Let $p \neq 2$ and $n \in \mathbb{N}$. From Proposition 37.1 it follows that

$$\Gamma_p(n+1)\, \Gamma_p(-n) = (-1)^{n+1-[n/p]}$$

With $n-1$ in place of n this becomes

$$\Gamma_p(n)\, \Gamma_p(1-n) = (-1)^{n-[(n-1)/p]} = (-1)^{n-p[(n-1)/p]}$$

Now let $n = a_0 + a_1 p + \ldots$ in base p. If $a_0 \neq 0$ then $[(n-1)/p] = [(a_0 - 1 + a_1 p + \ldots)/p] = a_1 + a_2 p + \ldots$ and $n - p[(n-1)/p] = a_0 = l(n)$. If $a_0 = 0$ then $n - 1 = (p-1) + b_1 p + \ldots$ in base p, $[(n-1)/p] = b_1 + b_2 p +$

. . . so that $n - p[(n-1)/p] = 1 + (p-1) = p = l(n)$. We get

$$\Gamma_p(n)\,\Gamma_p(1-n) = (-1)^{l(n)}$$

and the first formula follows by continuity. The proof of the second formula is left to the reader.

Remark. The formulas for $\Gamma_p(x)\,\Gamma_p(1-x)$ form the p-adic analogue of the classical formula

$$(*) \qquad \Gamma(z)\,\Gamma(1-z) = \frac{\pi}{\sin \pi z}$$

Notice that in our case the function $x \mapsto \Gamma_p(x)\,\Gamma_p(1-x)$ is locally constant!

Substituting $z = \frac{1}{2}$ in (*) we obtain $\Gamma(\frac{1}{2})^2 = \pi$. It is therefore interesting to consider $\Gamma_p(\frac{1}{2})$. Obviously, $\Gamma_2(\frac{1}{2})$ does not make sense so let $p \neq 2$. Using the formula of Proposition 37.2 we get

$$\Gamma_p(\tfrac{1}{2})^2 = (-1)^{l(\frac{1}{2})}$$

Now $l(\frac{1}{2}) = l(\frac{1}{2}(p+1)) = \frac{1}{2}(p+1)$ so that

$$\Gamma_p(\tfrac{1}{2})^2 = \begin{cases} 1 \text{ if } p \equiv 3 \pmod 4 \\ -1 \text{ if } p \equiv 1 \pmod 4 \end{cases}$$

Thus, the role played by $\pi = 3.14\ldots$ is taken over by one of the numbers $1, -1$. Observe that the formula for $\Gamma_p(\frac{1}{2})^2$ yields a new proof for the existence of $\sqrt{-1}$ in $\mathbb{Q}_p$ if $p \equiv 1 \pmod 4$, see Exercise 27.E.

Exercise 37.A. Show that $\Gamma_p(\frac{1}{2})$ and $(-1)^{(p+1)/2}\,(\frac{1}{2}(p-1))!$ lie in the same coset of $p\mathbb{Z}_p$ and deduce that

(i) if $p \equiv 3 \pmod 4$ then

$$\Gamma_p(\tfrac{1}{2}) = \begin{cases} 1 \text{ if } (\frac{1}{2}(p-1))! \equiv 1 \pmod p \\ -1 \text{ if } (\frac{1}{2}(p-1))! \equiv -1 \pmod p \end{cases}$$

(ii) if $p \equiv 1 \pmod 4$ and $i \in \mathbb{Q}_p$, $i^2 = -1$ then

$$\Gamma_p(\tfrac{1}{2}) = \begin{cases} i \text{ if } (\frac{1}{2}(p-1))! \equiv -\bar{i} \pmod p \\ -i \text{ if } (\frac{1}{2}(p-1))! \equiv \bar{i} \pmod p \end{cases}$$

Exercise 37.B. Give alternative (and short) proofs of the formulas of Proposition 37.2 as follows. For $x \in \mathbb{Z}_p$ set $f(x) := \Gamma_p(x)\,\Gamma_p(1-x)$ and $g(x) := (-1)^{l(x)}$ if $p \neq 2$; $g(x) := (-1)^{\sigma_1(x)+1}$ if $p = 2$. Show that

$$\frac{f(x+1)}{f(x)} = \frac{g(x+1)}{g(x)} = \begin{cases} -1 \text{ if } |x|_p = 1 \\ 1 \text{ if } |x|_p < 1 \end{cases}$$

Use induction to show that $f(n) = g(n)$ for all $n \in \mathbb{N}$. Conclude that $f = g$.

Exercise 37.C. Let $m \in \mathbb{N}$ be not divisible by p. Prove that

$$\prod_{j=0}^{m-1} \Gamma_p\left(\frac{j}{m}\right) \in \begin{cases} \{1, -1\} & \text{if } m \text{ is odd} \\ \{\Gamma_p(\frac{1}{2}), -\Gamma_p(\frac{1}{2})\} & \text{if } m \text{ is even} \end{cases}$$

38. The p-adic Gauss-Legendre multiplication formula

The classical multiplication formula is

$$\Gamma(z)\,\Gamma\left(z+\frac{1}{m}\right)\Gamma\left(z+\frac{2}{m}\right)\ldots\Gamma\left(z+\frac{m-1}{m}\right) = G_m(z)\,\Gamma(mz)$$

where

$$G_m(z) := (2\pi)^{(m-1)/2}\, m^{1/2-mz} \qquad (m \in \{2, 3, 4, \ldots\})$$

To arrive at a similar p-adic formula we have to suppose that m is not divisible by p since otherwise $\Gamma_p(x + 1/m)$ is not defined. For such $m \in \mathbb{N}$, $m \geqslant 2$ we set

$$(*)\,\Gamma_p(x)\,\Gamma_p\left(x+\frac{1}{m}\right)\Gamma_p\left(x+\frac{2}{m}\right)\ldots\Gamma_p\left(x+\frac{m-1}{m}\right) = G_m(x)\,\Gamma_p(mx)\ (x \in \mathbb{Z}_p)$$

Formula (*) defines the functions G_m; we would like to know more about them. Let $f(x)$ be the left hand expression of (*). Then

$$f\left(x+\frac{1}{m}\right) = f(x)\,\Gamma_p(x+1)\,\Gamma_p(x)^{-1} = f(x)\,h_p(x)$$

so that also

$$G_m\left(x+\frac{1}{m}\right)\Gamma_p\left(m\left(x+\frac{1}{m}\right)\right) = G_m(x)\,\Gamma_p(mx)\,h_p(x)$$

Since $\Gamma_p(m(x+1/m)) = \Gamma_p(mx+1) = \Gamma_p(mx)\,h_p(mx)$ we have

$$G_m\left(x+\frac{1}{m}\right)h_p(mx) = G_m(x)\,h_p(x)$$

According to the definition of h_p

$$\lambda(x) := h_p(x)\,h_p(mx)^{-1} = \begin{cases} m^{-1} & \text{if } |x|_p = 1 \\ 1 & \text{if } |x|_p < 1 \end{cases}$$

and we have

$$G_m\left(x+\frac{1}{m}\right) = G_m(x)\,\lambda(x) \qquad (x \in \mathbb{Z}_p)$$

Hence

$$G_m\left(\frac{n}{m}\right) = G_m(0)\prod_{j=0}^{n-1}\lambda\left(\frac{j}{m}\right) \qquad (n \in \mathbb{N})$$

By the definition of λ we have

$$\prod_{j=0}^{n-1}\lambda\left(\frac{j}{m}\right) = \prod_{j=0}^{n-1}\lambda(j) = (m^{-1})^{\mu(n)}$$

where $\mu(n)$ is the number of elements of $\{1, 2, \ldots, n-1\}$ that are not divisible by p, i.e.

$$\mu(n) = n - 1 - \left[\frac{n-1}{p}\right]$$

We get

$$G_m\left(\frac{n}{m}\right) = G_m(0)\, m^{-\mu(n)}$$

After replacing n by mn we find

$$G_m(n) = G_m(0)\, m^{-\mu(nm)} \qquad (n \in \mathbb{N})$$

and a first version of the p-adic multiplication formula

$$\Gamma_p(n)\,\Gamma_p\left(n+\frac{1}{m}\right)\ldots\Gamma_p\left(n+\frac{m-1}{m}\right) = \left(\prod_{j=0}^{m-1}\Gamma_p\left(\frac{j}{m}\right)\right)\Gamma_p(mn)\, m^{-\mu(nm)} \qquad (n \in \mathbb{N})$$

We wish to transform it in such a way that it becomes valid for $x \in \mathbb{Z}_p$ instead of $n \in \mathbb{N}$. Of course we know already that $n \mapsto m^{-\mu(nm)}$ can be interpolated. For any $j \in \mathbb{N}$ we have (with l as in Proposition 37.2)

$$\mu(j) = j - 1 - \left[\frac{j-1}{p}\right] = l(j) - 1 + (p-1)\cdot l_1(j)$$

where l_1 is defined by the formula $x = l(x) + pl_1(x)$ $(x \in \mathbb{Z}_p)$. We see that

$$m^{\mu(j)} = m^{l(j)-1}\cdot(m^{p-1})^{l_1(j)}$$

Observe now that l is locally constant, integer valued, that $m^{p-1} \in \mathbb{Q}_p^+$ and that by Theorem 32.4 the expression $(m^{p-1})^s$ makes sense for each $s \in \mathbb{Z}_p$. So the function

$$x \mapsto m^{l(mx)-1}\cdot(m^{p-1})^{l_1(xm)} \qquad (x \in \mathbb{Z}_p)$$

is a continuous extension of $n \mapsto m^{\mu(nm)}$. We have proved the following.

THEOREM 38.1. (p-adic multiplication formula) *For each $x \in \mathbb{Z}_p$, let $l(x) \in \{1, 2, \ldots p\}$ be such that $|x - l(x)|_p < 1$. Further, let $l_1(x) = p^{-1}(x - l(x))$ $(x \in \mathbb{Z}_p)$. Then for $m > 1$, m not divisible by p*

$$\prod_{j=0}^{m-1}\Gamma_p\left(x+\frac{j}{m}\right) = \left(\prod_{j=0}^{m-1}\Gamma_p\left(\frac{j}{m}\right)\right) m^{1-l(mx)}\cdot(m^{p-1})^{-l_1(mx)}\,\Gamma_p(mx) \qquad (x \in \mathbb{Z}_p)$$

(Observe that in Exercise 37.C it is shown that $\left(\prod_{j=0}^{m-1}\Gamma_p\left(\frac{j}{m}\right)\right)^4 = 1$.)

If we compare the p-adic and the classical multiplication formulas for the case $m = 2$ the resemblance shall become apparent. The classical formula for $m = 2$ reads

$$\Gamma(2z) = 2^{2z-1}\,\Gamma(\tfrac{1}{2})^{-1}\,\Gamma(z)\,\Gamma(z+\tfrac{1}{2})$$

Suppose $p \neq 2$, $m = 2$ and let $n \in \mathbb{N}$. The 'first version' of the p-adic formula says

$$\Gamma_p(n)\,\Gamma_p(n+\tfrac{1}{2}) = \Gamma_p(0)\,\Gamma_p(\tfrac{1}{2})\,\Gamma_p(2n)\,2^{-\mu(2n)}$$
$$= \Gamma_p(\tfrac{1}{2})\,\Gamma_p(2n)\,2^{-(2n-1)}\,2^{[(2n-1)/p]}$$

so that

$$\Gamma_p(2n) = 2^{2n-1}\,\Gamma_p(\tfrac{1}{2})^{-1}\,\Gamma_p(n)\,\Gamma_p(n+\tfrac{1}{2})\,2^{-[(2n-1)/p]}\ (n \in \mathbb{N})$$

39. Some other formulas involving Γ_p

In this section we consider a few elementary p-adic limits in which Γ_p plays a role.

PROPOSITION 39.1. (Elementary formulas concerning Γ_p) *Let $n \in \mathbb{N}$ and let s_n be the sum of the digits of $n = \sum_{j=0}^{s} a_j p^j$ ($a_s \neq 0$) in base p* (see Lemma 25.5). *Then*

(i) $\Gamma_p(n+1) = (-1)^{n+1}\,\dfrac{n!}{[n/p]!\;p^{[n/p]}}$

(ii) $\Gamma_p(p^n) = (-1)^p\,\dfrac{p^n!}{p^{n-1}!\;p^{p^{n-1}}}$

(iii) $n! = (-1)^{n+1-s}\,(-p)^{(n-s_n)/(p-1)}\displaystyle\prod_{j=0}^{s}\Gamma_p\left(\left[\frac{n}{p^j}\right]+1\right)$

(iv) $p^n! = (-1)^p(-p)^{(p^n-1)/(p-1)}\displaystyle\prod_{j=0}^{n}\Gamma_p(p^j)$

Proof. (i) and (ii) are straightforward consequences of the definition of Γ_p. Formulas (iii) and (iv) follow from (i) and (ii) by induction.

From (i)-(iv) we can derive certain interesting relations. We shall see that Γ_p has something to do with the behaviour of the terms $x^n/n!$ of the exponential function on the 'boundary' $\{x \in \mathbb{C}_p : |x|_p = p^{1/(1-p)}\}$ of its region of convergence. Of course for such x the sequence $1, x, x^2/2!, \ldots$ does not tend to zero, but has it convergent subsequences? We have $|x^n/n!|_p = p^{s_n/(1-p)}$ (Lemma 25.5) so in order to have any chance we should choose a subsequence $n_1, n_2, \ldots$ for which s_{n_j} is eventually constant; let us try $n_j = p^j$ ($j \in \mathbb{N}$). Then $s_{n_j} = 1$ for all j. The following question makes sense.

For which $x \in \mathbb{C}_p$, $|x|_p = p^{1/(1-p)}$ does $\lim_{n\to\infty} x^{p^n}/p^n!$ exist?

PROPOSITION 39.2. *Let $x \in \mathbb{C}_p$, $|x|_p = p^{1/(1-p)}$. Then $\lim_{n\to\infty} x^{p^n}/p^n!$ exists if and only if $x = a\tau$ where $a \in \mathbb{C}_p^+$ and $\tau^{p-1} = -p$. In that case we have*

$$\lim_{n\to\infty} \frac{x^{p^n}}{p^n!} = (-1)^p \tau \prod_{j=0}^{\infty} \Gamma_p(p^j)^{-1}$$

Proof. First, notice that since $\Gamma_p(0) = 1$ the product $c_p := \prod_{j=0}^{\infty} \Gamma_p(p^j)^{-1}$ is well defined. Now let $x = a\tau$ where $a \in \mathbb{C}_p^+$ and $\tau^{p-1} = -p$. Then $\lim_{n\to\infty} a^{p^n} = 1$ (Theorem 32.2). By formula (iv) of Proposition 39.1

$$p^n! = (-1)^p \tau^{p^n-1} \prod_{j=0}^{n} \Gamma_p(p^j)$$

so that $\lim_{n\to\infty} \tau^{p^n}/p^n! = (-1)^p \tau c_p$. Then

$$\lim_{n\to\infty} \frac{x^{p^n}}{p^n!} = \lim_{n\to\infty} a^{p^n} \cdot \lim_{n\to\infty} \frac{\tau^{p^n}}{p^n!} = (-1)^p \tau c_p$$

Conversely, suppose $\lim_{n\to\infty} x^{p^n}/p^n! = a \in \mathbb{C}_p$. Choose any $(p-1)$th root θ of $-p$. Then $|\theta|_p = |x|_p$ so that $x = y\theta$ where $|y|_p = 1$. By the first part of the proof $\lim_{n\to\infty} \theta^{p^n}/p^n!$ exists so that also $\lim_{n\to\infty} y^{p^n} = \beta$ must exist. β is a $(p-1)$th root of unity. Set $a := \beta^{-1}y$, $\tau := \beta\theta$. Then $x = a\tau$; $\lim_{n\to\infty} a^{p^n} = 1$ so that $a \in \mathbb{C}_p^+$ (Theorem 32.2) and τ is a $(p-1)$th root of $-p$.

Other relations we leave as an exercise. Let $c_p := \prod_{j=0}^{\infty} \Gamma_p(p^j)^{-1}$.

Exercise 39.A. Show that $|c_p + 1|_p < 1$.

Exercise 39.B. Show that $\lim_{n\to\infty} p^{n+1}!/(p^n!)^p = -pc_p^{p-1}$ in $\mathbb{Q}_p$.

Exercise 39.C. Prove that for $s \in \mathbb{N}$

$$(1+p+\ldots+p^s)\,\Gamma_p(1+p+\ldots+p^s) = \binom{1+p+\ldots+p^s}{p^s} \prod_{j=0}^{s} \Gamma_p(p^j)$$

and obtain the following p-adic limit

$$\lim_{s\to\infty} \binom{1+p+\ldots+p^s}{p^s} = (1-p)^{-1}\,\Gamma_p((1-p)^{-1})\,c_p$$

Exercise 39.D. (van Hamme) Let $m, n \in \mathbb{N}$, $m > n$. Prove that for $j \in \mathbb{N}$

$$\binom{mp^j}{np^j} = \binom{mp^{j-1}}{np^{j-1}} \cdot \frac{\Gamma_p(mp^j)}{\Gamma_p(np^j)\,\Gamma_p((m-n)p^j)}$$

and obtain the p-adic limit

$$\lim_{j \to \infty} \binom{mp^j}{np^j} = \binom{m}{n} \prod_{j=1}^{\infty} \frac{\Gamma_p(mp^j)}{\Gamma_p(np^j)\,\Gamma_p((m-n)p^j)}$$

Remark. In Part 2 of Chapter 3 we shall return to the subject of p-adic gamma functions.

PART 3: ANALYTIC FUNCTIONS

In Section 25 we mentioned already some facts on power series and analytic functions. In this part we go into the subject a little deeper and treat some examples. However, our approach shall remain 'naive' and elementary. In particular we shall avoid the use of formal power series. For a more 'sophisticated' treatment, see Amice (1975), Koblitz (1977), Koblitz (1980) and the references given there.

40. Convergence of power series

Let $a_0, a_1, \ldots \in K$. Recall that the power series $\Sigma\, a_n x^n$ converges for $|x| < \rho$, diverges for $|x| > \rho$ and either converges or diverges on the whole of $\{x \in K : |x| = \rho\}$. Here ρ, the radius of convergence of $\Sigma\, a_n x^n$, was defined by $\rho := (\overline{\lim}_{n \to \infty} \sqrt[n]{|a_n|})^{-1}$. In this section we shall prove that the convergence of $\Sigma\, a_n x^n$ on proper subdiscs of $B_0(\rho^-)$ is stronger than pointwise or uniform. But first we consider – in the form of exercises – a few properties of ρ itself to warn the reader that one has to be a little careful when dealing with the radius of convergence.

Exercise 40.A. (On the definition of the radius of convergence) Let $a_0, a_1, \ldots \in K$ and let ρ be the radius of convergence of $\Sigma\, a_n x^n$.
(i) Let the valuation on K be dense. Show that there exists a unique $\tau \in [0, \infty]$ such that $\Sigma\, a_n x^n$ converges on $\{x \in K : |x| < \tau\}$, diverges on $\{x \in K : |x| > \tau\}$ and that $\tau = \rho$.
(ii) Let the valuation of K be discrete. Suppose $0 < \rho < \infty$. Show that there are *infinitely many* $\tau \in (0, \infty)$ such that $\Sigma\, a_n x^n$ converges on $\{x \in K : |x| < \tau\}$, diverges on $\{x \in K : |x| > \tau\}$. Prove that there is a unique $\tau_0 \in |K^\times|$ such that $\Sigma\, a_n x^n$ converges on $\{x \in K : |x| \leqslant \tau_0\}$, diverges on $\{x \in K : |x| > \tau_0\}$ but that τ_0 may be unequal to ρ.
(iii) (Justification of choice of ρ, see Remark 1 following Theorem 22.7) Let τ be such that $\Sigma\, a_n x^n$ converges on $\{x \in L : |x| < \tau\}$ and diverges on $\{x \in L : |x| > \tau\}$ for any (complete) valued extension L of K. Show that $\tau = \rho$.

Exercise 40.B. For each $\rho \in [0, \infty]$ there exists a power series in K whose radius of convergence is precisely ρ. Prove this.

Exercise 40.C. (Convergence of $\Sigma\, a_n x^n$ and its formal derivative $\Sigma\, n a_n x^{n-1}$) Prove the following. Recall that k is the residue class field of K.
(i) If $\mathrm{char}(k) = 0$ then a power series and its formal derivative have the same region of convergence.
(ii) If $K \supset \mathbb{Q}_p$ then a power series and its formal derivative have equal radii of convergence yet the regions of convergence may differ.
(iii) If $\mathrm{char}(K) = p$ then for every ρ_1, ρ_2, $(0 < \rho_1 \leqslant \rho_2 < \infty)$ there is a power series with radius of convergence ρ_1, whereas its formal derivative has radius of convergence ρ_2.

Exercise 40.D. Let ρ be the radius of convergence of a power series $\Sigma\, a_n x^n$ in K. Prove $\rho = \sup\{\tau \in [0, \infty]: \lim_{n \to \infty} |a_n|\, \tau^n = 0\} = \sup\{\tau \in [0, \infty]: \sup_n |a_n|\, \tau^n \neq \infty\}$. Show by means of examples that each of the following cases may occur. (i) $\sup_n |a_n|\, \rho^n = \infty$ and $\rho < \infty$, (ii) $\sup_n |a_n|\, \rho^n < \infty$ but not $\lim_{n \to \infty} |a_n|\, \rho^n = 0$, (iii) $\lim_{n \to \infty} |a_n|\, \rho^n = 0$.

Next, we turn to convergence. To get a first impression consider the following exercise.

Exercise 40.E. (Convergence of power series of log and exp) Let $K = \mathbb{C}_p$.
(i) Recall that $\Sigma\, x^n/n$ converges on $B_0(1^-)$ and diverges elsewhere. Show that $-\log(1-x) = \Sigma\, x^n/n$ is not bounded on $B_0(1^-)$ and that the partial sums of $\Sigma\, x^n/n$ do not converge uniformly on $B_0(1^-)$.
(ii) The result of (i) may not surprise you. In that case show that the power series of exp does not converge uniformly on E.

DEFINITION 40.1. Let $n \in \mathbb{N} \cup \{0\}$, let X be a nonempty subset of K without isolated points. A sequence $f_1, f_2, \ldots$ in $C^n(X \to K)$ (Definition 29.1) *converges to* $f \in C^n(X \to K)$ *in the* C^n*-sense* ($f = \lim_{j \to \infty} f_j$ *in* C^n) *if* $\lim_{j \to \infty} \|f - f_j\|_n = 0$. (See Exercise 29.C.)

It is shown in Corollary 29.11 that analytic functions are C^∞. In this context we have the following.

THEOREM 40.2. *Let f be an analytic function defined on a disc $V \subset K$.*
(i) If K has discrete valuation or if V has the form $B_a(r)$ for some $r \in |K^\times|$ then for each $n \in \mathbb{N} \cup \{0\}$ the function $\Phi_n f$ is bounded and uniformly continuous on V^{n+1}. For each $b \in V$ the Taylor series of f at b converges in the C^n-sense to f.
(ii) If K has a dense valuation and if V has the form $B_a(r^-)$ for some $r \in |K^\times|$ then the conclusion of (i) *holds if V is replaced by a proper subdisc of V.*

This theorem is an immediate consequence of the proof of Corollary 29.11, Exercise 29.D, and

LEMMA 40.3. *Let $r \in |K^\times|$ and let $f(x) = \sum_{j=0}^{\infty} a_j x^j$ for $x \in B_0(r)$. Then for each $n \in \mathbb{N} \cup \{0\}$ the partial sums $m \mapsto \sum_{j=0}^{m} a_j x^j$ converge to f in the sense of C^n.*

Proof. Let $\epsilon > 0, n \in \mathbb{N} \cup \{0\}$. Notice that $\lim_{m\to\infty} |a_m| r^m = 0$; there is an N with $|a_m| r^m < \epsilon r^n$ for $m > N$. Using the estimates of $\Phi_n(x^j)$ of Corollary 29.11 and writing $f_m(x) = \sum_{j=0}^{m} a_j x^j$ we arrive at

$$\|\Phi_n(f - f_m)\|_\infty = \|\sum_{j=m+1}^{\infty} a_j \Phi_n(x^j)\|_\infty \leqslant r^{-n} \sup_{j>m} |a_j| r^j \leqslant \epsilon$$

for $m > N$. It follows that $\lim_{m\to\infty} \|\bar{\Phi}_n(f - f_m)\|_\infty = 0$ for each n, hence $\lim_{m\to\infty} \|f - f_m\|_n = 0$ for each n.

41. Substitution of power series

Are compositions of analytic functions analytic? Stated more precisely, let f, g be analytic functions defined on discs V, W respectively and let $f(V) \subset W$. Does it follow that $g \circ f$ is analytic? The following simple example shows that the answer is 'no'.

EXAMPLE 41.1. (Two analytic functions whose composition is not analytic) *The formulas*

$$f(x) = x^p - x$$
$$g(x) = (1-x)^{-1} = \sum_{n=0}^{\infty} x^n$$

define analytic functions f and g on $\mathbb{Z}_p$, $p\mathbb{Z}_p$ respectively for which $f(\mathbb{Z}_p) \subset p\mathbb{Z}_p$ but $g \circ f$ is not analytic.

Proof. For any $x \in \mathbb{Z}_p$ we have $\bar{x}^p - \bar{x} = 0$ in $\mathbb{F}_p$, hence $x^p - x \in p\mathbb{Z}_p$ and $g \circ f$ is defined. Suppose it is analytic. Then we have $a_0, a_1, \ldots \in \mathbb{Q}_p$ such that

$$g(f(x)) = (1-(x^p - x))^{-1} = a_0 + a_1 x + \ldots \quad (x \in \mathbb{Z}_p)$$

The substitution $x = 1$ leads to $\lim_{n\to\infty} a_n = 0$ so that $\sum a_n x^n$ converges not only on $\mathbb{Z}_p$ but on the unit disc of $\mathbb{C}_p$. Set

$$h(x) := (1-(x^p - x))\left(\sum_{n=0}^{\infty} a_n x^n\right) - 1 \qquad (x \in \mathbb{C}_p, |x|_p \leqslant 1)$$

$h+1$, being a product of analytic functions, is analytic. Then so is h. But h vanishes on $\mathbb{Z}_p$ so by Exercise 25.D it is identically zero on $\{x \in \mathbb{C}_p : |x|_p \leqslant 1\}$. On the other hand, if $z \in \mathbb{C}_p$ is a solution of $z^p - z - 1 = 0$ then $|z|_p = 1$ and $h(z) = -1$, a contradiction. It follows that $g \circ f$ is not analytic on $\mathbb{Z}_p$.

Exercise 41.A. Show that $g \circ f$ is locally analytic of order 1 (see Definition 25.3).

We shall prove that compositions of locally analytic functions are locally analytic (Theorem 41.3). If K is not too small then compositions of analytic functions are analytic as we will see in Theorem 42.4. In its turn, this fact shall lead to a satisfactory condition on analytic functions f and g in order that, for general K, their composition $g \circ f$ is analytic (Corollary 42.5). The following lemma supplies the necessary facts.

LEMMA 41.2. (Substitution of power series) *Let $\Sigma\, a_n x^n$ and $\Sigma\, b_n x^n$ be power series in K. Suppose that $\Sigma\, b_n x^n$ converges for some $x \in K$. Set $\tau := \max_{n \geqslant 0} |b_n|\, |x|^n$, and assume $\lim_{n \to \infty} |a_n|\, \tau^n = 0$. Then*

$$h(x) := \sum_{n=0}^{\infty} a_n \left(\sum_{j=0}^{\infty} b_j x^j \right)^n$$

is well defined and $h(x) = \Sigma_{m=0}^{\infty}\, c_m x^m$ where

$$c_m = \sum_{n=0}^{\infty} a_n b_m^{(n)}$$

$$b_0^{(0)} = 1,\ b_m^{(0)} = 0\ (m > 1),\ b_m^{(n)} = \sum_{j_1 + \ldots + j_n = m} b_{j_1} b_{j_2} \ldots b_{j_n}\ (n \in \mathbb{N})$$

Proof. Set $y := \Sigma_{j=0}^{\infty}\, b_j x^j$. Then $|y| \leqslant \max_j |b_j|\, |x|^j = \tau$. Hence $\lim_{n \to \infty} a_n y^n = 0$ so the series $\Sigma\, a_n y^n$ converges and $h(x)$ is well defined. By Exercise 23.M

$$(*) \qquad \left(\sum_{j=0}^{\infty} b_j x^j \right)^n = \sum_{m=0}^{\infty} b_m^{(n)} x^m \qquad (n \in \{0, 1, \ldots\})$$

where the $b_m^{(n)}$ are as above. So $h(x) = \displaystyle\sum_{n=0}^{\infty} a_n \left(\sum_{m=0}^{\infty} b_m^{(n)} x^m \right)$. To see that we may interchange the order of summation, set $c_{mn} := a_n b_m^{(n)} x^m$ and observe that from (*) it follows that $\lim_{m \to \infty} c_{mn} = 0$ for each n. Further, we have for $n \geqslant 1$

$$|c_{mn}| = |a_n| \left| \sum_{j_1 + \ldots + j_n = m} b_{j_1} x^{j_1} b_{j_2} x^{j_2} \ldots b_{j_n} x^{j_n} \right| \leqslant |a_n| \tau^n$$

It follows that $\lim_{n \to \infty} c_{mn} = 0$ uniformly in m. By Exercise 23.B we then have

$$h(x) = \sum_{n=0}^{\infty} \sum_{m=0}^{\infty} c_{mn} = \sum_{m=0}^{\infty} \sum_{n=0}^{\infty} c_{mn} = \sum_{m=0}^{\infty} \left(\sum_{n=0}^{\infty} a_n b_m^{(n)} \right) x^m$$

which proves the lemma.

THEOREM 41.3. *Let f, g be locally analytic functions defined on convex sets V, W respectively. If $f(V) \subset W$ then $g \circ f$ is locally analytic. The Taylor series of $g \circ f$ at any point of V can be computed by formal substitution.*
Proof. It suffices to prove the following statement. If $\Sigma a_n x^n$ converges for $|x| < \rho$, if $\Sigma b_n x^n$ has a positive radius of convergence and if $|b_0| < \rho$ then there is a $\rho_1 > 0$ such that $x \mapsto \Sigma_{n=0}^{\infty} a_n (\Sigma_{j=0}^{\infty} b_j x^j)^n$ is the sum of a power series for $|x| \leqslant \rho_1$. To prove it, choose $\rho_2 > 0$ such that $\lim_{n \to \infty} |b_n| \rho_2^n = 0$, choose $M > \max(\rho, \max_n |b_n| \rho_2^n)$ and set $\rho_1 := \rho_2 \rho M^{-1}$. Then for $|x| \leqslant \rho_1$ and $n \in \{1, 2, \ldots\}$

$$|b_n x^n| \leqslant |b_n| \rho_2^n \rho^n M^{-n} < M M^{n-1} \rho M^{-n} = \rho$$

It was given that $|b_0| < \rho$, so we have $\max_n |b_n| \rho_1^n < \rho$. Now apply the preceding lemma for any x with $|x| \leqslant \rho_1$ and $\tau := \max_n |b_n x^n|$.

42. The maximum principle

The maximum principle for complex analytic functions states that if f is an analytic function on the disc $\{ z \in \mathbb{C} : |z| \leqslant r \}$ then $|f|$ attains its maximum at a point of the boundary, i.e.

$$\max_{|z| \leqslant r} |f(z)| = \max_{|z| = r} |f(z)|$$

In this section we shall prove an ultrametric version of this statement for K not locally compact. (In the next section we consider the locally compact case.)

LEMMA 42.1. *Let K be not locally compact, let $f : B_0(1) \to K$ be an analytic function given by its power series*

$$f(x) = \sum_{n=0}^{\infty} a_n x^n \qquad (|x| \leqslant 1)$$

Then $\sup \{ |f(x)| : |x| \leqslant 1 \} = \max \{ |a_n| : n \geqslant 0 \}$. *If the valuation of K*

is dense then

$$\sup\{|f(x)| : |x| \leqslant 1\} = \sup\{|f(x)| : |x| < 1\}$$

If the residue class field k of K is infinite then

$$\sup\{|f(x)| : |x| \leqslant 1\} = \sup\{|f(x)| : |x| = 1\} = \max\{|f(x)| : |x| = 1\}$$

Proof. The power series $\Sigma\, a_n x^n$ converges at $x = 1$, so $\lim_{n\to\infty} a_n = 0$ and $\sup\{|a_n| : n \geqslant 0\} = \max\{|a_n| : n \geqslant 0\} < \infty$. By the strong triangle inequality we have (without any assumption on K) $\sup\{|f(x)| : |x| \leqslant 1\} \leqslant \max\{|a_n| : n \geqslant 0\}$. To prove the opposite inequality for non-locally compact K we may assume that $\max\{|a_n| : n \geqslant 0\} = 1$ (multiply f by a suitable constant). By Corollary 12.2 either k is infinite or $|K^\times|$ is dense. Thus we distinguish two cases.

(i) Suppose k is infinite. Let $x \mapsto \bar{x}$ be the quotient map $B_0(1) \to k$. Since $\lim_{n\to\infty} a_n = 0$ we have $\bar{a}_n = 0$ for large n so that the function $t \mapsto \Sigma_{n=0}^{\infty} \bar{a}_n t^n$ is a nonzero polynomial function on k; it has only finitely many zeros. Now k is infinite so there is $s \in k$ for which $\Sigma_{n=0}^{\infty} \bar{a}_n s^n \neq \bar{0}$. Choose $b \in K$ for which $\bar{b} = s$. Then $\Sigma_{n=0}^{\infty} a_n b^n \neq \bar{0}$, i.e. $|\Sigma_{n=0}^{\infty} a_n b^n| = 1$. So we have found

$$\sup\{|f(x)| : |x| \leqslant 1\} = \max\{|f(x)| : |x| = 1\} = \max\{|a_n| : n \geqslant 0\}$$

(ii) Suppose $|K^\times|$ is dense. We have trivially

$$\sup\{|f(x)| : |x| < 1\} \leqslant \sup\{|f(x)| : |x| \leqslant 1\} \leqslant \max\{|a_n| : n \geqslant 0\} = 1$$

and we are done if we can produce a $b \in B_0(1^-)$ such that $|f(b)|$ is close to 1. If $|a_0| = 1$ we can choose $b = 0$, so assume $N := \min\{j : |a_j| = 1\} > 0$. Let $0 < \epsilon < 1$, $\max(|a_0|, |a_1|, \ldots, |a_{N-1}|) \leqslant 1 - \epsilon$. We claim that if $b \in K$, $1 - \epsilon < |b^N| < 1$ then $|f(b)| \geqslant 1 - \epsilon$. Indeed, we have

$$\left|\sum_{j=0}^{N-1} a_j b^j\right| \leqslant \max(|a_0|, |a_1|, \ldots, |a_{N-1}|) \leqslant 1 - \epsilon < |b^N|$$

$$|a_N b^N| = |b^N|$$

$$\left|\sum_{j=N+1}^{\infty} a_j b^j\right| \leqslant \sup_{j>N} |a_j|\,|b^j| \leqslant \sup_{j>N} |b^j| \leqslant |b|^{N+1} < |b^N|$$

By the isosceles triangle principle, $|f(b)| = \left|\Sigma_{j=0}^{\infty} a_j b^j\right| = |b^N| > 1 - \epsilon$ and the lemma is proved. See also Exercise 42.B.

From Lemma 42.1 one obtains several interesting corollaries for analytic functions defined on discs. First we consider 'closed' discs.

THEOREM 42.2. *Let K be not locally compact, let* $r \in |K^\times|$.

(i) ('Maximum principle') *Let f be an analytic function on* $B_0(r)$ *given by*

its power series $f(x) = \sum_{n=0}^{\infty} a_n x^n$ $(x \in B_0(r))$.
If the valuation of K is dense then

$$\sup\{|f(x)| : |x| \leqslant r\} = \sup\{|f(x)| : |x| < r\} = \max\{|a_n| r^n : n \geqslant 0\} < \infty$$

If the residue class field of K is infinite then

$$\max\{|f(x)| : |x| \leqslant r\} = \max\{|f(x)| : |x| = r\} = \max\{|a_n| r^n : n \geqslant 0\} < \infty$$

(ii) (Uniform closedness) *If* $f_1, f_2, \ldots$ *are analytic on* $B_0(r)$ *and if* $f := \lim_{n \to \infty} f_n$ *uniformly on* $B_0(r)$ *then* f *is analytic.*

Proof. To prove (i) choose $a \in K$ such that $|a| = r$ and apply Lemma 42.1 to the function $x \mapsto f(ax)$ $(x \in B_0(1))$. The same trick reduces (ii) to the case $r = 1$. Let A be the space of all analytic functions on $B_0(1)$ normed by $f \mapsto \|f\|_\infty = \sup\{|f(x)| : x \in B_0(1)\}$. Using Lemma 42.1 we conclude that the formula

$$f \mapsto (a_0, a_1, \ldots) \qquad \left(f(x) = \sum_{n=0}^{\infty} a_n x^n, x \in B_0(1)\right)$$

defines an isometrical isomorphism between A and c_0. By Exercise 13.A the latter is a Banach space over K. Then so is A and (ii) is proved.

Next, we consider analytic functions on an 'open' disc $B_0(r^-)$ where $r \in (0, \infty)$. We assume the valuation to be dense since otherwise we are in the situation of Theorem 42.2. Let

$$f(x) = \sum_{n=0} a_n x^n \qquad (|x| < r)$$

For any $d \in |K^\times|, d < r$ we have by Theorem 42.2

$$\sup\{|f(x)| : |x| \leqslant d\} = \sup\{|a_n| d^n : n \geqslant 0\}$$

From this it follows easily that

$$\sup\{|f(x)| : |x| < r\} = \sup\{|a_n| r^n : n \geqslant 0\}$$

(Notice that these suprema may be infinite.) The formula

$$f \mapsto (a_0, a_1, \ldots) \qquad \left(f(x) = \sum_{n=0}^{\infty} a_n x^n, |x| \leqslant d\right)$$

defines an isometrical isomorphism between the space of all bounded analytic functions on $B_0(d)$ with the supremum norm and the space of all sequences $(a_0, a_1, \ldots)$ for which

$$\|(a_0, a_1, \ldots)\| := \sup\{|a_n| d^n : n \geqslant 0\} < \infty$$

The latter space is easily seen to be a K-Banach space. Then so is the first. We have proved the following theorem.

THEOREM 42.3. *Let the valuation of K be dense, let $r \in (0, \infty)$.*
(i) ('Supremum principle') *Let f be an analytic function on $B_0(r^-)$ given by its power series $f(x) = \sum_{n=0}^{\infty} a_n x^n$ $(x \in B_0(r^-))$. Then*

$$\sup \{|f(x)| : |x| < r\} = \sup \{|a_n| r^n : n \geqslant 0\}$$

(ii) (Uniform closedness) *If $f_1, f_2, \ldots$ are analytic on $B_0(r^-)$ and if $f := \lim_{n \to \infty} f_n$ uniformly on $B_0(r^-)$ then f is analytic.*

We now use the 'maximum principle' to prove that compositions of analytic functions are again analytic if, for example, K is algebraically closed. (Compare Example 41.1.)

THEOREM 42.4. *Let the residue class field k of K be infinite. Let D_1 and D_2 be discs in K, let $f : D_1 \to K$ and $g : D_2 \to K$ be analytic and let $f(D_1) \subset D_2$. Then $g \circ f$ is analytic.*

Proof. It suffices to consider the case where D_1 and D_2 are discs containing 0. Let $f(x) = \sum_{n=0}^{\infty} b_n x^n$ $(x \in D_1)$ and $g(x) = \sum_{n=0}^{\infty} a_n x^n$ $(x \in D_2)$. We shall apply Lemma 41.2 to these power series in $x \in D_1$, and $\tau := \max_{n \geqslant 0} |b_n|\, |x|^n$. By the 'maximum principle'

$$\max \{|f(z)| : |z| \leqslant |x|\} = \tau$$

so there is $z \in D_1$ such that $|f(z)| = \tau$. Then $\lim_{n \to \infty} |a_n|\, \tau^n = \lim_{n \to \infty} |a_n|\, |f(z)|^n = 0$ since $f(z) \in D_2$. Hence the conditions of Lemma 41.2 are satisfied and we may conclude that there are $c_0, c_1, \ldots$ in K only depending on the coefficients $a_0, a_1, \ldots$ and $b_0, b_1, \ldots$ such that $g \circ f(x) = \sum_{n=0}^{\infty} c_n x^n$.

COROLLARY 42.5. *Let K be arbitrary, let D_1, D_2, be discs in K, let $f : D_1 \to K$ and $g : D_2 \to K$ be analytic and let $f(D_1) \subset D_2$. Let the power series of f and g be $\sum b_n x^n$, $\sum a_n x^n$ respectively. Let L be the completion of the algebraic closure of K. Suppose there are discs $\tilde{D}_1, \tilde{D}_2$ in L such that $\tilde{f}(x) := \sum_{n=0}^{\infty} b_n x^n$, $\tilde{g}(x) := \sum_{n=0}^{\infty} a_n x^n$ converge on $\tilde{D}_1, \tilde{D}_2$ respectively, and $\tilde{D}_1 \supset D_1, \tilde{D}_2 \supset D_2$. If $\tilde{f}(\tilde{D}_1) \subset \tilde{D}_2$ then $g \circ f$ is analytic.*

THEOREM 42.6. (Ultrametric Liouville theorem) *Let K be not locally compact. Then a bounded analytic function $K \to K$ is constant.*

Proof. Let $f(x) = \sum_{n=0}^{\infty} a_n x^n$ $(x \in K)$. By Theorem 42.2 we have for each $r \in |K^\times|$

$$\sup \{|a_n| r^n : n \geqslant 0\} = \sup \{|f(x)| : |x| \leqslant r\} \leqslant \|f\|_\infty$$

Hence $a_1 = a_2 = \ldots = 0$ and $f(x) = a_0$ for all $x \in K$.

Exercise 42.A. Verify that the functions f, g of Example 41.1 do not satisfy the conditions of Corollary 42.5.

Exercise 42.B. Let K be not locally compact, let $f : B_0(1) \to K$ be analytic. Show that $\sup \{|f(x)| : |x| \leqslant 1\} = \sup \{|f(x)| : |x| = 1\}$. (Hint. Let $|K^\times|$ be dense. Consider $g(x) := f(x+1)$ and apply Lemma 42.1 to g.) Does $\max \{|f(x)| : |x| = 1\}$ exist?

43. Failure of the maximum principle for locally compact K

Almost everything we have proved in Section 42 becomes a falsity if we take K to be locally compact. First of all, we have seen (Example 41.1) that compositions of analytic functions need not be analytic if $K = \mathbb{Q}_p$. In this section we shall construct, for a locally compact K, a bounded non-constant analytic function $K \to K$ destroying the locally compact version of the ultrametric Liouville Theorem 42.6 and prove Kaplansky's theorem which is an ultrametric translation of Weierstrass' approximation theorem. As a consequence, for locally compact K a uniform limit of analytic functions need not be analytic, in contrast to Theorems 42.2 (ii) and 42.3 (ii).

EXAMPLE 43.1. *Let K be locally compact. Then there exists a bounded nonconstant analytic function $K \to K$.*

Proof. Let $a_1, a_2, \ldots \in K$, $\lim_{j\to\infty} a_j = 0$. Then

$$(*) \qquad f(x) := \prod_{j=1}^{\infty} (1 + a_j x)$$

exists for all $x \in K$. By Exercise 23.N (ii) there are $c_0, c_1, \ldots$ such that $f(x) = \sum_{j=0}^{\infty} c_j x^j$ for all $x \in K$, i.e. f is analytic. We proceed to show that for a proper choice of $a_1, a_2, \ldots f$ is bounded. Let $1 = \epsilon_0, \epsilon_1, \ldots$ be positive numbers. We prove the existence of a null sequence $a_1, a_2, \ldots$ in K and an increasing sequence $n_0, n_1, \ldots$ of natural numbers such that for all $m \in \{0, 1, 2, \ldots\}$

$$\left| \prod_{j=1}^{n_m} (1 + a_j x) \right| \leqslant \epsilon_s \qquad (|x| = |\pi|^{-s}, 0 \leqslant s \leqslant m)$$

(for π see the beginning of this chapter). First take $m = 0$. Choose $n_0 := 1$ and $a_1 := 1$. Then clearly $|1 + a_1 x| \leqslant 1 = \epsilon_0$ for all $x \in B_0(1)$. Suppose we have chosen $n_0, n_1, \ldots, n_{m-1}$ and $a_1, a_2, \ldots a_{n_{m-1}}$ such that

$$\left|\prod_{j=1}^{n_{m-1}} (1+a_jx)\right| \leqslant \epsilon_s \qquad (|x| = |\pi|^{-s}, 0 \leqslant s \leqslant m-1)$$

Let $\tau := \max\left\{\left|\prod_{j=1}^{n_{m-1}} (1+a_jx)\right| : |x| \leqslant |\pi|^{-m}\right\}$. Choose a positive number r such that $r < \min(1, \epsilon_m \tau^{-1}) |\pi|^{-m}$. The clopen set $\{x \in K : |x| = |\pi|^{-m}\}$ is a disjoint union of balls of the form $B_{b_i}(r)$ $(i = 1, \ldots, t)$. Then $|b_i| = |\pi|^{-m}$ for all i. Define

$$a_{n_{m-1}+i} := b_i^{-1} \qquad (i = 1, \ldots, t)$$

$$n_m := n_{m-1} + t$$

We claim that for $0 \leqslant s \leqslant m$ and $|x| = |\pi|^{-s}$

$$\left|\prod_{j=1}^{n_m} (1+a_jx)\right| \leqslant \epsilon_s$$

Indeed, if $|x| = |\pi|^{-m}$ then for $j \in \{n_{m-1}+1, \ldots, n_m\}$ we have $|a_jx| = 1$ so that $|1+a_jx| \leqslant 1$. Among $n_{m-1}+1, \ldots, n_m$ there is one j for which $|x+b_j| \leqslant r$, i.e. $|1+a_jx| \leqslant r\,|\pi|^m$ and we get

$$\left|\prod_{j=1}^{n_m} (1+a_jx)\right| \leqslant \tau \left|\prod_{j=n_{m-1}+1}^{n_m} (1+a_jx)\right| \leqslant \tau r |\pi|^m \leqslant \epsilon_m$$

If $|x| = |\pi|^{-s}$ and $0 \leqslant s < m$ then for $j \in \{n_{m-1}+1, \ldots, n_m\}$ we have $|a_jx| = |\pi|^m\, |\pi|^{-s} < 1$ so that $|1+a_jx| = 1$. Together with the induction hypothesis this yields

$$\left|\prod_{j=1}^{n_m} (1+a_jx)\right| = \left|\prod_{j=1}^{n_{m-1}} (1+a_jx)\right| \leqslant \epsilon_s$$

and we have proved what we announced. Our function f, formally defined by (*), satisfies

$$|f(x)| = \lim_{m\to\infty} \left|\prod_{j=1}^{n_m} (1+a_jx)\right| \leqslant \epsilon_s \; (|x| = |\pi|^{-s}, s \in \{0, 1, \ldots\})$$

If we choose $\epsilon_0, \epsilon_1, \ldots$ to be a bounded sequence and, say, $\epsilon_1 < 1$ then f is bounded, nonconstant. This finishes the proof.

COROLLARY 43.2. *Let K be locally compact. There exists an analytic function $f: K \to K$ with $f(0) = 1$ and $\lim_{|x|\to\infty} f(x) = 0$.*

Proof. In the proof of above, choose $\epsilon_n := 1/n$ for all $n \in \mathbb{N}$.

THEOREM 43.3. (Kaplansky) *Let X be a compact subset of K, let $f \in C(X \to K)$ and $\epsilon > 0$. Then there is a polynomial function $P : K \to K$ such that $|P(x) - f(x)| < \epsilon$ for all $x \in X$.*

Proof. By Theorem 26.2 it suffices to solve the problem for a locally constant function f. By compactness there is a $\delta \in (0, 1)$ such that f is constant on each of the (finitely many) balls in X of radius δ. So we may even assume that f is the characteristic function of a ball (in X) of radius δ. Without loss of generality, let $0 \in X$, $f(0) = 1$. Choose $c_1, \ldots, c_m \in X$ such that $X \subset B_0(\delta) \cup B_{c_1}(\delta) \cup \ldots \cup B_{c_m}(\delta)$, where $B_0(\delta), B_{c_1}(\delta), \ldots, B_{c_m}(\delta)$ are pairwise disjoint and $|c_1| \leqslant |c_2| \ldots \leqslant |c_m|$. Then $\delta < |c_1|$; choose $s \in \mathbb{N}$ such that $(\delta/|c_1|)^s < \epsilon$. Inductively we shall define integers $n_1, \ldots, n_m$ such that the polynomial function P defined by

$$P(x) = \prod_{j=1}^{m} (1 - (c_j^{-1} x)^s)^{n_j} \qquad (x \in K)$$

does the job, i.e. we prove that for a suitable choice of $n_1, \ldots, n_m$ we have $|P(x) - 1| < \epsilon$ for $x \in B_0(\delta)$ and $|P(x)| < \epsilon$ for $x \in B_{c_1}(\delta) \cup \ldots \cup B_{c_m}(\delta)$. First let $x \in B_0(\delta)$. Then $1 - (c_j^{-1} x)^s \in B_1(\epsilon)$ for all j. Since $B_1(\epsilon)$ is a multiplicative group (at least if $\epsilon < 1$) we may conclude that $P(x) \in B_1(\epsilon)$, i.e. $|P(x) - 1| < \epsilon$. This result does not depend on the choice of $n_1, \ldots, n_m$. Now let $x \in B_{c_i}(\delta)$ for some $i \in \{1, 2, \ldots, m\}$. Then $|x - c_i| \leqslant \delta$ and $|x| = |c_i|$ so that

$$|1 - c_j^{-s} x^s| \leqslant \max(1, |x/c_j|^s) \leqslant |c_i/c_j|^s \qquad (j < i)$$

$$|1 - c_i^{-s} x^s| \leqslant |1 - c_i^{-1} x| \leqslant \delta\, |c_i^{-1}| \leqslant \delta\, |c_1^{-1}|$$

$$|1 - c_j^{-s} x^s| \leqslant \max(1, |x/c_j|^s) \leqslant 1 \qquad (j > i)$$

In order that $|P(x)| < \epsilon$ we need

$$(*) \qquad \left|\frac{c_i}{c_1}\right|^{sn_1} \left|\frac{c_i}{c_2}\right|^{sn_2} \cdots \left|\frac{c_i}{c_{i-1}}\right|^{sn_{i-1}} \left(\frac{\delta}{|c_1|}\right)^{n_i} < \epsilon$$

If $n_1, \ldots, n_{i-1}$ are already chosen we can, since $\delta/|c_1| < 1$, always choose n_i such that (*) holds.

Remarks.

1. Kaplansky's theorem holds for arbitrary K whereas Weierstrass' approximation theorem (Let X be a compact subset of $\mathbb{R}$. Then a continuous $f : X \to \mathbb{R}$ can uniformly be approximated by polynomial functions.) becomes a falsity if we replace $\mathbb{R}$ by $\mathbb{C}$.
2. Kaplansky's theorem, applied for $X = B_0(1) \subset K$ where K is locally compact, shows that the limit of a uniformly convergent sequence of analytic

functions (polynomials) on $B_0(1)$ is in general not analytic. But it has applications that are more important. We shall meet them in Chapter 3.
3. For a generalization of Kaplansky's theorem see Appendix A.4.

44. exp and log

In the remaining sections of this chapter we shall consider some special elementary (locally) analytic functions. We start with the basic properties of the exponential function and the logarithm introduced in Section 25. Recall that $K^+ = B_1(1^-)$ and that E is the region of convergence of $\Sigma\, x^n/n!$.
IN THIS SECTION WE ASSUME $\mathrm{char}(K) = 0$

PROPOSITION 44.1.
(i) $\exp(x+y) = (\exp x)(\exp y)$ $(x, y \in E)$.
(ii) $\exp'(x) = \exp x$ $(x \in E)$.
(iii) exp *is an isometry of E onto $1+E$.*
(i)' $\log xy = \log x + \log y$ $(x, y \in K^+)$.
(ii)' $\log' x = x^{-1}$ $(x \in K^+)$.
(iii)' log *maps* $1+E$ *isometrically onto* E.
(iv) $\log(\exp x) = x$, $\exp(\log y) = y$ $(x \in E, y \in 1+E)$.
Proof. Most of these statements occur in previous exercises so we leave a proof of properties (i), (ii), (iii), (i)', (ii)' to the reader. It is easily verified that log maps $1+E$ into E, so that log exp and exp log $|\, 1+E$ are well defined. By Corollary 42.5 they are analytic. Differentiation yields (iv). Finally (iii)' follows from (iii) and (iv).

Something to be curious about is the behaviour of the logarithm on that part of its domain that does not meet the range of exp, i.e. the set $K^+ \backslash 1+E$. Of course, this set is empty if $\mathrm{char}(k) = 0$ and also if $K = \mathbf{Q}_p$ for $p \neq 2$ (Exercise 25.F). In $\mathbf{Q}_2$ we have $-1 \in \mathbf{Q}_2^+ \backslash 1+E$ so $\log(-1)$ exists and is in fact 0 since $2\log(-1) = \log(-1)^2 = \log 1 = 0$. Hence log has nontrivial zeros in $\mathbf{Q}_2^+$. This leads to the following.

PROPOSITION 44.2. (Zeros of the logarithm) *Let* $x \in K^+$. *Then* $\log x = 0$ *if and only if x is a root of unity. More specifically we have the following.*
(i) *If* $\mathrm{char}(k) = 0$ *then* log *has no zeros on* K^+ *except* 1.
(ii) *If* $K \supset \mathbf{Q}_p$ *then* $x \in K^+$, $\log x = 0$ *if and only if* $x^{p^n} = 1$ *for some* $n \in \{0, 1, 2, \ldots\}$.
Proof. (i) follows from Proposition 44.1 (iii)' and the fact that $1+E = K^+$. For (ii) it suffices to show that $\log x = 0$, $x \in K^+$ implies $x^{p^n} = 1$ for some n. By Theorem 32.2 we have $\lim_{n\to\infty} x^{p^n} = 1$ so $x^{p^n} \in 1+E$ for some n.

Now $0 = p^n \log x = \log x^{p^n}$. By the injectivity of log on $1 + E$ we get $x^{p^n} = 1$.

PROPOSITION 44.3. ((Un)boundedness of the logarithm)
(i) *If* $\text{char}(k) = 0$ *or if* $|K^\times|$ *is discrete then* log *is bounded on* K^+.
(ii) *If* $K \supset \mathbb{Q}_p$, $|K^\times|$ *is dense then* log *is unbounded on* K^+. *If, in addition, K is algebraically closed then* $\log : K^+ \to K$ *is surjective.*
Proof. If $\text{char}(k) = 0$ then $|x^n/n| < 1$ for all $n \in \mathbb{N}$, $x \in B_0(1^-)$ so that log is bounded by 1. If $|K^\times|$ is discrete we have $x \in B_0(1^-) \Leftrightarrow x \in B_0(|\pi|)$, hence $|\log(1-x)| \leqslant \sup\{|\pi|^n/|n| : n \geqslant 1\} \leqslant \sup\{n\,|\pi|^n : n \geqslant 1\} < \infty$. This establishes (i). To prove (ii) we apply the 'supremum principle' (Theorem 42.3(i)). We get $\sup\{|\log(1-x)| : |x| < 1\} = \sup\{|1/n|_p : n \geqslant 1\} = \infty$, i.e. log is unbounded. If K is also algebraically closed, let $a \in K$. We prove the existence of an $x \in K^+$ for which $\log x = a$. In fact, $p^n a \in E$ for some $n \in \mathbb{N}$. By Proposition 44.1 (iii)' there is a $y \in 1 + E$ such that $\log y = p^n a$. Now choose $x \in K$ for which $x^{p^n} = y$. By Corollary 32.3, $x \in K^+$ and $\log x$ is defined. Now $p^n \log x = \log y = p^n a$. Hence, $\log x = a$.

COROLLARY 44.4. $\log : \mathbb{C}_p^+ \to \mathbb{C}_p$ *is surjective. Its set of zeros is precisely* Γ_r (see Section 33).

Exercise 44.A. For which $a \in \mathbb{C}_p^+$ is the formula $a^x = \exp(x \log a)$ true for all $x \in \mathbb{Z}_p$? (See Section 32.)

Exercise 44.B. Let $K = \mathbb{C}_p$. Show that the domain of the function $x \mapsto \exp(\log x)$ is precisely the set of products $\Gamma_r(1 + E)$.

**Exercise* 44.C. Let $a \in \mathbb{C}_p^+$. Use the fact that $a^{p^n} = \exp(p^n \log a)$ for large n to find a proof of

$$\lim_{n \to \infty} \frac{a^{p^n} - 1}{p^n} = \log a$$

which is different from the one of Exercise 32.H.

45. Extensions of exp and log

We shall extend the functions $\exp : E \to 1 + E$ and $\log : \mathbb{C}_p^+ \to \mathbb{C}_p$ to homomorphisms $\text{EXP} : \mathbb{C}_p \to \mathbb{C}_p^+$ and $\text{LOG} : \mathbb{C}_p^\times \to \mathbb{C}_p$ respectively. The function LOG is particularly interesting since it is unique after prescribing LOG p (Corollary 45.10). The heart of the construction is Lemma 45.3 on the extension of homomorphisms between groups.

DEFINITION 45.1. An abelian group Y (written multiplicatively) is *divisible* if for each $y \in Y$, $n \in \mathbb{N}$ there is $x \in Y$ such that $x^n = y$.

The additive group $\mathbb{C}_p$ is divisible and so is, by the algebraic closedness of $\mathbb{C}_p$, the multiplicative group $\mathbb{C}_p^\times$. The next lemma shows that also $\mathbb{C}_p^+$ is divisible.

LEMMA 45.2. $\mathbb{C}_p^+ = \{x \in \mathbb{C}_p : |1-x|_p < 1\}$ *is divisible.*

Proof. Let $y \in \mathbb{C}_p^+$, $n \in \mathbb{N}$. We prove the existence of an $x \in \mathbb{C}_p^+$ for which $x^n = y$. Write $n = p^j m$ where $j, m \in \{0, 1, \ldots\}$ and m is not divisible by p. By algebraic closedness there is a $z \in \mathbb{C}_p$ for which $z^{p^j} = y$; by Corollary 32.3 we have $z \in \mathbb{C}_p^+$. We have to find an $x \in \mathbb{C}_p^+$ for which $x^m = z$. But this is easily done by applying Hensel's lemma (Theorem 27.6) to the function $x \mapsto x^m - z$ with $a = 1$.

LEMMA 45.3. (Extension of homomorphisms) *Let G, Y be abelian groups and let Y be divisible. Let X be a subgroup of G and let $a : X \to Y$ be a homomorphism. Then a can be extended to a homomorphism $\tilde{a} : G \to Y$.*

Proof. Thanks to Zorn's lemma we may assume that G is generated by X and $\{y\}$ for some $y \in G \backslash X$. Let us write G additively and Y multiplicatively. We distinguish two cases.

(i) $ny \notin X$ for all $n \in \mathbb{N}$. For every $g \in G$ there are unique $n \in \mathbb{Z}$ and $x \in X$ such that $g = ny + x$. Choose any $z \in Y$ and define

$$\tilde{a}(g) := z^n a(x)$$

One easily establishes that $\tilde{a}$ is as required. (Observe that in this part we have not used the divisibility of Y.)

(ii) $ny \in X$ for some $n \in \mathbb{N}$, $n > 1$. Let $m := \min\{n \in \mathbb{N} : ny \in X\}$. It is easy to prove that for every $g \in G$ there are unique $n \in \{0, 1, \ldots, m-1\}$ and $x \in X$ such that $g = ny + x$. Since Y is divisible there is $z \in Y$ such that $z^m = a(my)$. Define

$$\tilde{a}(g) := z^n a(x)$$

It is an elementary exercise to show that $\tilde{a}$ is a well-defined homomorphism extending a.

Remark. The above extension $\tilde{a}$ is, in general, not unique.

COROLLARY 45.4. *The functions* $\exp : E \to 1 + E$ *and* $\log : \mathbb{C}_p^+ \to \mathbb{C}_p$ *can be extended to continuous homomorphisms* $\mathbb{C}_p \to \mathbb{C}_p^+$ *and* $\mathbb{C}_p^\times \to \mathbb{C}_p$ *respectively.*

DEFINITION 45.5. Let EXP : $\mathbb{C}_p \to \mathbb{C}_p^+$ be a homomorphism extending exp and let LOG : $\mathbb{C}_p^\times \to \mathbb{C}_p$ be a homomorphism extending log.

PROPOSITION 45.6. EXP : $\mathbb{C}_p \to \mathbb{C}_p^+$ *is an extension of* exp *with the following properties.*

(i) EXP *is analytic on the additive cosets of* E, *hence locally analytic.*

(ii) $\mathrm{EXP}(x+y) = (\mathrm{EXP}\, x)\,(\mathrm{EXP}\, y)$ $(x, y \in \mathbb{C}_p)$.

(iii) $\log \mathrm{EXP}\, x = x$ $(x \in \mathbb{C}_p)$.

(iv) EXP *is injective.*

(v) $\mathrm{EXP}' = \mathrm{EXP}$.

Proof. For $a \in \mathbb{C}_p$, $x \in E$ we have $\mathrm{EXP}(a+x) = (\mathrm{EXP}\, a)\,(\exp x)$ and (i), (v) follow. To prove (iii), let $x \in \mathbb{C}_p$. Then $p^n x \in E$ for some $n \in \mathbb{N}$ and $p^n x = \log \exp p^n x = \log(\mathrm{EXP}\, x)^{p^n} = p^n \log \mathrm{EXP}\, x$. Finally, (iv) is a consequence of (iii).

PROPOSITION 45.7. LOG : $\mathbb{C}_p^\times \to \mathbb{C}_p$ *is an extension of* log *with the following properties.*

(i) LOG *is analytic on the multiplicative cosets of* $\mathbb{C}_p^+$ (*i.e. on the 'sides of 0', see Definition* 24.4).

(ii) $\mathrm{LOG}\, xy = \mathrm{LOG}\, x + \mathrm{LOG}\, y$ $(x, y \in \mathbb{C}_p^\times)$.

(iii) $\mathrm{LOG}\,\mathrm{EXP}\, x = x$ $(x \in \mathbb{C}_p)$.

(iv) $\mathrm{LOG}'\, x = 1/x$ $(x \in \mathbb{C}_p^\times)$.

Proof. As $\mathrm{EXP}\, x \in \mathbb{C}_p^+$ for all $x \in \mathbb{C}_p$ we have $\mathrm{LOG}\,\mathrm{EXP}\, x = \log \mathrm{EXP}\, x = x$ by (iii) of Proposition 45.6. Let $a \in \mathbb{C}_p^\times$. If $x \in \mathbb{C}_p$, $|x-a|_p < |a|_p$ then $|xa^{-1}-1|_p < 1$ so that $\mathrm{LOG}\, x = \mathrm{LOG}\, a + \log xa^{-1}$. From this the properties (i) and (iv) follow easily.

**Exercise* 45.A. (On EXP LOG) Define the function $h : \mathbb{C}_p^\times \to \mathbb{C}_p^\times$ by

$$h(x) := x/\mathrm{EXP}\,\mathrm{LOG}\, x \qquad (x \in \mathbb{C}_p^\times)$$

(i) Show that $h(xy) = h(x)h(y)$ and $|h(x)|_p = |x|_p$ for all $x, y \in \mathbb{C}_p^\times$.

(ii) Let $G := \{\mathrm{EXP}\, x : x \in \mathbb{C}_p\}$. Prove that $G \supset 1 + E$ is an open multiplicative subgroup of $\mathbb{C}_p^+$ and that G does not contain a root of unity except 1.

(iii) Prove that $\{x \in \mathbb{C}_p^\times : h(x) = 1\} = G$. Deduce that h is locally constant.

(iv) Prove that $\{h(x) : x \in \mathbb{C}_p^\times\} = \{x \in \mathbb{C}_p^\times : \mathrm{LOG}\, x = 0\}$.

Now let j be the restriction of h to $\{x \in \mathbb{C}_p : |x|_p = 1\}$. Show that $\{x \in \mathbb{C}_p : |x|_p = 1, \mathrm{LOG}\, x = 0\}$ equals the group Γ of roots of unity of $\mathbb{C}_p$. Conclude that j maps $\{x \in \mathbb{C}_p : |x|_p = 1\}$ onto Γ.

(v) Use the formula $x = j(x)\,\mathrm{EXP}\,\mathrm{LOG}\, x$ $(|x|_p = 1)$ to show that the group $\{x \in \mathbb{C}_p : |x|_p = 1\}$ is the direct product of Γ and G.

(vi) Let $\omega_p : \{x \in \mathbb{C}_p : |x|_p = 1\} \to \Gamma_u$ be the Teichmüller character (see Definition 33.3). Prove that $\omega_p \circ j = j \circ \omega_p = \omega_p$.

(vii) Let $p \neq 2$. Prove that for $x \in \mathbb{Z}_p \backslash p\mathbb{Z}_p$ we have $j(x) = \omega_p(x)$ so that

$$\exp \mathrm{LOG}\, x = x\, \omega_p(x^{-1}) \qquad (x \in \mathbb{Z}_p \backslash p\mathbb{Z}_p)$$

For the uniqueness property of LOG we need the following extension of Definition 4.3.

DEFINITION 45.8. For $x \in \mathbb{C}_p^\times$ set $\mathrm{ord}_p(x) := r$ if $|x|_p = p^{-r}$.

ord_p is locally constant, its values are rational numbers and

$$\mathrm{ord}_p(xy) = \mathrm{ord}_p(x) + \mathrm{ord}_p(y) \qquad (x, y \in \mathbb{C}_p^\times)$$

so that ord_p, just like LOG, is a homomorphism $\mathbb{C}_p^\times \to \mathbb{C}_p$.

THEOREM 45.9. *Let $f, g : \mathbb{C}_p^\times \to \mathbb{C}_p$ be two extensions of* $\log : \mathbb{C}_p^+ \to \mathbb{C}_p$ *in the sense of Corollary* 45.4. *Then there is $c \in \mathbb{C}_p$ such that $f(x) = g(x) + c\, \mathrm{ord}_p(x)$ for all $x \in \mathbb{C}_p^\times$.*

Proof. First, let $x \in \mathbb{C}_p$, $|x|_p = 1$. Then $\bar{x}$ is a root of unity in the residue class field of $\mathbb{C}_p$ so $x^n \in \mathbb{C}_p^+$ for some $n \in \mathbb{N}$. Then $f(x) = f(x^n)/n = (\log x^n)/n = g(x^n)/n = g(x)$. We see that $f = g$ on $\{x \in \mathbb{C}_p : |x|_p = 1\}$. Now let $x \in \mathbb{C}_p^\times$. Then $|x|_p = p^{-r}$ where $\mathrm{ord}_p(x) = r = t/n$ for some $t, n \in \mathbb{Z}$ so that $x^n = p^t y$ where $|y|_p = 1$. By the first part $f(y) = g(y)$ so

$$f(x^n) = f(p^t) + f(y) = tf(p) + f(y)$$
$$g(x^n) = g(p^t) + g(y) = tg(p) + f(y)$$

Hence, $f(x) - g(x) = (f(x^n) - g(x^n))/n = (f(p) - g(p))t/n = c\, \mathrm{ord}_p(x)$ where $c := f(p) - g(p)$.

COROLLARY 45.10. *There is precisely one function $f : \mathbb{C}_p^\times \to \mathbb{C}_p$ such that*
(i) *f is an extension of* $\log : \mathbb{C}_p^+ \to \mathbb{C}_p$
(ii) *$f(xy) = f(x) + f(y)$ for all $x, y \in \mathbb{C}_p^\times$*
(iii) *$f(p) = 0$.*

Proof. Let LOG be as in Definition 45.5. The function $x \mapsto \mathrm{LOG}\, x - \mathrm{ord}_p(x)\, \mathrm{LOG}\, p$ satisfies (i), (ii), (iii). The uniqueness follows from the previous theorem.

DEFINITION 45.11. The function f of Corollary 45.10 is the *Iwasawa logarithm* on $\mathbb{C}_p^\times$ and denoted $\log_p$.

We state some properties of $\log_p$.

THEOREM 45.12. (Properties of the Iwasawa logarithm) *Besides the ones of Proposition* 45.7, $\log_p$ *has the following properties.*
(i) *Let $x \in \mathbb{C}_p^\times$. Then $\log_p x = 0$ if and only if for some $n \in \mathbb{N}$ the element*

x^n is an integral power of p. In particular, if $|x|_p = 1$ then $\log_p x = 0$ if and only if x is a root of unity.

(ii) *Let $x \in \mathbb{Q}_p^\times$. Then $\log_p x = 0$ if and only if $x = p^n y$ where $n \in \mathbb{Z}$, y is a root of unity.*

(iii) *Let $x \in \mathbb{Q}_p$, $|x|_p = 1$. Then*

$$\log_p x = \frac{1}{1-p} \sum_{n=1}^{\infty} \frac{(1-x^{p-1})^n}{n}$$

(iv) *Let $a \in \mathbb{C}_p^\times$. Then*

$$\log_p x = \log_p a + \sum_{n=1}^{\infty} (-1)^{n+1} \frac{a^{-n}(x-a)^n}{n} \qquad (|x-a|_p < |a|_p)$$

Proof. For $x \in \mathbb{C}_p^\times$ there is $n \in \mathbb{N}$ such that $x^n = p^m a$ for some $m \in \mathbb{Z}, a \in \mathbb{C}_p^+$. If $\log_p x = 0$ then $\log_p a = 0$ so, by Corollary 44.4, a is a root of unity. Hence some power of x is an integral power of p. The rest of (i) is obvious, (ii) is a consequence of (i). To prove (iii) observe that $|x^{p-1} - 1|_p < 1$ so that we can use power series expansion for $\log_p x = (p-1)^{-1} \log_p x^{p-1} = (p-1)^{-1} \log_p(1-(1-x^{p-1}))$. Finally we prove (iv). If $|x-a|_p < |a|_p$ then $|xa^{-1} - 1|_p < 1$ so that $\log_p(xa^{-1}) = \log_p(1-(1-xa^{-1})) = -\sum_{n=1}^{\infty} (1-xa^{-1})^n/n$. From this (iv) follows.

The following property of $\log_p$ is also of interest.

THEOREM 45.13.

(i) *Let L be a closed subfield of $\mathbb{C}_p$, $\mathbb{Q}_p \subset L \subset \mathbb{C}_p$. Then $\log_p$ maps $L^\times$ into L.*

(ii) *Let $\sigma : \mathbb{C}_p \to \mathbb{C}_p$ be a continuous automorphism. Then*

$$\log_p \sigma(x) = \sigma(\log_p x) \qquad (x \in \mathbb{C}_p^\times)$$

Proof. (i) Let $x \in L^\times$. There is $n \in \mathbb{N}$ such that $x^n = p^m a$ for some $m \in \mathbb{Z}$, $a \in \mathbb{C}_p^+$. Since $p \in L$ we have $a = p^{-m} x^n \in L$, so $a \in L^+$. The power series of $\log_p$ tells us that $\log_p$ maps L^+ into L, so $\log_p x = (\log_p x^n)/n = (\log_p a)/n \in L$.

(ii) By Exercise 9.D the inverse σ^{-1} of σ is continuous; $|x|_p < 1$ if and only if $|\sigma(x)|_p < 1$. Using this and the power series of $\log_p$ on $\mathbb{C}_p^+$ we arrive at

$$\sigma^{-1} \log_p \sigma(x) = \log_p x \qquad (x \in \mathbb{C}_p^+)$$

Thus $x \mapsto \sigma^{-1} \log_p \sigma(x)$ $(x \in \mathbb{C}_p^\times)$ is a homomorphism extending $\log_p$ and having the value 0 at p. By Corollary 45.10

$$\sigma^{-1} \log_p \sigma(x) = \log_p x \qquad (x \in \mathbb{C}_p^\times)$$

Remark. Without proof we mention that for a continuous automorphism

σ of $\mathbb{C}_p$ it is in general not true that $\sigma \circ \mathrm{EXP} = \mathrm{EXP} \circ \sigma$; neither do we have a uniqueness theorem for EXP in the style of Corollary 45.10.

Exercise 45.B. Discuss the existence of $\lim_{|x|_p \to \infty} (\log_p x)/x$ and $\lim_{x \to 0} x \log_p x$ in $\mathbb{Q}_p$ and $\mathbb{C}_p$.

Exercise 45.C. (Extension of Exercise 44.C) Show that

$$\log_p a = \lim_{n \to \infty} \frac{a^{p^{n!}} - \omega_p(a)}{\omega_p(a)\, p^{n!}} \qquad (a \in \mathbb{C}_p,\ |a|_p = 1)$$

(where ω_p is the Teichmüller character) and also that

$$\log_p a = \sum_{n=1}^{\infty} (-1)^{n+1} \frac{\omega_p(a^{-n})\,(a - \omega_p(a))^n}{n} \qquad (a \in \mathbb{C}_p,\ |a|_p = 1)$$

Exercise 45.D. Let $p \equiv 3 \pmod 4$, let $L := \mathbb{Q}_p(\sqrt{-1})$, choose $i \in L$ such that $i^2 = -1$. Let $x + iy \in L$ $(x, y \in \mathbb{Q}_p)$, let $\log_p (x + iy) = u + iv$ $(u, v \in \mathbb{Q}_p)$. Show that $\log_p (x - iy) = u - iv$.

Exercise 45.E. (p-adic analogue of $e = 2.718\ldots$?) In this exercise, let $e_p := \mathrm{EXP}\, 1 \in \mathbb{C}_p$. Then e_p is a pth root of $\exp p \in \mathbb{Q}_p$ (if $p \neq 2$). However, which root it is actually, depends on the choice of EXP. Show that $e_p \notin \mathbb{Q}_p$. Since $\log_p$ seems to be more 'canonical' than EXP we may look at set $E_p := \{x \in \mathbb{C}_p : \log_p x = 1\}$. Describe this set.

In the last six exercises we consider differentiable homomorphisms between the additive and multiplicative groups of $\mathbb{C}_p$.

Exercise 45.F. (Extension of differentiable homomorphisms) Prove the following.
(i) Let U be an open additive subgroup of $\mathbb{C}_p$ and let $f : U \to \mathbb{C}_p$ be a differentiable homomorphism. Then f can be extended to a homomorphism $\tilde{f} : \mathbb{C}_p \to \mathbb{C}_p$ and $\tilde{f}$ is differentiable.
(ii) Let U be an open additive subgroup of $\mathbb{C}_p$ and let $g : U \to \mathbb{C}_p^\times$ be a differentiable homomorphism. Then $g(U) \subset \mathbb{C}_p^+$, g can be extended to a homomorphism $\tilde{g} : \mathbb{C}_p \to \mathbb{C}_p^+$ and $\tilde{g}$ is differentiable.
(iii) Let V be an open multiplicative subgroup of $\mathbb{C}_p^\times$ and let $h : V \to \mathbb{C}_p$, $j : V \to \mathbb{C}_p^\times$ be differentiable homomorphisms. Then h, j can be extended to homomorphisms $\tilde{h} : \mathbb{C}_p^\times \to \mathbb{C}_p$ and $\tilde{j} : \mathbb{C}_p^\times \to \mathbb{C}_p^\times$ respectively. These extensions are differentiable.

Exercise 45.G. Let $f : \mathbb{C}_p \to \mathbb{C}_p$ be a differentiable function such that $f(x + y) = f(x) + f(y)$ for all $x, y \in \mathbb{C}_p$. Prove that there is a constant c such that

$f(x) = cx$ $(x \in \mathbb{C}_p)$.

Exercise 45.H. Let $h : \mathbb{C}_p^\times \to \mathbb{C}_p$ be a differentiable function such that $h(xy) = h(x) + h(y)$ for all $x, y \in \mathbb{C}_p^\times$. Show that

$$h(x) = a \log_p x + b \operatorname{ord}_p x \qquad (x \in \mathbb{C}_p^\times)$$

for some $a, b \in \mathbb{C}_p$ and that in fact $a = h'(1)$, $b = h(p)$.

Exercise 45.I. Let $g : \mathbb{C}_p \to \mathbb{C}_p^\times$ be a differentiable function such that $g(x + y) = g(x)\, g(y)$ for all $x, y \in \mathbb{C}_p$.
(i) Prove that $g'(0) = 0$ implies local constantness of g.
(ii) Use (i) to show the existence of a constant $c \in \mathbb{C}_p$ and a locally constant homomorphism $\alpha : \mathbb{C}_p \to \mathbb{C}_p^\times$ such that

$$g(x) = \operatorname{EXP}(cx)\, \alpha(x) \qquad (x \in \mathbb{C}_p)$$

Exercise 45.J. Let $j : \mathbb{C}_p^\times \to \mathbb{C}_p^\times$ be a differentiable function such that $j(xy) = j(x)\, j(y)$ for all $x, y \in \mathbb{C}_p^\times$. Show that there exist $c \in \mathbb{C}_p$ and a locally constant homomorphism $\beta : \mathbb{C}_p^\times \to \mathbb{C}_p^\times$ such that

$$j(x) = \operatorname{EXP}(c \log_p x)\ \beta(x) \qquad (x \in \mathbb{C}_p^\times)$$

From the four exercises above we may conclude that 'differentiable homomorphisms are locally analytic'. In Exercise 50.H we shall prove the existence of continuous nowhere differentiable homomorphisms of each of the four types, where the ground field is $\mathbb{C}_p$. From the next exercise we may conclude that continuous homomorphisms of $\mathbb{Q}_p$ are automatically locally analytic.

Exercise 45.K. Let $f : \mathbb{Z}_p \to \mathbb{C}_p$ be continuous and let $f(x + y) = f(x)\, f(y)$ for all $x, y \in \mathbb{Z}_p$. Show that there is an $a \in \mathbb{C}_p^+$ for which $f(x) = a^x$ $(x \in \mathbb{Z}_p)$.

46. Trigonometric functions

In this section we shall have a closer look at the functions sin, cos, tan, arctan. For the inverses of sin, cos see Section 49. For simplicity we shall work in $\mathbb{C}_p$ where $p \neq 2$. The reader is asked to investigate the case $p = 2$ and also the case where $\mathbb{C}_p$ is replaced by a general non-archimedean field K for which $\operatorname{char}(K) = 0$.

Let us fix, in Section 46, an element $i \in \mathbb{C}_p$ such that $i^2 = -1$. The proof of the following proposition is left to the reader. (See Definition 25.7.)

PROPOSITION 46.1 *Let $x, y \in E$. Then*
(i) $\sin x = (\exp ix - \exp(-ix))/2i$, $\cos x = (\exp ix + \exp(-ix))/2$.

(ii) $\exp ix = \cos x + i \sin x$, $\sin^2 x + \cos^2 x = 1$.
(iii) $\sin(x+y) = \sin x \cos y + \cos x \sin y$,
$\cos(x+y) = \cos x \cos y - \sin x \sin y$.
(iv) $\sin' x = \cos x$, $\cos' x = -\sin x$.
(v) $|\sin x|_p = |x|_p$, $|\cos x|_p = |\exp ix|_p = 1$.
$|\sin x - x|_p < |x|_p$, $|\cos x - 1|_p < |x|_p$ $(x \neq 0)$.
$|\sin x - \sin y|_p = |x-y|_p$, $|\cos x - \cos y|_p \leqslant |x-y|_p$.

PROPOSITION 46.2. sin *is an increasing function of E onto E.*
Proof. Let r be a real number, $0 < r < p^{1/(1-p)}$. It suffices to prove that the restriction of sin to $B_0(r)$ is an increasing function onto $B_0(r)$. Let $x, y \in B_0(r), x \neq y$. Then

$$\left|\frac{\sin x - \sin y}{x-y} - 1\right|_p \leqslant \sup_{n \geqslant 3} \left|\frac{x^n - y^n}{x-y}\frac{1}{n!}\right|_p \leqslant \sup_{n \geqslant 3} r^{n-1}|n!|_p^{-1}$$

There is an $a > 1$ such that $r = p^{a/(1-p)}$. For $n \geqslant 3$ we have $r^{n-1}|n!|_p^{-1} = p^{\mu(n)}$ where (with s_n as in Lemma 25.5)

$$\mu(n) = (a(n-1) + s_n - n)/(1-p)$$

Now $a(n-1) + s_n - n \geqslant (a-1)(n-1) \geqslant 2(a-1)$ so that

$$\sup\left\{\left|\frac{\sin x - \sin y}{x-y} - 1\right|_p : x, y \in B_0(r), x \neq y\right\} \leqslant p^{2(a-1)/(1-p)} < 1$$

We see that sin is increasing on $B_0(r)$. By Lemma 27.4 sin maps $B_0(r)$ onto $B_0(r)$, and the proposition is proved.

We turn to cos. It is an even function, hence not injective in a neighbourhood of 0.

PROPOSITION 46.3. cos *maps E onto* $1 + E^2 = \{y \in \mathbb{C}_p : |1-y|_p < p^{2/(1-p)}\}$. *The restriction of* cos *to a side of* 0 *within E is a scalar multiple of an isometry.*
Proof. We have the formula $\cos x = 1 - 2\sin^2 \frac{1}{2}x$ $(x \in E)$. If x runs through E then so does $\frac{1}{2}x$ and, as we have seen in Proposition 46.2, $\sin\frac{1}{2}x$. Therefore the range of cos is $\{y \in \mathbb{C}_p : y = 1 - 2x^2 \text{ for some } x \in E\} = \{y \in \mathbb{C}_p : \frac{1}{2}(1-y) \text{ is a square of an element of } E\} = \{y \in \mathbb{C}_p : 1 - y \in E^2\} = 1 + E^2$.

Let $a \in E$, $a \neq 0$ and let x, y be in the side of 0 determined by a, i.e. $|x-a|_p < |a|_p$, $|y-a|_p < |a|_p$. Then $|x+y-2a|_p < |a|_p$ so that $|x+y|_p = |a|_p$. The formula

$$\cos x - \cos y = -2\sin\tfrac{1}{2}(x+y)\sin\tfrac{1}{2}(x-y)$$

yields

$$|\cos x - \cos y|_p = |x+y|_p\ |x-y|_p = |a|_p\ |x-y|_p$$

and we are done.

**Exercise* 46.A. Show that cos, restricted to $\{x \in \mathbb{C}_p : |x-a|_p < |a|_p\}$ $(a \in E,\ a \neq 0)$ is monotone of type $-\operatorname{sgn} a$ and maps $B_a(|a|_p^-)$ onto $B_{\cos a}(|a^2|_p^-)$.

PROPOSITION 46.4. *The function* $\tan : E \to \mathbb{C}_p$ *defined by*

$$\tan x := \frac{\sin x}{\cos x} \qquad (x \in E)$$

is analytic.

Proof. The function $\log_p \cos$ is well defined and analytic by Theorem 42.4. Then so is its derivative $-\tan$.

In Definition 25.7 we introduced the function arctan on $B_0(1^-)$. van Hamme (informal communication) suggested the following natural extension of this function by means of a formula which is well known in 'classical' analysis.

DEFINITION 46.5. (van Hamme)

$$\arctan x := \frac{1}{2i}\log_p\left(\frac{1+ix}{1-ix}\right) \qquad (x \in \mathbb{C}_p,\ x \notin \{i, -i\})$$

The proofs of the following formulas are left to the reader.

PROPOSITION 46.6. (Properties of arctan)

(i) $\arctan x = x - x^3/3 + x^5/5 - \ldots$ $(|x|_p < 1)$.

(ii) $\arctan(\tan x) = x$, $\tan(\arctan x) = x$ $(x \in E)$.

(iii) $\arctan' x = 1/(1+x^2)$ $(x \notin \{i, -i\})$.

(iv) $\arctan(-x) = -\arctan x$ $(x \notin \{i, -i\})$.

(v) $\arctan 1 = 0$ (!)

(vi) $\arctan x + \arctan(1/x) = 0$ $(x \notin \{0, i, -i\})$.

(vii) $\arctan x + \arctan y = \arctan((x+y)/(1-xy))$ $(xy \neq 1,\ x, y \notin \{i, -i\})$.

(viii) $\lim_{|x| \to \infty} \arctan x = 0$ (!)

(ix) $\log_p(x+iy) = \frac{1}{2}\log_p(x^2+y^2) + i\arctan(y/x)$ $(x^2+y^2 \neq 0,\ x \neq 0)$.

Exercise 46.B. (Local analyticity of arctan) Of course arctan, being the composition of locally analytic functions, is locally analytic. In this exercise we determine the maximal discs on which it is analytic. Let $a \in \mathbb{C}_p$, $a \notin \{i, -1\}$. The largest disc containing a on which arctan is defined is $D_a := \{x \in$

$\mathbb{C}_p : |x-a|_p < \min(|a-i|_p, |a+i|_p)\}$. Show that arctan is, indeed, analytic on D_a. (Hint. Write $2i(\arctan x - \arctan a) = \log_p((1+ix)/(1+ia)) - \log_p((1-ix)/(1-ia))$.)

Exercise 46.C. Say something sensible about the zeros of arctan.

47. $(1+x)^a$

We shall consider the function $x \mapsto (1+x)^a := \sum_{n=0}^{\infty} \binom{a}{n} x^n$ where a is an element of K. First we establish some properties of $\binom{a}{n}$. These shall play an important role in the following chapter.

THROUGHOUT SECTION 47 WE ASSUME char$(K) = 0$

DEFINITION 47.1. We define the symbol $\binom{x}{n}$ ($x \in K$, $n \in \{0, 1, \ldots\}$) by $\binom{x}{0} := 1$ and

$$\binom{x}{n} := \frac{x(x-1)\ldots(x-n+1)}{n!} \qquad (n \in \mathbb{N})$$

PROPOSITION 47.2. (Properties of $\binom{x}{n}$)

(i) $x \mapsto \binom{x}{n}$ *is a polynomial function of degree* n. *If* $j \in \{0, 1, 2, \ldots\}$ *and* $j < n$ *then* $\binom{j}{n} = 0$; $\binom{n}{n} = 1$.

(ii) *If* $f : K \to K$ *is a polynomial function of degree* n *and if* $f(j) = 0$ *for* $j \in \{0, 1, 2, \ldots\}$, $j < n$ *and* $f(n) = 1$ *then* $f(x) = \binom{x}{n}$ *for all* $x \in K$.

(iii) *For all* $x, y \in K$, $n \in \{0, 1, 2, \ldots\}$

$$\binom{x+y}{n} = \sum_{j=0}^{n} \binom{x}{j}\binom{y}{n-j}$$

In particular

$$\binom{x+1}{n} = \binom{x}{n} + \binom{x}{n-1} \qquad (n \in \{1, 2, \ldots\})$$

(iv) *For all* $x \in K$, $n \in \{0, 1, 2, \ldots\}$

$$\binom{x}{n+1} = \frac{x-n}{n+1}\binom{x}{n} = \frac{x}{n+1}\binom{x-1}{n}$$

(v) *If* $n, m \in \{0, 1, 2, \ldots\}$ *then* $\binom{m}{n} = \dfrac{m!}{n!(m-n)!}$ *is an integer if* $n \leqslant m$. *For* $x \in \mathbb{Z}_p$ *we have*

$$\left|\binom{x}{n}\right|_p \leqslant 1$$

(vi) *Let* s_j *denote the sum of the p-adic digits of* $j = a_0 + a_1 p + \ldots \in \{0, 1, 2, \ldots\}$ (see Lemma 25.5). *For* $m, n \in \{0, 1, 2, \ldots\}$, $n \leqslant m$ *we have*

$$\left|\binom{m}{n}\right|_p = p^{\lambda(m,n)}$$

where $\lambda(m, n) := (s_n + s_{m-n} - s_m)/(1-p)$.

Proof. We restrict the proof to a few comments. (i), (ii), (iv), (v) are straightforward. To prove (iii), first let $x, y \in \mathbb{N}$. Then $\binom{x+y}{n}$ is the coefficient of X^n in $(1+X)^{x+y}$ whereas $\sum_{j=0}^{n} \binom{x}{j}\binom{y}{n-j}$ is the coefficient of X^n in $(1+X)^x (1+X)^y$. The polynomial functions in $x, y \in K$ given by either side of the equality sign of (iii) coincide on $\mathbb{N} \times \mathbb{N}$, hence on $K \times K$. (vi) is a consequence of Lemma 25.5.

In Chapter 3 we shall need some inequalities concerning $\binom{x}{n}$ where $x \in \mathbb{Z}_p$.

DEFINITION 47.3. Let $j \in \mathbb{N}$ be written in base p

$$j = a_0 + a_1 p + \ldots + a_s p^s \qquad (a_s \neq 0)$$

Then set

$$j_- := a_0 + a_1 p + \ldots + a_{s-1} p^{s-1}$$

PROPOSITION 47.4. *For* $x, y \in \mathbb{Z}_p, j \in \mathbb{N}$ *we have*

$$\left|\binom{x}{j} - \binom{y}{j}\right|_p \leqslant |x-y|_p\, |j-j_-|_p^{-1}$$

Proof. It suffices to prove the inequality for $x = y + p^n$ for some $n \in \mathbb{N}$. In that case $(x-y)^{-1}\left(\binom{x}{j} - \binom{y}{j}\right) = p^{-n}\left\{\binom{y+p^n}{j} - \binom{y}{j}\right\} = p^{-n} \sum_{s=0}^{j-1} \binom{y}{s} \cdot \binom{p^n}{j-s} = \sum_{s=0}^{j-1} \binom{y}{s} (j-s)^{-1} \binom{p^n-1}{j-s-1}$. It follows that

$$\left|\binom{x}{j} - \binom{y}{j}\right|_p \leqslant |x-y|_p \cdot \max_{0 \leqslant s \leqslant j-1} |1/(j-s)|_p = |x-y|_p\, |j-j_-|_p^{-1}.$$

COROLLARY 47.5. *Let* $n \in \{0, 1, 2, \ldots\}$ *and* $x, y \in \mathbb{Z}_p$. *Then*

$$|x-y|_p \leqslant p^{-n} \textit{ implies } \left|\binom{x}{j} - \binom{y}{j}\right|_p < 1 \qquad (j = 0, 1, \ldots p^n - 1)$$

Proof. If $1 \leqslant j < p^n$ then $j - j_- = ap^s$ where $a \in \{1, 2, \ldots, p-1\}$ and $s < n$, so $|j-j_-|_p > p^{-n}$. Now apply Proposition 47.4.

Exercise 47.A. Compute $|\binom{p}{j}|_p$ $(j \in \{0, 1, \ldots, p-1\})$; $\left|\binom{p^n-1}{j}\right|_p$ $(j \in \{0, 1, \ldots, p^n - 1\})$; $|\binom{2p}{p}|_p$; $|\binom{p^n}{pj}|_p$ $(j, n \in \{0, 1, 2, \ldots\})$.

Exercise 47.B. Show that $|\binom{n}{j}|_p \geqslant |n - n_-|_p \geqslant 1/n$ $(1 \leqslant j \leqslant n)$. (Hint. Let $n = a_0 + a_1 p + \ldots + a_s p^s$ $(a_s \neq 0)$. Show that $s_j + s_{n-j} \leqslant s_n + s(p-1)$.)

Now we come to the subject of this section. In Exercise 31.I you were asked to prove the identity

$$(1+x)^a = \sum_{n=0}^{\infty} \binom{a}{n} x^n \qquad (a \in \mathbb{Z}_p, x \in \mathbb{C}_p, |x|_p < 1)$$

where $a \mapsto (1+x)^a$ was obtained by interpolation of $n \mapsto (1+x)^n$ ($n \in \mathbb{N}$). The proof is not hard: at both sides of the equality sign we have continuous functions of a coinciding on $\{1, 2, \ldots\}$.

In this section we consider $(1+x)^a$ as a function of x rather than a and try to give a meaning to $(1+x)^a$ in a more general context.

LEMMA 47.6. *Let $K \supset \mathbb{Q}_p$, $a \in \mathbb{Z}_p$, $a \notin \{0, 1, 2, \ldots\}$. The region of convergence of $\sum \binom{a}{n} x^n$ is $B_0(1^-)$.*

Proof. If $|x| < 1$ then $\lim_{n\to\infty} |\binom{a}{n} x^n| \leqslant \lim_{n\to\infty} |x|^n = 0$. If $a \in \mathbb{Z}_p \setminus \{0, 1, 2, \ldots\}$ then it has a p-adic expansion $a = \sum_{j=0}^{\infty} a_{n_j} p^{n_j}$ where $n_0 < n_1 < \ldots$ and $a_{n_j} \neq 0$ for each j. For $n \in \{0, 1, 2, \ldots\}$ set

$$a_n := \sum_{j=0}^{n} a_{n_j} p^{n_j}$$

From Proposition 47.2 (vi) it follows that $\left|\binom{a_m}{a_n}\right|_p = 1$ if $m > n$. By continuity $\left|\binom{a}{a_n}\right|_p = 1$ for all n. So we see that $|\binom{a}{n}|_p = 1$ for infinitely many n so that the power series $\sum \binom{a}{n} x^n$ does not converge for $x = 1$, which finishes the proof.

Exercise 47.C. From the above proof it follows that for $a \in \mathbb{Z}_p$ we have $a \notin \{0, 1, 2, \ldots\}$ if and only if $|\binom{a}{n}|_p = 1$ for infinitely many n. Show that $|\binom{a}{n}|_p = 1$ for *all* $n \in \mathbb{N} \cup \{0\}$ if and only if $a = -1$ (for a short proof consider the identity $\binom{a}{n+1} = \binom{a}{n} \frac{a-n}{n+1}$ for $n+1$ a large power of p).

Exercise 47.D. It follows from Lemma 47.6 that $\overline{\lim}_{n\to\infty} \sqrt[n]{|\binom{a}{n}|_p} = 1$ for $a \in \mathbb{Z}_p \setminus \{0, 1, 2, \ldots\}$. Show that

$$\lim_{n\to\infty} \sqrt[n]{\left|\binom{\frac{1}{2}}{n}\right|_p} = \begin{cases} 1 \text{ if } p \neq 2 \\ 4 \text{ if } p = 2 \end{cases}$$

(One should not think that $\overline{\lim}_{n\to\infty} \sqrt[n]{|\binom{a}{n}|_p}$ exists for all $a \in \mathbb{Z}_p$, see Exercise 67.C.)

DEFINITION 47.7. Let $a \in K$. We define $r_a \in (0, \infty)$ in the following way.
(i) If $K \supset \mathbb{Q}_p$ then

$$r_a := \begin{cases} 1 \text{ if } a \in \mathbb{Z}_p \\ p^{1/(1-p)} \text{ if } |a| \leqslant 1, a \in K \setminus \mathbb{Z}_p \\ |a|^{-1} p^{1/(1-p)} \text{ if } |a| > 1 \end{cases}$$

(ii) If char$(k) = 0$ then

$$r_a := \begin{cases} 1 \text{ if } |a| \leqslant 1 \\ |a|^{-1} \text{ if } |a| > 1 \end{cases}$$

THEOREM 47.8. *The power series* $\Sigma \binom{a}{n} x^n$ *converges for* $|x| < r_a$.

Proof. (i) Let $K \supset \mathbb{Q}_p$. It suffices to consider the case $a \notin \mathbb{Z}_p$. If $|a| \leqslant 1$ then $|a(a-1)\ldots(a-n+1)| \leqslant 1$ so that $\overline{\lim}_{n \to \infty} \sqrt[n]{|\binom{a}{n}|} \leqslant \overline{\lim}_{n \to \infty} |n!|_p^{-1} = p^{-1/(1-p)}$ and the radius of convergence is $\geqslant p^{1/(1-p)}$. If $|a| > 1$ then $|a(a-1)\ldots(a-n+1)| = |a|^n$ and $\overline{\lim}_{n \to \infty} \sqrt[n]{|\binom{a}{n}|} = |a| p^{-1/(1-p)}$ and the radius of convergence is $|a|^{-1} p^{1/(1-p)}$.
(ii) Let char$(k) = 0$ and let $|a| \leqslant 1$. As the elements $0, 1, 2, \ldots$ of K are equidistant, there is at most one j for which $|a-j| < 1$. If $n \neq j$ then $|a-n| = 1$. Hence $|\binom{a}{n}|$ is a positive constant for large n (if $a \notin \{0, 1, 2, \ldots\}$). We see that $\Sigma \binom{a}{n} x^n$ converges if $|x| < 1$. If $|a| > 1$ then $|a(a-1)\ldots(a-n+1)| = |a|^n$ and $|n!| = 1$. It follows that $\Sigma \binom{a}{n} x^n$ converges on $\{x \in K : |x| < |a|^{-1}\}$.

DEFINITION 47.9. Let x, $a \in K$, $|1-x| < r_a$. Then

$$x^a := \sum_{n=0}^{\infty} \binom{a}{n} (x-1)^n$$

Observe that this definition extends the one given in Theorem 32.4. The following theorem shows that x^a behaves like a power function.

THEOREM 47.10. *Let* x, $a \in K$, $|1-x| < r_a$. *Then* $x^a \in K^+$ *and* $\log x^a = a \log x$. *We have even* $x^a = \exp(a \log x)$ *except when* $(K \supset \mathbb{Q}_p) \wedge (a \in \mathbb{Z}_p) \wedge (|1-x| \geqslant p^{1/(1-p)})$.

Proof. Inspection of all cases yields $x^a \in K^+$. Set $f(x) := x^a$. Elementary computation shows that $xf'(x) = af(x)$. The derivative of $x \mapsto \log x^a$ is in consequence $x \mapsto a/x$, a property that is shared by $x \mapsto a \log x$. By Corollary 42.5 the function $x \mapsto \log x^a$ is analytic, so $\log x^a - a \log x$ is a constant which is easily seen to be 0. If not $(K \supset \mathbb{Q}_p) \wedge (a \in \mathbb{Z}_p) \wedge (|1-x| \geqslant p^{1/(1-p)})$ then inspection of all cases yields $x^a \in 1 + E$ so that $x^a = \exp \log x^a = \exp(a \log x)$.

Exercise 47.E. Discuss the formulas $x^a x^\beta = x^{a+\beta}$, $(x^a)^\beta = x^{a\beta}$.

Exercise 47.F. Let $x \in K^+$ and $n \in \mathbb{N}$, not divisible by p if $K \supset \mathbb{Q}_p$. Use Theorems 32.4 and 47.10 to show that

$$\sqrt[n]{x} := x^{1/n}$$

is the unique positive nth root of x in K. Show that in $\mathbb{Q}_3$ we have $\sqrt{16} = 4$ but $\sqrt{25} = -5$ (!)

48. The Artin-Hasse exponential

As an application of the theory of the preceding sections we shall introduce a modification of the exponential function in $\mathbb{C}_p$, the 'Artin-Hasse exponential' E_p (see Theorem 48.1 below). The basic idea is similar to the one that played a role when we constructed the p-adic gamma function, namely to remove the 'bad terms' in a product (see Section 35). The same principle works in a way for the construction of a p-adic zeta function (see Section 61).

The *Möbius function* $\mu : \mathbb{N} \to \mathbb{Z}$ is given by

$$\mu(n) := \begin{cases} (-1)^j & \text{if } n \text{ is a product of } j \text{ distinct primes} \\ 0 & \text{if } n \text{ is divisible by a square} \neq 1 \end{cases}$$

We have the following formula from elementary number theory.

$$\sum_{d \mid n} \mu(d) = \begin{cases} 1 \text{ if } n = 1 \\ 0 \text{ if } n \in \{2, 3, \ldots\} \end{cases}$$

(the sum is taken over all divisors d of n).

THEOREM 48.1. *If* $x \in E \subset \mathbb{C}_p$ *then*

$$(*) \qquad \exp x = \prod_{n=1}^{\infty} (1 - x^n)^{-\mu(n)/n}$$

The formula

$$E_p(x) = \prod_{n=1}^{\infty}{}' (1 - x^n)^{-\mu(n)/n}$$

defines an analytic function on $\{x \in \mathbb{C}_p : |x|_p < 1\}$. *We have*

$$E_p(x) = \exp\left(x + \frac{x^p}{p} + \frac{x^{p^2}}{p^2} + \ldots\right) \qquad (x \in E)$$

but this formula does not hold for all $x \in B_0(1^-)$.

Before starting the proof we make a few comments. It is not hard to carry out a calculation showing that (*) holds as an identity of formal power series over $\mathbb{Q}$ (take log at both sides). But we prefer to avoid the use of formal power series; also we are interested in interpreting (*) as an identity between genuine functions. This leads to the following proof presented in the form of easy exercises.

Exercise 48.A. Let $n \in \mathbb{N}$. Show that $\mu(n)/n \in \mathbb{Z}_p$ if n is not divisible by p and that $\mu(n)/n \in p^{-1}\mathbb{Z}_p$ otherwise. Use this fact to show that

$$x \mapsto f_n(x) := (1-x^n)^{-\mu(n)/n}$$

is a well-defined analytic function, on $\{x \in \mathbb{C}_p : |x|_p < 1\}$ if n is not divisible by p, on E if p divides n.

Exercise 48.B. Show that $f_n(x) = \exp \log_p f_n(x)$ $(x \in E,\ n \in \mathbb{N})$ so that for each $N \in \mathbb{N}$

$$\prod_{n=1}^{N} f_n(x) = \exp\left(\sum_{n=1}^{N} \sum_{j=1}^{\infty} \frac{\mu(n)}{n}\frac{x^{nj}}{j}\right) \qquad (x \in E)$$

Use unconditional convergence of the double series to arrive at formula (*) of Theorem 48.1.

Exercise 48.C. Use Exercise 48.A. to show that

$$E_p(x) := \sum_{n=1}^{\infty} (1-x^n)^{-\mu(n)/n}$$

is well defined for $x \in B_0(1^-)$. Use uniform convergence to show that E_p is analytic on $B_0(1^-)$.

Exercise 48.D. Show that $x \in E$ implies $E_p(x) = \exp \log_p E_p(x)$ and carry out a computation similar to the one of Exercise 48.B to show that $E_p(x) = \exp(x + x^p/p + x^{p^2}/p^2 + \ldots)$ for $x \in E$. Finish the proof of Theorem 48.1.

49. arcsin and arccos

IN SECTION 49 WE WORK IN $\mathbb{C}_p, p \neq 2$.

We shall define inverses of sin and cos.

Recall that for each $a \in \mathbb{C}_p^+$ the element $\sqrt{a} := a^{\frac{1}{2}} = \sum_{n=0}^{\infty} \binom{1/2}{n}(a-1)^n$ is the unique positive square root of a. To find an inverse for sin, let $x = \sin y$ where $y \in E$. Then also $x \in E$ and $\exp(iy) - \exp(-iy) = 2ix$. To express y as a function of x we solve the quadratic equation $z^2 - 1 = 2ixz$ and obtain the two possibilities

$$\exp(iy) = ix \pm \sqrt{1-x^2}$$

However, $\exp(iy) \in \mathbb{C}_p^+$ and $ix - \sqrt{1-x^2} \in -\mathbb{C}_p^+$ so that the only choice is $\exp(iy) = ix + \sqrt{1-x^2}$, hence $iy = \log_p(ix + \sqrt{1-x^2})$. This expression makes sense for all $x \in B_0(1^-)$, which leads to the following definition.

DEFINITION 49.1. (van Hamme)

$$\arcsin x = \frac{1}{i} \log_p(ix + \sqrt{1-x^2}) \qquad (x \in \mathbb{C}_p, |x|_p < 1)$$

The proof of the following proposition is left to the reader.

PROPOSITION 49.2. (Properties of arcsin)

(i) $\sin(\arcsin x) = \arcsin(\sin x) = x$ $(x \in E)$.
(ii) $\arcsin(-x) = -\arcsin x = (1/i) \log_p(ix - \sqrt{1-x^2})$ $(|x|_p < 1)$.
(iii) arcsin *is analytic on* $\{x \in \mathbb{C}_p : |x|_p < 1\}$.
(iv) $\arcsin' x = (1-x^2)^{-\frac{1}{2}}$ $(|x|_p < 1)$.
(v) $\arcsin x = \sum_{n=0}^{\infty} \binom{-\frac{1}{2}}{n} (-1)^n x^{2n+1}/(2n+1)$ $(|x|_p < 1)$.
(vi) arcsin *maps* $B_0(1^-)$ *onto* $\mathbb{C}_p$ (!)
(vii) *Let* $x \in \mathbb{C}_p$, $|x|_p < 1$. *Then* $\arcsin x = 0$ *if and only if* $x = (\theta - \theta^{-1})/2i$ *for some* $\theta \in \Gamma_r$ (see Section 33).

A similar procedure leads to an inverse of cos.

DEFINITION 49.3. Let $a \in E$, $a \neq 0$.

$$\arccos_a(x) := \frac{1}{i} \log_p (x + ia\sqrt{a^{-2}(1-x^2)}) \quad (x \in B_{\sqrt{1-a^2}}(|a^2|_p^-))$$

We show that the definition is meaningful. If $|x - \sqrt{1-a^2}|_p < |a^2|_p$ then x is positive so that $|x + \sqrt{1-a^2}|_p = 1$ and $|x^2 - (1-a^2)|_p < |a^2|_p$. In other words, $a^{-2}(1-x^2)$ is positive and $\sqrt{a^{-2}(1-x^2)}$ is defined. One easily checks that $x + ia\sqrt{a^{-2}(1-x^2)} \neq 0$. Further, observe that $|\cos a - \sqrt{1-a^2}|_p = |\sqrt{1-\sin^2 a} - \sqrt{1-a^2}|_p \leqslant |a^2 - \sin^2 a|_p = |(a+\sin a)\cdot(a - \sin a)|_p < |a|^2$. Thus the domain of $\arccos_a$ may be written as $B_{\cos a}(|a^2|_p^-)$, which is precisely the range of cos, restricted to $B_a(|a|_p^-)$, see Exercise 46.A. With all this the proof of the following proposition is easy.

PROPOSITION 49.4. (Inverse of cos) *Let* $a \in E$, $a \neq 0$.

(i) cos, *restricted to* $B_a(|a|_p^-)$ *is monotone of type* $-\operatorname{sgn} a$. *Its range is the domain of* $\arccos_a$.
(ii) $\cos(\arccos_a x) = x$ $(|x - \cos a|_p < |a|_p^2)$, $\arccos_a(\cos x) = x$ $(x \in E)$.
(iii) $\arccos_a$ *is analytic*.
(iv) $\arccos_a' x = -a^{-1}\sqrt{a^2(1-x^2)^{-1}}$ $(|x - \cos a|_p < |a|_p^2)$.
(v) *If* $b \in E$, $|b - a|_p < |a|_p$ *then* $\arccos_b = \arccos_a$.

3
Functions on $\mathbb{Z}_p$

PART 1: MAHLER'S BASE AND p-ADIC INTEGRATION

Kaplansky's theorem (Theorem 43.3) shows that a continuous function : $\mathbb{Z}_p \to \mathbb{C}_p$ can uniformly be approximated by polynomial functions. In this part we shall go one step further and ask whether there exists a system $e_1, e_2, \ldots$ of polynomial functions on $\mathbb{Z}_p$ such that for each $f \in C(\mathbb{Z}_p \to \mathbb{C}_p)$ there are unique $\lambda_1, \lambda_2, \ldots \in K$ such that $f = \sum_{n=1}^{\infty} \lambda_n e_n$ uniformly. It will turn out that there are many choices for $e_1, e_2, \ldots$ among which the Mahler base $\binom{x}{n}$ $(n = 0, 1, 2, \ldots)$ seems to be the most important one.

EXCEPT FOR SECTION 50 WE SUPPOSE THROUGHOUT CHAPTER 3 THAT $K \supset \mathbb{Q}_p$.

50. Orthogonal bases in Banach spaces

In order to have a theoretical framework at our disposal we raise the level of abstraction a little by considering orthogonal bases in arbitrary K-Banach spaces. The following resembles somewhat the theory of bases in Hilbert spaces.

THROUGHOUT SECTION 50 $(E, \| \ \|)$ IS A K-BANACH SPACE

DEFINITION 50.1. Let $x, y \in E$. We write $x \perp y$ if 0 is a best approximation of x in $Ky := \{\lambda y : \lambda \in K\}$. In other words, $x \perp y$ if and only if

$$(*) \qquad \inf \{ \|x - \lambda y\| : \lambda \in K \} = \|x\|$$

(Without harm (*) may be replaced by the condition $\|x - \lambda y\| \geqslant \|x\|$ for all $\lambda \in K$.)

A similar definition can be given for elements of Banach spaces over $\mathbb{R}$ or $\mathbb{C}$. However, if the space is not a Hilbert space the relation $\perp$ may fail to be symmetric. A crucial property in the ultrametric theory is that $\perp$ is always symmetric.

PROPOSITION 50.2. *Let $x, y \in E$. If $x \perp y$ then $y \perp x$.*
Proof. Let $x \perp y$, $\lambda \in K^\times$. Then $\|y - \lambda x\| = |\lambda| \, \|x - \lambda^{-1} y\| \geqslant |\lambda| \, \|x\| = \|\lambda x\|$. By Lemma 13.4 we also have $\|y - \lambda x\| \geqslant \|y\|$. It follows that $y \perp x$.

DEFINITION 50.3. (Orthogonality)
(i) Let $x \in E$ and let $D_1, D_2 \subset E$. We write $x \perp D_2$ if $x \perp d$ for all $d \in D_2$. We write $D_1 \perp D_2$ if $d_1 \perp d_2$ for all $d_1 \in D_1, d_2 \in D_2$.
(ii) $\{x_1, x_2, \ldots\} \subset E$ is an *orthogonal set* if for each $n \in \mathbb{N}$

$$x_n \perp [\![x_1, x_2, \ldots, x_{n-1}, x_{n+1}, \ldots]\!]$$

where $[\![\quad]\!]$ indicates the K-linear span. An orthogonal set $\{x_1, x_2, \ldots\}$ is *orthonormal* if $\|x_n\| = 1$ for all $n \in \mathbb{N}$.

Exercise 50.A. Give an example of a nonorthogonal set $\{e_1, e_2, e_3\} \subset K^2$ for which, however, $e_1 \perp e_2$, $e_1 \perp e_3$, $e_2 \perp e_3$.

Exercise 50.B. Set $\|E\| := \{\|x\| : x \in E\}$, $|K| = \{|\lambda| : \lambda \in K\}$.
(i) Suppose $\|E\| = |K|$. Prove that for an orthogonal set $\{x_1, x_2, \ldots\}$ of nonzero elements one can find $\lambda_1, \lambda_2, \ldots$ in K such that $\{\lambda_1 x_1, \lambda_2 x_2, \ldots\}$ is orthonormal.
(ii) Show that the assumption $\|E\| = |K|$ made in (i) is not superfluous.

PROPOSITION 50.4. *Let $x_1, x_2, \ldots \in E$.*
(i) $\{x_1, x_2, \ldots\}$ *is orthogonal if and only if* $\{x_1, x_2, \ldots, x_n\}$ *is orthogonal for each $n \in \mathbb{N}$.*
(ii) $\{x_1, x_2, \ldots, x_n\}$ *is orthogonal if and only if for each* $\lambda_1, \lambda_2, \ldots, \lambda_n \in K$

$$(*) \qquad \Big\|\sum_{j=1}^{n} \lambda_j x_j\Big\| = \max\{|\lambda_j| \, \|x_j\| : 1 \leqslant j \leqslant n\}$$

(iii) $\{x_1, x_2, \ldots, x_n\}$ *is orthogonal if and only if for each* $\lambda_1, \lambda_2, \ldots, \lambda_n \in K$

$$(**) \qquad \Big\|\sum_{j=1}^{m} \lambda_j x_j\Big\| \geqslant |\lambda_m| \, \|x_m\| \qquad (m \in \{2, 3, \ldots, n\})$$

Proof. (i) is immediate. To prove (ii), let $\{x_1, x_2, \ldots, x_n\}$ be an orthogonal set, and let $j \in \{1, \ldots, n\}$. Then $\lambda_j x_j \perp [\![x_1, x_2, \ldots, x_{j-1}, x_{j+1}, \ldots, x_n]\!]$ so that

$$\Big\|\lambda_j x_j + \sum_{m \neq j} \lambda_m x_m\Big\| \geqslant \|\lambda_j x_j\|$$

Thus

$$\left\| \sum_{j=1}^{n} \lambda_j x_j \right\| \geqslant \max_j |\lambda_j| \, \|x_j\|$$

The opposite inequality is trivial. Conversely, assume (*) for all $\lambda_1, \lambda_2, \ldots, \lambda_n \in K$; we prove that $x_1 \perp [\![x_2, \ldots, x_n]\!]$. For any $v = \sum_{j=2}^{n} \lambda_j x_j \in [\![x_2, \ldots, x_n]\!]$ we have $\|x_1 - v\| = \max\{\|x_1\|, \|\lambda_2 x_2\|, \ldots, \|\lambda_n x_n\|\} \geqslant \|x_1\|$ and (ii) is proved. To prove (iii) it suffices to check that (**) implies (*). We have $\| \sum_{j=1}^{n} \lambda_j x_j \| \geqslant \|\lambda_n x_n\|$. By Lemma 13.4 $\| \sum_{j=1}^{n} \lambda_j x_j \| \geqslant \| \sum_{j=1}^{n-1} \lambda_j x_j \|$ and by (**) the latter is $\geqslant \|\lambda_{n-1} x_{n-1}\|$. Downward induction yields (*).

**Exercise* 50.C. If $\{x_1, x_2, \ldots\}$ is an orthonormal set then $x_1, x_2, \ldots$ are linearly independent. Prove this.

Exercise 50.D. Let $U_1, U_2, \ldots, U_n$ be mutually disjoint nonempty clopen subsets of K. Show that their K-valued characteristic functions form an orthonormal set in $BC(K \to K)$ (Definition 22.2). Is the disjointness condition necessary?

DEFINITION 50.5. Let $e_1, e_2, \ldots$ be nonzero elements of E. The system $e_1, e_2, \ldots$ is an *orthogonal* (*orthonormal*) *base* of E if
(i) $\{e_1, e_2, \ldots\}$ is an orthogonal (orthonormal) set
(ii) for each $x \in E$ there are $\lambda_1, \lambda_2, \ldots \in K$ such that $x = \sum_{n=1}^{\infty} \lambda_n e_n$.

Exercise 50.E. (i) Show that $(1, 0, 0, \ldots), (0, 1, 0, \ldots), \ldots$ is an orthonormal base of c_0 (see Exercise 13.A).
(ii) Prove the following. If $e_1, e_2, \ldots$ is an orthonormal base of E and if $\sigma: \mathbb{N} \to \mathbb{N}$ is a bijection then $e_{\sigma(1)}, e_{\sigma(2)}, \ldots$ is an orthonormal base of E.

PROPOSITION 50.6. *Let $e_1, e_2, \ldots$ be an orthonormal base of E. Let $x = \sum_{n=1}^{\infty} \lambda_n e_n \in E$ for some $\lambda_1, \lambda_2, \ldots \in K$. Then the following are true.*
(i) $\lim_{n \to \infty} \lambda_n = 0$.
(ii) $\|x\| = \max\{|\lambda_n| : n \in \mathbb{N}\}$.
(iii) *If also $x = \sum_{n=1}^{\infty} \mu_n e_n$ $(\mu_1, \mu_2, \ldots \in K)$ then $\mu_n = \lambda_n$ for all n.*
Proof. (i) follows from the convergence of $\sum \lambda_n e_n$. To prove (ii) set $x_m = \sum_{n=1}^{m} \lambda_n e_n$ $(m \in \mathbb{N})$ and observe that $\|x\| = \lim_{m \to \infty} \|x_m\| = \lim_{m \to \infty} \max\{|\lambda_1|, |\lambda_2|, \ldots, |\lambda_m|\} = \max\{|\lambda_n| : n \in \mathbb{N}\}$. Finally, application of (ii) to $0 = \sum_{n=1}^{\infty} (\lambda_n - \mu_n) e_n$ yields $|\lambda_n - \mu_n| = 0$, i.e. $\mu_n = \lambda_n$ for all n.

The following theorem is quite important. The analogy with the Hilbert space theory is obvious.

THEOREM 50.7. *Let $e_1, e_2, \ldots$ be an orthonormal set whose K-linear span is dense in E. Then $e_1, e_2, \ldots$ is an orthonormal base of E.*

Proof. Define a map $A : c_0 \to E$ by the formula

$$A(\lambda_1, \lambda_2, \ldots) = \sum_{n=1}^{\infty} \lambda_n e_n$$

Then A is K-linear. From $\| \sum_{n=1}^{\infty} \lambda_n e_n \| \leqslant \max_n |\lambda_n|$ it follows that A is continuous (Proposition 13.5). Proposition 50.4 tells us that the restriction of A to $c_{00} := \{(\lambda_1, \lambda_2, \ldots) \in c_0 : \lambda_n = 0 \text{ for large } n\}$ is an isometry. But c_{00} is dense in c_0 so that A itself is an isometry and, by consequence, $A(c_0)$ is a Banach space. It contains a dense subspace of E. Hence, $A(c_0) = E$. According to Definition 50.5 the system $e_1, e_2, \ldots$ is an orthonormal base of E.

**Exercise* 50.F. Prove the following. If $\{e_1, e_2, \ldots\}$ is an orthogonal subset of E consisting of nonzero elements and if its linear K-linear span is dense in E then $e_1, e_2, \ldots$ is an orthogonal base of E.

The next theorem is not essential for the sequel, yet we shall include it because it yields a positive solution to the ultrametric Schauder base problem for a locally compact scalar field K.

THEOREM 50.8. *Let E be a separable Banach space over a locally compact non-archimedean valued field K. Then E has an orthogonal base.*

Proof. There exists a countable subset $\{x_1, x_2, \ldots\}$ of E such that for each n, $\dim E_n = n$ $\quad (E_n := [\![x_1, x_2, \ldots, x_n]\!])$ and $\cup_n E_n$ is dense in E. We now orthogonalize the system $x_1, x_2, \ldots$ as follows. Let $e_1 := x_1$ and for $n \in \mathbb{N}$, let $e_{n+1} := x_{n+1} - v_n$, where v_n is a best approximation of x_{n+1} in E_n. (The existence of such best approximations follows from the local compactness of K and Theorem 13.3.) Our set $e_1, e_2, \ldots$ satisfies $e_{n+1} \perp [\![e_1, \ldots, e_n]\!]$ for all $n \in \mathbb{N}$. Hence, for each $n \geqslant 2$ and $\lambda_1, \ldots, \lambda_n \in K$ we have $\| \sum_{j=1}^{n} \lambda_j e_j \| = \| \lambda_n e_n + \sum_{j=1}^{n-1} \lambda_j e_j \| \geqslant \| \lambda_n e_n \|$, and $e_1, e_2, \ldots$ are orthogonal by Proposition 50.4 (iii). As $E_n = [\![x_1, \ldots, x_n]\!] = [\![e_1, \ldots, e_n]\!]$ we have that $[\![e_1, e_2, \ldots]\!]$ is dense in E. Now apply Exercise 50.F.

COROLLARY 50.9 *($C(\mathbb{Z}_p \to \mathbb{Q}_p)$, $\| \; \|_\infty$) has an orthogonal base.*

Proof. The polynomial functions with rational coefficients form a countable dense subset of $C(\mathbb{Z}_p \to \mathbb{Q}_p)$ (Theorem 43.3).

COROLLARY 50.10. $(\mathbb{C}_p, |\ |_p)$, *as a Banach space over* $\mathbb{Q}_p$, *has an orthogonal base.*
Proof. $\mathbb{C}_p$ is separable (Corollary 17.2(iv)).

Exercise 50.G. Show that $C(\mathbb{Z}_p \to \mathbb{Q}_p)$ has an orthonormal base, but $\mathbb{C}_p$ (as a Banach space over $\mathbb{Q}_p$) has not.

Exercise 50.H. (On nondifferentiable continuous homomorphisms in $\mathbb{C}_p$) Use Corollary 50.10 to show the existence of a nonzero continuous $\mathbb{Q}_p$-linear map $f : \mathbb{C}_p \to \mathbb{Q}_p$. Show that such an f is nowhere differentiable (as a map $\mathbb{C}_p \to \mathbb{C}_p$) and that EXP $\circ f$ is a nowhere differentiable homomorphism $\mathbb{C}_p \to \mathbb{C}_p^\times$. Find continuous nowhere differentiable homomorphisms $\mathbb{C}_p^\times \to \mathbb{C}_p^\times$ and $\mathbb{C}_p^\times \to \mathbb{C}_p$. (See Exercises 45.F-K.)

Remark. The results of this section can be generalized. For a thorough treatment of orthogonal bases see van Rooij (1978).

51. The Mahler base of $C(\mathbb{Z}_p \to K)$

We denote the functions $x \mapsto \binom{x}{n}$ $(x \in \mathbb{Z}_p)$ by $\binom{*}{n}$ $(n \in \{0, 1, \ldots\})$. Recall that we assume $K \supset \mathbb{Q}_p$, and that $C(\mathbb{Z}_p \to K)$ is normed by $\|\ \|_\infty$.

THEOREM 51.1. *The functions* $\binom{*}{0}, \binom{*}{1}, \binom{*}{2}, \ldots$ *form an orthonormal base* (the 'Mahler base') *of* $C(\mathbb{Z}_p \to K)$. *In other words we have* (i), (ii) *below.*
(i) *Let* $f \in C(\mathbb{Z}_p \to K)$. *Then there exist unique elements* $a_0, a_1, \ldots$ *of* K (the 'Mahler coefficients' of f) *such that*

$$f(x) = \sum_{n=0}^{\infty} a_n \binom{x}{n} \qquad (x \in \mathbb{Z}_p)$$

(the 'Mahler expansion' of f). *The series converges uniformly and*

$$\|f\|_\infty = \max\{|a_n| : n \in \{0, 1, 2, \ldots\}\}$$

(ii) *If* $a_0, a_1, \ldots$ *is a null sequence in* K *then* $x \mapsto \sum_{n=0}^{\infty} a_n \binom{x}{n}$ *defines a continuous function* $\mathbb{Z}_p \to K$.
Proof. By Proposition 47.2 the function $\binom{*}{n}$ is a polynomial of degree n. Hence, each polynomial function $\mathbb{Z}_p \to K$ can be expressed as a (finite) K-linear combination of $\binom{*}{0}, \binom{*}{1}, \ldots$ By Kaplansky's theorem (Theorem 43.3) the K-linear span of $\binom{*}{0}, \binom{*}{1}, \ldots$ is dense in $C(\mathbb{Z}_p \to K)$. For each $m, n \in \{0, 1, 2, \ldots\}$ we have that $\binom{m}{n}$ is an integer, $\binom{n}{n} = 1$ so, by continuity, $\|\binom{*}{n}\|_\infty = 1$ for all n. To prove orthogonality let $n \in \mathbb{N} \cup \{0\}$ and $a_0, a_1, \ldots, a_n \in K$. For each m, $0 \leqslant m \leqslant n$ we have

$$\|a_n\binom{*}{n} + a_{n-1}\binom{*}{n-1} + \ldots + a_m\binom{*}{m}\|_\infty \geq \left|\sum_{j=m}^{n} a_j\binom{m}{j}\right| = |a_m|$$

which shows (Proposition 50.4 (iii)) that $\binom{*}{0}, \binom{*}{1}, \ldots$ is an orthonormal base of $C(\mathbb{Z}_p \to K)$.

Remark. For other (more elementary) proofs of Theorem 51.1, see Exercises 52.E and 52.G.

Exercise 51.A. ($1, *, *^2, \ldots$ is not an orthonormal base) The simplest polynomial function of degree n is $x \mapsto x^n$ and one may wonder why in Theorem 51.1 we consider $\binom{*}{n}$ rather then $*^n$. Let X be the unit disc of a complete non-archimedean valued field K. Define $f_0, f_1, \ldots : X \to L$ by

$$f_n(x) = x^n \qquad (x \in X, n \in \{0, 1, 2, \ldots\})$$

and prove the following.
(i) The $f_0, f_1, \ldots$ are linearly independent and $\|f_n\|_\infty = 1$ for all n.
(ii) If L is locally compact then the L-linear span of $f_0, f_1, \ldots$ is dense in $C(X \to L)$ but $\{f_0, f_1, \ldots\}$ is not an orthogonal set.
(iii) If L is not locally compact then $\{f_0, f_1, \ldots\}$ is an orthonormal set but its L-linear span is not dense in $BC(X \to L)$.
Conclude that $f_0, f_1, \ldots$ is never an orthonormal base of $BC(X \to L)$.

**Exercise* 51.B. Do certain noncontinuous functions $f : \mathbb{Z}_p \to K$ admit a representation

$$(*) \qquad f(x) = \sum_{n=0}^{\infty} a_n\binom{x}{n} \qquad (x \in \mathbb{Z}_p)$$

where $a_0, a_1, \ldots \in K$? (If this is the case then obviously the convergence is not uniform and $a_0, a_1, \ldots$ is not a null sequence.) Show that the answer is negative by proving that *if $f : \mathbb{Z}_p \to K$ is a function for which there is a pointwise representation of the form* (*) *then f is continuous* (Hint. Consider $f(-1)$.)

Exercise 51.C. (Sequel to the previous exercise) Show that for $x \in \mathbb{Z}_p \setminus \{-1\}$

$$\sum_{n=0}^{\infty} p^{-n}\binom{x}{p^{2n}-1} = \sum_{n=0}^{\infty} \frac{p^n}{x+1}\binom{x+1}{p^{2n}}$$

and obtain an unbounded continuous function $f : \mathbb{Z}_p \setminus \{-1\} \to \mathbb{Q}_p$ and $a_0, a_1, \ldots \in \mathbb{Q}_p$ such that

$$f(x) = \sum_{n=0}^{\infty} a_n\binom{x}{n} \qquad (x \in \mathbb{Z}_p \setminus \{-1\})$$

Exercise 51.D. (Interpolation polynomials) Let $f \in C(\mathbb{Z}_p \to K)$ have the Mahler expansion $\sum_{n=0}^{\infty} a_n\binom{x}{n}$. The mth *interpolation polynomial* f_m of f ($m \in \{0, 1, 2, \ldots\}$) is given by

$$f_m(x) = \sum_{n=0}^{m} a_n\binom{x}{n} \qquad (x \in \mathbb{Z}_p)$$

(i) Show that f_m is the unique polynomial function P of degree $\leqslant m$ for which $P(n) = f(n)$ ($n = 0, 1, \ldots, m$).
(ii) Show that f_m is a best approximation of f in the set of all polynomial functions of degree $\leqslant m$.

Remark. One can construct other orthonormal bases of $C(\mathbb{Z}_p \to K)$ by generalizing the procedure used to define the Mahler base as follows. Let $s_0, s_1, \ldots$ be mutually distinct elements of $\mathbb{Z}_p$. Define the functions $P_0, P_1, \ldots$ by $P_0(x) := 1$ for all $x \in \mathbb{Z}_p$ and

$$P_n(x) := \frac{(x-s_0)(x-s_1)\ldots(x-s_{n-1})}{(s_n-s_0)(s_n-s_1)\ldots(s_n-s_{n-1})} \qquad (x \in \mathbb{Z}_p)$$

for $n \geqslant 1$. Observe that P_n is a polynomial of degree n, that $P_n(s_j) = 0$ for $j < n$ and $P_n(s_n) = 1$. (The choice $s_n := n$ for all n yields $P_n = \binom{x}{n}$.) Now *suppose* that $\|P_n\|_\infty = 1$ for all n. Then we use the trick in the proof of Theorem 51.1

$$\left\| \sum_{j=m}^{n} a_j P_j \right\|_\infty \geqslant \left| \sum_{j=m}^{n} a_j P_j(s_m) \right| = |a_m| = |a_m| \, \|P_m\|_\infty$$

($m, n \in \{0, 1, 2, \ldots\}$, $m \leqslant n$, $a_m, \ldots, a_n \in K$) and conclude that $P_0, P_1, \ldots$ is an orthonormal set (whose K-linear span is dense in $C(\mathbb{Z}_p \to K)$), hence an orthonormal base. However, the assumption $\|P_n\|_\infty = 1$ restricts the possible choices of $s_0, s_1, \ldots$ To see that there is still a lot of freedom, start with an arbitrary $s_0 \in \mathbb{Z}_p$. Choose $s_1 \in \mathbb{Z}_p$ such that $\max\{|x - s_0|_p : x \in \mathbb{Z}_p\} = |s_1 - s_0|_p$ and define

$$P_1(x) := \frac{x - s_0}{s_1 - s_0} \qquad (x \in \mathbb{Z}_p)$$

Then $\|P_1\|_\infty = 1$. Next choose $s_2 \in \mathbb{Z}_p$ such that $\max\{|(x-s_0)(x-s_1)|_p : x \in \mathbb{Z}_p\} = |(s_2 - s_0)(s_2 - s_1)|_p$ and define

$$P_2(x) := \frac{(x-s_0)(x-s_1)}{(s_2-s_0)(s_2-s_1)}$$

Then $\|P_2\|_\infty = 1$. Going on this way we obtain an orthonormal base $P_0, P_1,$

... of $C(\mathbb{Z}_p \to K)$. Constructions like this also work for more general domains. The study of these generalized bases goes beyond the scope of this book. For a good background account we refer to Y. Amice : Interpolation p-adique. *Bull. Soc. Math. France* 92(1964), 117-80.

52. The Mahler coefficients. Examples

Given an $f \in C(\mathbb{Z}_p \to K)$ its Mahler coefficients are uniquely determined. We shall prove a formula that expresses these coefficients in terms of values of f.

For $f \in C(\mathbb{Z}_p \to K)$ we set

$$(L_1 f)(x) := f(x+1) \qquad (x \in \mathbb{Z}_p)$$
$$\Delta f := L_1 f - f$$

(In difference calculus one mostly uses the symbol E rather than L_1.) Thus, $(\Delta f)(x) = f(x+1) - f(x)$ $(x \in \mathbb{Z}_p)$. The operators L_1 and $\Delta = L_1 - I$ (I is the identity) map $C(\mathbb{Z}_p \to K)$ into $C(\mathbb{Z}_p \to K)$. Suppose $f \in C(\mathbb{Z}_p \to K)$ has the Mahler expansion

$$f = \sum_{n=0}^{\infty} a_n \binom{*}{n}$$

Then

$$f(x+1) = \sum_{n=0}^{\infty} a_n \binom{x+1}{n} \qquad (x \in \mathbb{Z}_p)$$

Now

$$\binom{x+1}{n} = \begin{cases} \binom{x}{n} + \binom{x}{n-1} & \text{if } n \geqslant 1 \\ 1 & \text{if } n = 0 \end{cases}$$

We find that

$$f(x+1) = a_0 + \sum_{n=1}^{\infty} a_n \binom{x}{n} + \sum_{n=1}^{\infty} a_n \binom{x}{n-1}$$

$$= f(x) + \sum_{n=0}^{\infty} a_{n+1} \binom{x}{n}$$

Hence,

$$\Delta f = \sum_{n=0}^{\infty} a_{n+1} \binom{*}{n}$$

It follows that for $k \in \mathbb{Z}, k \geqslant 0$

$$\Delta^k f = \sum_{n=0}^{\infty} a_{n+k} \binom{x}{n}$$

so that

$$(\Delta^k f)\ (0) = a_k$$

Now

$$\Delta^k = (L_1 - I)^k = \sum_{j=0}^{k} L_1^j (-1)^{k-j} \binom{k}{j} = \sum_{j=0}^{k} (-1)^{k-j} \binom{k}{j} L_j$$

where

$$(L_j f)\ (x) := f(x+j) \qquad (x \in \mathbb{Z}_p)$$

We see that $(\Delta^k f)\ (0) = \sum_{j=0}^{k} (-1)^{k-j} \binom{k}{j} f(j)$ and we have proved the following.

THEOREM 52.1. *Let $f \in C(\mathbb{Z}_p \to K)$ have the Mahler expansion $\sum_{n=0}^{\infty} a_n \binom{x}{n}$. Then the coefficients a_n can be reconstructed from f by*

$$a_n = \sum_{j=0}^{n} (-1)^{n-j} \binom{n}{j} f(j) \qquad (n = 0, 1, 2, \ldots)$$

We now consider a few examples.

EXAMPLE 52.2. *Let $a \in \mathbb{C}_p^+$. Then $a^x = \sum_{n=0}^{\infty} (a-1)^n \binom{x}{n}$ $(x \in \mathbb{Z}_p)$. In particular,* $\exp(ax) = \sum_{n=0}^{\infty} (\exp a - 1)^n \binom{x}{n}$ $(x \in \mathbb{Z}_p)$ *for $a \in E$.*

Proof. $a^m = (a - 1 + 1)^m = \sum_{n=0}^{\infty} (a-1)^n \binom{m}{n}$ for $m \in \mathbb{N}$. Now use continuity. If $a \in E$ then $\exp(ax) = (\exp a)^x$ $(x \in \mathbb{Z}_p)$, which proves the second part.

EXAMPLE 52.3. (Mahler coefficients of sin and cos) *Let $p \neq 2, a \in p\mathbb{Z}_p$. Then* $\sin ax = \sum_{n=0}^{\infty} a_n \binom{x}{n}$, $\cos ax = \sum_{n=0}^{\infty} b_n \binom{x}{n}$ $(x \in \mathbb{Z}_p)$ *where for each n*

$$\begin{aligned}
a_{2n} &= (-1)^n 2^{2n} (\sin \tfrac{1}{2} a)^{2n} \sin na \\
a_{2n+1} &= (-1)^n 2^{2n+1} (\sin \tfrac{1}{2} a)^{2n+1} \cos(n + \tfrac{1}{2}) a \\
b_{2n} &= (-1)^n 2^{2n} (\sin \tfrac{1}{2} a)^{2n} \cos na \\
b_{2n+1} &= (-1)^{n+1} 2^{2n+1} (\sin \tfrac{1}{2} a)^{2n+1} \sin(n + \tfrac{1}{2}) a
\end{aligned}$$

Proof. $\sin ax = (\exp iax - \exp(-iax))/2i = \sum_{n=0}^{\infty} a_n \binom{x}{n}$ where $2ia_n =$

$(\exp ia-1)^n - (\exp(-ia)-1)^n$. Now $\exp ia - 1 = \exp\frac{1}{2}ia(\exp\frac{1}{2}ia - \exp(-\frac{1}{2}ia))$ so that $(\exp ia-1)^n = \exp\frac{1}{2}in a\,(2i)^n \sin^n \frac{1}{2}a$, etc.

To find an expression for the Mahler coefficients of the p-adic gamma function Γ_p introduced in Section 35 we could take the formula $a_n = \Sigma_{j=0}^{n} (-1)^{n-j}\binom{n}{j}\Gamma_p(j)$ of Theorem 52.1 but the following approach might be more appealing. First we have a lemma.

LEMMA 52.4. *Let $f:\mathbb{Z}_p \to K$ be a bounded function and let $a_n := \Sigma_{j=0}^{n} (-1)^{n-j}\binom{n}{j} f(j)$ $(n \in \{0,1,2,\ldots\})$. Then*

$$\sum_{n=0}^{\infty} f(n)\frac{x^n}{n!} = (\exp x)\sum_{n=0}^{\infty} a_n\frac{x^n}{n!} \qquad (x \in E)$$

Proof. $\exp(-x)\,\Sigma_{n=0}^{\infty} f(n)\,x^n/n! = \Sigma_{n=0}^{\infty}(-x)^n/n!\ \Sigma_{n=0}^{\infty} f(n)\,x^n/n! = \Sigma_{n=0}^{\infty}\ \Sigma_{j=0}^{n}(-1)^{n-j} f(j)/(j!(n-j)!)x^n = \Sigma_{n=0}^{\infty}(\Sigma_{j=0}^{n}(-1)^{n-j}\binom{n}{j}f(j))\,x^n/n! = \Sigma_{n=0}^{\infty} a_n x^n/n!$.

EXAMPLE 52.5. (Mahler coefficients of Γ_p) *Let*

$$\Gamma_p(x+1) = \sum_{n=0}^{\infty} a_n\binom{x}{n} \qquad (x \in \mathbb{Z}_p)$$

$$\exp\left(x+\frac{x^p}{p}\right)\cdot\frac{1-x^p}{1-x} = \sum_{n=0}^{\infty} b_n x^n \qquad (x \in E)$$

Then $a_n = (-1)^{n+1}\,n!\,b_n$ for all n.

Proof. We shall apply Lemma 52.4 to $x \mapsto \Gamma_p(x+1)$. Define $g(x) := \Sigma_{n=0}^{\infty}\Gamma_p(n+1)x^n/n!$ $(x \in E)$ then

$$g(x) = \sum_{j=0}^{p-1}\sum_{m=0}^{\infty}\frac{\Gamma_p(mp+j+1)}{(mp+j)!}\,x^{mp+j}$$

By formula (i) of Proposition 39.1

$$\Gamma_p(mp+j+1) = (-1)^{mp+j+1}\frac{(mp+j)!}{p^m m!}$$

so that

$$g(x) = \sum_{j=0}^{p-1}\sum_{m=0}^{\infty}(-1)^{mp+j+1}\left(\frac{x^p}{p}\right)^m\frac{1}{m!}\,x^j = -\exp\frac{(-x)^p}{p}\sum_{j=0}^{p-1}(-x)^j$$

$$g(-x) = -\exp\left(\frac{x^p}{p}\right)\frac{1-x^p}{1-x}$$

By the lemma we have $g(-x)=\exp(-x)\ \Sigma_{n=0}^{\infty}\ (-1)^n a_n \frac{x^n}{n!}$. Hence

$$\exp\left(x+\frac{x^p}{p}\right)\frac{1-x^p}{1-x} = \sum_{n=0}^{\infty} (-1)^{n+1} a_n \frac{x^n}{n!}$$

and we are done.

EXAMPLE 52.6. (Mahler coefficients of the indefinite sum) *Let* $f=\Sigma_{n=0}^{\infty} a_n\binom{*}{n}\in C(\mathbb{Z}_p\to K)$. *Then its indefinite sum* Sf (Definition 34.3) *has the Mahler expansion* $\Sigma_{n=1}^{\infty} a_{n-1}\binom{*}{n}$.

Proof. Set $Sf=\Sigma_{n=0}^{\infty} b_n\binom{*}{n}$. Then $b_0=0$. Using the formulas $Sf(x+1)-Sf(x)=f(x)$ and $\binom{x+1}{n}=\binom{x}{n}+\binom{x}{n-1}$ we can write

$$\sum_{n=0}^{\infty} a_n\binom{*}{n}=f=\sum_{n=1}^{\infty} b_n\binom{*+1}{n}-\sum_{n=1}^{\infty} b_n\binom{*}{n}=\sum_{n=1}^{\infty} b_n\binom{*}{n-1}$$

$$=\sum_{n=0}^{\infty} b_{n+1}\binom{*}{n}$$

It follows that $b_n=a_{n-1}$ for $n\in\mathbb{N}$.

We conclude this section with a number of exercises concerning the Mahler base.

Exercise 52.A. Let $f\in C(\mathbb{Z}_p\to K)$ and let $\Sigma_{n=0}^{\infty} a_n\binom{*}{n}$ be its Mahler expansion. Show that $\max(|a_0|, |a_1|, \ldots, |a_n|) = \max(|f(0)|, |, |f(1)|, \ldots, |f(n)|)$ $(n\in\mathbb{N}\cup\{0\})$.

Exercise 52.B. Let $f\in C(\mathbb{Z}_p\to K)$ and let $\Sigma_{n=0}^{\infty} a_n\binom{*}{n}$ be its Mahler expansion. Show that $xf(x)=\Sigma_{n=1}^{\infty} n(a_n+a_{n-1})\binom{x}{n}$ $(x\in\mathbb{Z}_p)$.

**Exercise* 52.C. (On the Mahler coefficients of $x\mapsto x^n$) For $m\in\mathbb{N}\cup\{0\}$ set

$$x^m = \sum_{n=0}^{\infty} a_{nm}\binom{x}{n} \qquad (x\in\mathbb{Z}_p)$$

(The $a_{nm}/n!$ are known as the *Stirling numbers* of the second kind.) Let $n, m\in\{0, 1, 2, \ldots\}$. Prove the following.

(i) $a_{00}=1; a_{0m}=a_{n0}=0\ (m, n\neq 0); a_{nm}=0$ if $n>m$.

(ii) $a_{nm}=\Sigma_{j=0}^{n}(-1)^{n-j}\binom{n}{j}j^m$.

(iii) a_{nm} is an integer, divisible by $n!$

(iv) Set

$$f_{nm}(x) := (x\tfrac{d}{dx})^m (x-1)^n \qquad (x\in\mathbb{Z}_p)$$

Then $a_{nm} = f_{nm}(1)$.
(v) $a_{n,m+1} = n(a_{nm} + a_{n-1,m})$ $(n \geqslant 1)$
(vi) $a_{n+1,m} = \sum_{j=0}^{m-1} \binom{m}{j} a_{nj}$ $(m \geqslant 1)$

**Exercise* 52.D. (Mahler base for $C(\mathbb{Z}_p \times \mathbb{Z}_p \to K)$) Prove that the functions

$$(x, y) \mapsto \binom{x}{m}\binom{y}{n} \qquad (m, n \in \{0, 1, 2, \ldots\})$$

form an orthonormal base of $C(\mathbb{Z}_p \times \mathbb{Z}_p \to K)$. Show also that if $f \in C(\mathbb{Z}_p \times \mathbb{Z}_p \to K)$ has the Mahler expansion

$$f(x, y) = \sum_{m,n} a_{mn} \binom{x}{m}\binom{y}{n} \qquad ((x, y) \in \mathbb{Z}_p \times \mathbb{Z}_p)$$

then

$$a_{mn} = \sum_{i=0}^{m} \sum_{j=0}^{n} (-1)^{m+n-i-j} \binom{m}{i}\binom{n}{j} f(i, j) \; (m, n \in \{0, 1, 2, \ldots\})$$

**Exercise* 52.D. (Mahler base for $C(\mathbb{Z}_p \times \mathbb{Z}_p \to K)$) Prove that the functions of Theorem 51.1 that uses neither Kaplansky's theorem nor the notion of orthogonality. Let $f \in C(\mathbb{Z}_p \to K)$.
(i) For $n \in \{0, 1, 2, \ldots\}$ set $\rho_n := \sup\{|f(x) - f(y)| : |x - y|_p \leqslant p^{-n}\}$.
Show that $\lim_{n \to \infty} \rho_n = 0$.
(ii) From the formula for Δ^k proved in the beginning of this section it follows that for $x \in \mathbb{Z}_p$

$$(\Delta^n f)(x) = \sum_{j=0}^{n} (-1)^{n-j} \binom{n}{j} f(x + j) \qquad (n \in \{0, 1, 2, \ldots\})$$

For $n \geqslant 1$ we have $\sum_{j=0}^{n} (-1)^{n-j} \binom{n}{j} = 0$ so that

$$(\Delta^n f)(x) = \sum_{j=0}^{n} (-1)^{n-j} \binom{n}{j} (f(x + j) - f(x)) \qquad (n \in \mathbb{N})$$

(iii) Prove that $|\binom{p^n}{j}|_p = p^{-(n-s)}$ $(0 < j < p^n,\ s = \operatorname{ord}_p(j))$ and use this fact to show that for $n = 0, 1, 2, \ldots$ and $x \in \mathbb{Z}_p$

$$|(\Delta^{p^n} f)(x)| \leqslant \max_{0 \leqslant s \leqslant n} p^{-n+s} \rho_s$$

Conclude that $\lim_{n \to \infty} \Delta^{p^n} f = 0$ uniformly.
(iv) Prove that $\|\Delta g\|_\infty \leqslant \|g\|_\infty$ $(g \in C(\mathbb{Z}_p \to K))$. Combine this with (iii) to show that $\lim_{n \to \infty} \Delta^n f = 0$ uniformly.
(v) Set $a_n := (\Delta^n f)(0)$. Show that $\lim_{n \to \infty} a_n = 0$ and that $f(x) = \sum_{n=0}^{\infty}$

$a_n \binom{x}{n}$, first for $x \in \mathbb{N}$ and then, by continuity, for all $x \in \mathbb{Z}_p$. Now finish the proof of Theorem 51.1.

Exercise 52.F. (Preamble to the following exercise) Let $e_0, e_1, \ldots$ be an orthonormal base of $C(\mathbb{Z}_p \to \mathbb{Q}_p)$. Let $f_0, f_1, \ldots$ be in $C(\mathbb{Z}_p \to \mathbb{Q}_p)$ such that $\|f_n - e_n\|_\infty < 1$ for each n. Show that $f_0, f_1, \ldots$ is also an orthonormal base of $C(\mathbb{Z}_p \to \mathbb{Q}_p)$. Deduce that $e_0^p, e_1^p, \ldots$ is an orthonormal base of $C(\mathbb{Z}_p \to \mathbb{Q}_p)$. (In general it is not true that $e_0^j, e_1^j, \ldots$ is orthonormal if $1 < j < p$.)

Exercise 52.G. (On powers of $\binom{*}{n}$) In this exercise we shall prove the following result due to Caenepeel (informal communication). *For each $k \in \mathbb{N}$ the functions $\binom{*}{0}^k, \binom{*}{1}^k, \ldots$ form an orthonormal base of $C(\mathbb{Z}_p \to K)$.* Observe that the preceding exercise establishes the theorem for $k = p$ and $K = \mathbb{Q}_p$. The general case is less easy.
(i) Prove that $\binom{*}{0}^k, \binom{*}{1}^k, \ldots$ is an orthonormal set in $C(\mathbb{Z}_p \to K)$.
(ii) Show that the K-linear span of $C(\mathbb{Z}_p \to \mathbb{Q}_p)$ is dense in $C(\mathbb{Z}_p \to K)$ so that it suffices to show that the $\mathbb{Q}_p$-linear span of $\binom{*}{0}^k, \binom{*}{1}^k, \ldots$ is dense in $C(\mathbb{Z}_p \to \mathbb{Q}_p)$.
(iii) Let $n \in \{0, 1, 2, \ldots\}$. Use Corollary 47.5 to prove that if $|x - y|_p \leqslant p^{-n}$ then

$$\left|\binom{x}{j}^k - \binom{y}{j}^k\right|_p < 1 \qquad (j = 0, 1, \ldots, p^n - 1)$$

(iv) Let $n \in \{0, 1, 2, \ldots\}$. Define $e_0, e_1, \ldots, e_{p^n-1} : \mathbb{Z}_p \to \mathbb{Z}_p / p\mathbb{Z}_p = \mathbb{F}_p$ by

$$e_j(x) = \binom{x}{j}^k \pmod{p\mathbb{Z}_p}$$

Show that $e_0, e_1, \ldots, e_{p^n-1}$ is a base of the $\mathbb{F}_p$-linear space consisting of all functions $\mathbb{Z}_p \to \mathbb{F}_p$ that are constant on cosets of $p^n \mathbb{Z}_p$.
(v) Let $f \in C(\mathbb{Z}_p \to \mathbb{Q}_p)$, $\|f\|_\infty \leqslant 1$. Then there is a $\mathbb{Q}_p$-linear combination g_0 of $\binom{*}{0}^k, \binom{*}{1}^k, \ldots$ such that $\|f - g_0\|_\infty < 1$. Prove this first for a locally constant function f and then for a general f.
(vi) Apply (v) to $p^{-1}(f - g_0)$ to show the existence of a $\mathbb{Q}_p$-linear combination g_1 of $\binom{*}{0}^k, \binom{*}{1}^k, \ldots$ such that $\|f - g_0 - pg_1\|_\infty < 1/p$. Inductively, define $\mathbb{Q}_p$-linear combinations $g_0, g_1, \ldots$ of $\binom{*}{0}^k, \binom{*}{1}^k, \ldots$ such that for every $n \in \{0, 1, 2, \ldots\}$

$$\left\| f - \sum_{j=0}^{n} p^j g_j \right\|_\infty < p^{-n},$$

from which it follows that the $\mathbb{Q}_p$-linear span of $\binom{*}{0}^k, \binom{*}{1}^k, \ldots$ is dense in $C(\mathbb{Z}_p \to \mathbb{Q}_p)$.

Remarks on Exercise 52.G.
1. The above exercise furnishes a third proof of the fact that $\binom{*}{0}, \binom{*}{1}, \ldots$

is an orthonormal base of $C(\mathbb{Z}_p \to K)$. Observe that Kaplansky's theorem is not needed for the proof.
2. The proof of Exercise 52.G works also for the set $\binom{*}{0}^{k_0}, \binom{*}{1}^{k_1}, \ldots$ where $k_0, k_1, \ldots \in \mathbb{N}$. Thus, $\binom{*}{0}^{k_0}, \binom{*}{1}^{k_1}, \ldots$ *is an orthonormal base for* $C(\mathbb{Z}_p \to K)$.

Exercise 52.H. (Sequel to the previous exercise) For $n \in \{0, 1, 2, \ldots\}$ define

$$u_n(x) := \lim_{m \to \infty} \binom{x}{n}^{(p-1)p^m} \qquad (x \in \mathbb{Z}_p)$$

Show that $u_0, u_1, \ldots$ is an orthonormal base of $C(\mathbb{Z}_p \to K)$ consisting of characteristic functions of clopen sets.

Exercise 52.I. Let $a \in \mathbb{Z}_p$, $|a|_p = 1$. Show that $\binom{ax}{0}, \binom{ax}{1}, \binom{ax}{2}, \ldots$ is an orthonormal base of $C(\mathbb{Z}_p \to K)$.

Exercise 52.J. (Sequel to Exercise 34.E) In Exercise 34.E it was proved that if $f, g \in l^\infty$ can be interpolated then so can $f * g$. We shall sketch another proof which is less involved but uses the observations at the beginning of this section. For any $h : \mathbb{N} \cup \{0\} \to \mathbb{Q}_p$, let $h_1(n) := h(n+1)$ $(n \in \mathbb{N})$, $\Delta h := h_1 - h$.
(i) Prove that $h : \mathbb{N} \cup \{0\} \to \mathbb{Q}_p$ can be interpolated if and only if $\lim_{n \to \infty} \|\Delta^n h\|_\infty = 0$.
(ii) Let $f, g \in l^\infty$, $n \in \mathbb{N} \cup \{0\}$. Show that

$$\Delta^{n+1}(f * g) = f * \Delta^{n+1} g + \sum_{j=0}^{n} \Delta^j f_1 \, \Delta^{n-j} g(0)$$

(iii) Suppose that $f, g \in l^\infty$ can be interpolated. Use (i) and (ii) to show that $f * g$ can be interpolated.

53. Mahler's base for $C^1(\mathbb{Z}_p \to K)$

One of the interesting aspects of the Mahler expansion $f = \sum_{n=0}^{\infty} a_n \binom{*}{n}$ is the existence of a simple condition on the Mahler coefficients characterizing continuous differentiability of f. In fact, we shall prove (Theorem 53.5) that $f \in C^1(\mathbb{Z}_p \to K)$ if and only if $\lim_{n \to \infty} |a_n| n = 0$.

Recall that $\mathrm{Lip}_1(\mathbb{Z}_p \to K)$ (Definition 26.5) is a K-Banach space with respect to the norm

$$f \mapsto \|f\|_1 = \|f\|_\infty \vee \|\Phi_1 f\|_\infty$$

(Exercise 26.E) and that $C^1(\mathbb{Z}_p \to K)$ is a closed subspace (Exercise 27.C).

**Exercise* 53.A. Show that $\|f\|_1 = |f(0)| \vee \|\Phi_1 f\|_\infty$ ($f \in \mathrm{Lip}_1(\mathbb{Z}_p \to K)$).

We first characterize the Lipschitz functions by means of a condition on their Mahler coefficients.

LEMMA 53.1. *Let $f : \mathbb{Z}_p \backslash \{-1\} \to K$ be bounded and continuous. Suppose there are $a_0, a_1, \ldots \in K$ such that $f(x) = \sum_{n=0}^{\infty} a_n \binom{x}{n}$ for all $x \in \mathbb{Z}_p \backslash$* $x \in \mathbb{Z}_p \backslash \{-1\}$.

(i) $a_n = \sum_{j=0}^{n} (-1)^{n-j} \binom{n}{j} f(j)$ $(n \in \{0, 1, 2, \ldots\})$.

(ii) $\|f\|_\infty = \sup\{|a_n| : n \in \mathbb{N} \cup \{0\}\}$.

Proof. Induction yields (i). From (i) we infer $|a_n| \leqslant \max(|f(0)|, |f(1)|, \ldots, |f(n)|)$. Hence, $\sup_n |a_n| \leqslant \|f\|_\infty$. Then opposite inequality follows from $|f(x)| = |\sum_{n=0}^{\infty} a_n \binom{x}{n}| \leqslant \sup_n |a_n|$.

DEFINITION 53.2. Let $\gamma_0, \gamma_1, \ldots$ be the p-adic integers defined as follows. $\gamma_0 := 1$ and $\gamma_n := n - n_-$ (see Definition 47.3) for $n \in \mathbb{N}$.

LEMMA 53.3. *The numbers $\gamma_0, \gamma_1, \ldots$ satisfy the following.*

(i) $|\gamma_n|_p = \min(|1|_p, |2|_p, \ldots, |n|_p)$ $(n \in \mathbb{N})$.

(ii) $1/n \leqslant |\gamma_n|_p \leqslant p/n$ $(n \in \mathbb{N})$.

(iii) $|\gamma_n|_p / p \leqslant |\gamma_{n+1}|_p \leqslant |\gamma_n|_p$ $(n \in \{0, 1, 2, \ldots\})$.

Proof. All statements are direct consequences of the fact that if $n = a_0 + a_1 p + \ldots + a_s p^s (a_s \neq 0)$ in base p then $\gamma_n = a_s p^s$ so that $|\gamma_n|_p = p^{-s}$.

THEOREM 53.4. (Characterization of Lipschitz functions by Mahler coefficients) *Let $f \in C(\mathbb{Z}_p \to K)$ have Mahler expansion $\sum_{n=0}^{\infty} a_n \binom{x}{n}$. Then $f \in \mathrm{Lip}_1(\mathbb{Z}_p \to K)$ if and only if $\sup_n |a_n| n < \infty$. More precisely, for $f = \sum_{n=0}^{\infty} a_n \binom{x}{n} \in C(\mathbb{Z}_p \to K)$ we have the following.*

(i) $\|\Phi_1 f\|_\infty = \sup\{|a_n| |\gamma_n|_p^{-1} : n \in \mathbb{N}\}$.

(ii) $\|\Phi_1 f\|_\infty \leqslant \sup\{|a_n| n : n \in \mathbb{N}\} \leqslant p \|\Phi_1 f\|_\infty$.

(iii) *If $f \in \mathrm{Lip}_1(\mathbb{Z}_p \to K)$ then* $\|f\|_1 = \sup\{|a_n| |\gamma_n|_p^{-1} : n \in 0, 1, 2, \ldots\}$.

(iv) *The formula*

$$\|f\|_1^\sim := |a_0| \vee \sup\{|a_n| n : n \in \mathbb{N}\}$$

defines a norm $\| \ \|_1^\sim$ on $\mathrm{Lip}_1(\mathbb{Z}_p \to K)$ for which

$$\|f\|_1 \leqslant \|f\|_1^\sim \leqslant p \|f\|_1 \quad (f \in \mathrm{Lip}_1(\mathbb{Z}_p \to K))$$

Proof. We only prove (i). (The other statements are obvious consequences of (i) and Lemma 53.3.) For $x, y \in \mathbb{Z}_p$, $y \neq 0$ we have by Proposition 47.2 (iii)

$$\Phi_1 f(x+y,x) = y^{-1} \sum_{n=0}^{\infty} a_n \left(\binom{x+y}{n} - \binom{x}{n}\right) = \sum_{n=1}^{\infty} \sum_{j=0}^{n-1} a_n \binom{x}{j} y^{-1} \binom{y}{n-j}$$

Set $b_{nj} := a_n \binom{x}{j} y^{-1} \binom{y}{n-j} = (a_n/(n-j)) \binom{x}{j} \binom{y-1}{n-j-1}$ if $j < n$ and $b_{nj} := 0$ if $j \geqslant n$. Then $\lim_{j \to \infty} b_{nj} = 0$ for each n, $\lim_{n \to \infty} b_{nj} = 0$ uniformly in j so that by Exercises 23.A and 23.B the summation of the b_{nj} is unconditional. After the substitution $n = j+m+1$ we obtain $\Phi_1 f(x+y,x) = \sum_{n=1}^{\infty} \sum_{m=0}^{\infty} b_{nj} = \sum_{m=0}^{\infty} \sum_{j=0}^{\infty} b_{j+m+1,j}$. With $y+1$ in place of y we arrive at the following formula which is valid for all $x, y \in \mathbb{Z}_p (y \neq -1)$.

$$(*) \qquad \Phi_1 f(x+y+1,x) = \sum_{m=0}^{\infty} \sum_{j=0}^{\infty} \frac{a_{j+m+1}}{m+1} \binom{x}{j}\binom{y}{m}$$

$$= \sum_{j=0}^{\infty} \sum_{m=0}^{\infty} \frac{a_{j+m+1}}{m+1} \binom{y}{m}\binom{x}{j}$$

(We shall need the second equality later on.) Set

$$\tau := \sup_{m, j \geqslant 0} \left| \frac{a_{j+m+1}}{m+1} \right|$$

From (*) we have immediately $\|\Phi_1 f\|_\infty \leqslant \tau$. Now $\tau = \sup_{0 \leqslant j \leqslant n} |(n-j+1)^{-1} a_{n+1}| = \sup_{n \geqslant 0} |a_{n+1}| (\max |n+1|_p^{-1}, |n|_p^{-1}, \ldots, |1|_p^{-1}) = \sup_{n \geqslant 1} |a_n| |\gamma_n|_p^{-1}$ and we have proved

$$\|\Phi_1 f\|_\infty \leqslant \sup_{n \geqslant 1} |a_n| |\gamma_n|_p^{-1} = \tau$$

To prove the inequality $\|\Phi_1 f\|_\infty \geqslant \tau$ we may assume that $\Phi_1 f$ is bounded. Applying Lemma 53.1 to the function $y \mapsto \Phi_1 f(x+y+1,x)$ defined on $\mathbb{Z}_p \backslash \{-1\}$ we get, using (*)

$$\sup_{y \neq -1} |\Phi_1 f(x+y+1,x)| = \sup_{m \geqslant 0} \left| \sum_{j=0}^{\infty} \frac{a_{j+m+1}}{m+1} \binom{x}{j} \right|$$

The series $\sum_{j=0}^{\infty} (a_{j+m+1}/(m+1)) \binom{x}{j}$ converges for all $x \in \mathbb{Z}_p$ hence represents a continuous function, by Exercise 51.B, so that $\max \{ | \sum_{j=0}^{\infty} (a_{j+m+1}/(m+1)) \binom{x}{j} | : x \in \mathbb{Z}_p \} = \max_j |a_{j+m+1}/(m+1)|$. It follows that $\|\Phi_1 f\|_\infty = \sup_{m,j} |a_{j+m+1}/(m+1)| = \tau$. This finishes the proof.

THEOREM 53.5. (Characterization of C^1-functions by Mahler coefficients) *Let $f \in C(\mathbb{Z}_p \to K)$ have the Mahler expansion $\sum_{n=0}^{\infty} a_n \binom{*}{n}$. Then $f \in C^1(\mathbb{Z}_p \to K)$ if and only if $\lim_{n \to \infty} |a_n| n = 0$. More precisely, for $f = \sum_{n=0}^{\infty} a_n \binom{*}{n}$ we have the following.*

(i) *$f \in C^1(\mathbb{Z}_p \to K)$ if and only if* $\lim_{n\to\infty} |a_n|\, |\gamma_n|_p^{-1} = 0$.
(ii) *If $f \in C^1(\mathbb{Z}_p \to K)$ then* $\|f\|_1 = \max\{|a_n|\, |\gamma_n|_p^{-1} : n \in \{0, 1, 2, \ldots\}\}$.
(iii) *The functions* $\gamma_0\binom{*}{0}, \gamma_1\binom{*}{1}, \gamma_2\binom{*}{2}, \ldots$ *form an orthonormal base of* $C^1(\mathbb{Z}_p \to K)$.

Proof. (i) Let $f \in C^1(\mathbb{Z}_p \to K)$. Formula (*) in the proof of the preceding theorem tells us that

$$(*) \qquad \Phi_1 f(x+y+1, x) = \sum_{j=0}^{\infty} \sum_{m=0}^{\infty} (m+1)^{-1} a_{j+m+1} \binom{y}{m}\binom{x}{j}$$

$$(x, y \in \mathbb{Z}_p, y \neq -1)$$

Now $\Phi_1 f$ can be extended to a continuous function $\bar{\Phi}_1 f$ on $\mathbb{Z}_p \times \mathbb{Z}_p$. So there are (Exercise 52.D) $b_{jm} \in K$ with $\lim_{j+m\to\infty} b_{jm} = 0$ such that

$$(**) \qquad \bar{\Phi}_1 f(x+y+1, x) = \sum_{j=0}^{\infty} \sum_{m=0}^{\infty} b_{jm} \binom{y}{m}\binom{x}{j} \qquad (x, y \in \mathbb{Z}_p)$$

The expressions (*) and (**) depend continuously on x. For each $j \in \{0, 1, 2, \ldots\}$ we have therefore

$$\sum_{m=0}^{\infty} ((m+1)^{-1} a_{j+m+1} - b_{jm}) \binom{y}{m} = 0 \qquad (y \in \mathbb{Z}_p, y \neq -1)$$

Lemma 53.1 (ii) now says that $b_{jm} = (m+1)^{-1} a_{j+m+1}$ for all j, m. It follows that

$$\lim_{j+m\to\infty} (m+1)^{-1} a_{j+m+1} = 0$$

For $n \in \mathbb{N}$ we have $\sup_{j+m=n} |(m+1)^{-1} a_{j+m+1}| = |a_{n+1}| \max(|n+1|_p^{-1}, \ldots |1|_p^{-1}) = |a_{n+1}|\, |\gamma_{n+1}|_p^{-1}$. Hence $\lim_{n\to\infty} |a_n|\, |\gamma_n|_p^{-1} = 0$. Conversely, if $\lim_{n\to\infty} |a_n|\, |\gamma_n|_p^{-1} = 0$ then also $\lim_{j+m\to\infty} (m+1)^{-1} a_{j+m+1} = 0$ and a glance at the formula (*) tells us that $\Phi_1 f$ is the restriction of a continuous function on $\mathbb{Z}_p \times \mathbb{Z}_p$, i.e. $f \in C^1(\mathbb{Z}_p \to K)$.
(ii) In Theorem 53.4 (iii) we already proved that $\|f\|_1 = \sup_{n\geq 0} |a_n|\, |\gamma_n|_p^{-1}$. Since $\lim_{n\to\infty} |a_n|\, |\gamma_n|_p^{-1} = 0$ we may replace 'sup' by 'max'.
(iii) By (ii), $\|\gamma_n\binom{*}{n}\|_1 = 1$ for each $n \in \{0, 1, 2, \ldots\}$. Again from (ii), now applied for a finite linear combination of $\gamma_0\binom{*}{0}, \gamma_1\binom{*}{1}, \ldots$, the orthonormality of $\gamma_0\binom{*}{0}, \gamma_1\binom{*}{1}, \ldots$ follows. Finally, let $f = \sum_{j=0}^{\infty} a_j\binom{*}{j} \in C^1(\mathbb{Z}_p \to K)$. Set

$$f_n := \sum_{j=0}^{n} a_j\binom{*}{j} \qquad (n \in \mathbb{N})$$

Then $\|f-f_n\|_1 = \max_{m>n} |a_m| \, |\gamma_m|_p^{-1}$. It follows that $\lim_{n\to\infty} \|f-f_n\|_1 = 0$ so that

$$f = \sum_{n=0}^{\infty} a_n \binom{*}{n} = \sum_{n=0}^{\infty} a_n \gamma_n^{-1} \gamma_n \binom{*}{n}$$

where the convergence is with respect to the norm $\| \ \|_1$. According to Definition 50.5, $\gamma_0 \binom{*}{0}, \gamma_1 \binom{*}{1}, \ldots$ is an orthonormal base of $C^1(\mathbb{Z}_p \to K)$.

Exercise 53.B. By Theorem 53.4 the norm $\| \ \|_1^{\sim}$ on $C^1(\mathbb{Z}_p \to K)$ given by

$$\|f\|_1^{\sim} = |a_0| \vee \sup_{n \geq 1} |a_n| \, n$$

satisfies $\|f\|_1 \leq \|f\|_1^{\sim} \leq p \|f\|_1$ $(f \in C^1(\mathbb{Z}_p \to K))$. Let $f \in C^1(\mathbb{Z}_p \to \mathbb{Q}_p)$, $f \neq 0$. Show that $\|f\|_1$ is an integral power of p and that, in general, $\|f\|_1^{\sim}$ is not. Show that $\|f\|_1$ is the largest among the integral powers of p that are $\leq \|f\|_1^{\sim}$.

Exercise 53.C. (Best approximation in $C^1(\mathbb{Z}_p \to K)$) In Exercise 51.D it was stated that for $f = \sum_{n=0}^{\infty} a_n \binom{*}{n} \in C(\mathbb{Z}_p \to K)$ the function $f_m := \sum_{n=0}^{m} a_n \binom{*}{n}$ is a best approximation (with respect to $\| \ \|_\infty$) of f in the space of the polynomial functions of degree $\leq m$. Show that if $f \in C^1(\mathbb{Z}_p \to K)$ then f_m is a best approximation of f in the same space but with $\| \ \|_1$ instead of $\| \ \|_\infty$.

Exercise 53.D. (Derivatives versus Mahler base)
(i) Show that for $n \in \mathbb{N}$

$$\binom{x}{n}' = \sum_{j=0}^{n-1} \frac{(-1)^{n-j-1}}{n-j} \binom{x}{j}$$

(ii) Let $f \in C^1(\mathbb{Z}_p \to K)$ have the Mahler expansion $\sum_{n=0}^{\infty} a_n \binom{*}{n}$. Find the Mahler expansion of the derivative f' of f.

Theorem 53.5 has an important consequence.

THEOREM 53.6. *Let $f \in C^1(\mathbb{Z}_p \to K)$. Then its indefinite sum Sf is also in $C^1(\mathbb{Z}_p \to K)$ and*

$$\|f\|_1 \leq \|Sf\|_1 \leq p \, \|f\|_1$$

Proof. Let $f = \sum_{n=0}^{\infty} a_n \binom{*}{n}$. Then, by Example 52.6, $Sf = \sum_{n=1}^{\infty} a_{n-1} \binom{*}{n}$. Clearly, $\lim_{n\to\infty} |a_n| \, n = 0$ if and only if $\lim_{n\to\infty} |a_{n-1}| \, n = 0$. By Theorem 53.5, $Sf \in C^1(\mathbb{Z}_p \to K)$ and $\|f\|_1 = \max_{n \geq 0} |a_n| \, |\gamma_n|_p^{-1}$, $\|Sf\|_1 = \max_{n \geq 0} |a_n| \, |\gamma_{n+1}|_p^{-1}$. By Lemma 53.3 (ii), $\|f\|_1 \leq \|Sf\|_1 \leq p \, \|f\|_1$.

Exercise 53.E. Let $f \in C^1(\mathbb{Z}_p \to \mathbb{Q}_p)$, $f \neq 0$. Show that either $\|Sf\|_1 = \|f\|_1$ or $\|Sf\|_1 = p\,\|f\|_1$. Give examples showing that both cases do occur.

Exercise 53.F. If $f \in \mathrm{Lip}_1(\mathbb{Z}_p \to K)$ then $Sf \in \mathrm{Lip}_1(\mathbb{Z}_p \to K)$ and $\|f\|_1 \leqslant \|Sf\|_1 \leqslant p\,\|f\|_1$. Prove this.

Exercise 53.G. (Characterization of increasing functions by Mahler coefficients) Let $f = \sum_{n=0}^{\infty} a_n \binom{*}{n} \in C(\mathbb{Z}_p \to K)$. Show that f is increasing (Proposition 24.7) if and only if $a_1 \in K^+$ and $|a_n| < |\gamma_n|_p$ $(n \geqslant 2)$.

Exercise 53.H. Show that $\binom{*}{0}^2, \binom{*}{1}^2, \ldots$ (see Exercise 52.G) do not form on orthogonal base of $C^1(\mathbb{Z}_p \to K)$. (Hint. Suppose they do. Expand $f \in C^1(\mathbb{Z}_p \to K)$ and consider $f'(0)$.)

Exercise 53.I. Let $f \in C^1(\mathbb{Z}_p \to K)$. Show that $s \mapsto f_s$ $(s \in \mathbb{Z}_p)$ is a continuous mapping of $\mathbb{Z}_p$ into $C^1(\mathbb{Z}_p \to K)$ (here $f_s(x) := f(s+x)$ for all $s, x \in \mathbb{Z}_p$).

54. Mahler coefficients of C^n-functions

In the previous section we have seen how to characterize Lipschitz functions and C^1-functions in terms of Mahler coefficients (Theorems 53.4 and 53.5). In a similar way we shall characterize C^n-functions and analytic functions. Recall that $f : \mathbb{Z}_p \to K$ is a C^n-function if its nth order difference quotient $\Phi_n f$ can be extended to a continuous function $\bar{\Phi}_n f$ on $\mathbb{Z}_p^{n+1}$.

THEOREM 54.1. (Characterization of C^n-functions by Mahler coefficients) *Let* $n \in \mathbb{N}$, $f = \sum_{m=0}^{\infty} a_m \binom{*}{m} \in C(\mathbb{Z}_p \to K)$. *Then* $f \in C^n(\mathbb{Z}_p \to K)$ *if and only if* $\lim_{m \to \infty} |a_m|\, m^n = 0$.

Proof. From formula (*) in the proof of Theorem 53.4 we obtain

$$\Phi_1 f(x+y, y) = \sum_{j=0}^{\infty} \sum_{k=1}^{\infty} \frac{a_{j+k}}{k} \binom{x-1}{k-1} \binom{y}{j} \quad (x, y \in \mathbb{Z}_p, x \neq 0)$$

where the terms of the double series tend to 0 in the sense of Exercise 23.A. An induction process shows that if $x_1, x_2, \ldots, x_n, y$ are elements of $\mathbb{Z}_p$ such that $v := (x_1 + \ldots + x_n + y, x_1 + \ldots + x_{n-1} + y, \ldots, x_1 + x_2 + y, x_1 + y, y) \in \nabla^{n+1} \mathbb{Z}_p$ then

$$(*) \quad \Phi_n f(v) = \sum_{j=0}^{\infty} \sum_{k_1, \ldots, k_n = 1}^{\infty} \frac{a_{j+k_1+\ldots+k_n}}{k_n(k_n + k_{n-1}) \ldots (k_n + \ldots + k_1)} \cdot \binom{x_1 - 1}{k_1 - 1} \binom{x_2 - 1}{k_2 - 1} \cdots \binom{x_n - 1}{k_n - 1} \binom{y}{j}$$

where the terms occuring in the series tend to 0 on $(\mathbb{N} \cup \{0\}) \times \mathbb{N}^n$ in the sense of Exercise 23.A. We shall show that $f \in C^n(\mathbb{Z}_p \to K)$ is equivalent to

$$(**) \qquad \lim_{j+k_1+\ldots+k_n \to \infty} \frac{a_{j+k_1+\ldots+k_n}}{k_n(k_n+k_{n-1})\ldots(k_n+\ldots+k_1)} = 0$$

and that in its turn (**) is equivalent to

$$\lim_{m \to \infty} |a_m| \, m^n = 0$$

If (**) holds then the right-hand side of formula (*) defines a continuous function on $\mathbb{Z}_p^{n+1}$ and we see that $\Phi_n f$ can be extended to a continuous function on $\mathbb{Z}_p^{n+1}$, i.e. $f \in C^n(\mathbb{Z}_p \to K)$. Conversely, if $f \in C^n(\mathbb{Z}_p \to K)$ then there exist $b_{jk_1\ldots k_n} \in K$ such that $\lim_{j+k_1+\ldots+k_n \to \infty} b_{jk_1\ldots k_n} = 0$ and

$$\bar{\Phi}_n f(x_1+\ldots+x_n+y, \ldots, x_1+y, y) = \sum_{j=0}^{\infty} \sum_{k_1,\ldots,k_n=1}^{\infty} b_{jk_1\ldots k_n} \cdot \binom{x_1-1}{k_1-1} \cdots \binom{x_n-1}{k_n-1}\binom{y}{j}$$

for all $x_1, x_2, \ldots, x_n, y \in \mathbb{Z}_p$. Subtraction from (*) yields

$$(***) \qquad 0 = \sum_{j=0}^{\infty} \sum_{k_1,\ldots,k_n=1}^{\infty} c_{jk_1,\ldots k_n} \binom{x_1-1}{k_1-1} \cdots \binom{x_n-1}{k_n-1}\binom{y}{j}$$

for all $x_1, x_2, \ldots, x_n, y \in \mathbb{Z}_p$ for which $(x_1+\ldots+x_n+y, \ldots, x_1+y, y) \in \nabla^{n+1}\mathbb{Z}_p$, where $c_{jk_1\ldots k_n}$ equals

$$\frac{a_{j+k_1+\ldots+k_n}}{k_n(k_n+k_{n-1})\ldots(k_n+\ldots+k_1)} - b_{jk_1\ldots k_n}$$

For all $y \in \{0, 1, 2, \ldots\}$ and $x_1, \ldots, x_n \in \{1, 2, 3, \ldots\}$ we have that $(x_1+\ldots+x_n+y, x_1+\ldots+x_{n-1}+y, \ldots, x_1+y, y) \in \nabla^{n+1}\mathbb{Z}_p$. By substituting successively $y = 0, 1, 2, \ldots$ in (***) we arrive at

$$0 = \sum_{k_1,\ldots,k_n=1}^{\infty} c_{jk_1\ldots k_n} \binom{x_1-1}{k_1-1} \cdots \binom{x_n-1}{k_n-1} \qquad (j = 0, 1, 2, \ldots)$$

Successive substitution of $x_n = 1, 2, \ldots$ in the latter formula yields

$$0 = \sum_{k_1,\ldots,k_{n-1}=1}^{\infty} c_{jk_1\ldots k_n} \binom{x_1-1}{k_1-1} \cdots \binom{x_{n-1}-1}{k_{n-1}-1}$$

which formula holds for $j \in \{0, 1, 2, \ldots\}$, $k_n \in \{1, 2, \ldots\}$. Continuing this way we find that all coefficients $c_{jk_1 \ldots k_n}$ are zero, and (**) follows.

Now consider the term $a_{j+k_1+\ldots+k_n}/k_n(k_n+k_{n-1})\ldots(k_n+\ldots+k_1)$ of (**). We have (with $m = j + k_1 + \ldots + k_n$)

$$\begin{aligned} |k_n|_p &\geqslant 1/k_n \geqslant 1/m \\ |k_n + k_{n-1}|_p &\geqslant 1/(k_n + k_{n-1}) \geqslant 1/m \\ &\vdots \\ |k_n + \ldots + k_1|_p &\geqslant 1/m \end{aligned}$$

so that

$$\left| \frac{a_{j+k_1+\ldots+k_n}}{k_n(k_n+k_{n-1})\ldots(k_n+\ldots+k_1)} \right| \leqslant |a_m|\, m^n$$

Thus, if $\lim_{m\to\infty} |a_m|\, m^n = 0$ then we have (**). Conversely, suppose (**). Let $m \in \mathbb{N}$, $m \geqslant p^n$. Then m has the p-adic expansion $m = a_0 + a_1 p + \ldots + a_t p^t$ where $a_t \neq 0$ and $t \geqslant n$. Choose $k_1, \ldots, k_n, j$ such that

$$\begin{aligned} k_1 + \ldots + k_n + j &= m \\ k_1 + \ldots + k_n &= p^t \\ k_2 + \ldots + k_n &= p^{t-1} \\ &\vdots \\ k_n &= p^{t-n+1} \end{aligned}$$

Then

$$\left| \frac{a_m}{k_n(k_n+k_{n-1})\ldots(k_n+\ldots+k_1)} \right| = |a_m|\, p^{nt-n(n-1)/2} \geqslant |a_m|\, m^n\, p^{-n(n+1)/2}$$

From this inequality it follows easily that (**) implies $\lim_{m\to\infty} |a_m|\, m^n = 0$.

**Exercise* 54.A. Let $n \in \mathbb{N}$. Show the existence of positive constants c_1 and c_2 such that for all $f \in C^n(\mathbb{Z}_p \to K)$

$$c_1 \sup_{m \geqslant n} |a_m|\, m^n \leqslant \|\Phi_n f\|_\infty \leqslant c_2 \sup_{m \geqslant n} |a_m|\, m^n$$

Deduce that $\|\ \|_n$ is equivalent to $\|\ \|_n^{\sim}$ on $C^n(\mathbb{Z}_p \to K)$ where

$$\|f\|_n := \max_{0 \leqslant j \leqslant n} \|\Phi_j f\|_\infty$$

$$\|f\|_n^{\sim} := \sup_{m \geqslant 0} |a_m|\, m^n$$

COROLLARY 54.2. (Characterization of C^∞-functions by Mahler coefficients) *Let* $f = \sum_{m=0}^{\infty} a_m \binom{x}{m} \in C(\mathbb{Z}_p \to K)$. *Then* f *is a* C^∞*-function if and only if* $\lim_{m\to\infty} |a_m| m^n = 0$ *for each* $n \in \mathbb{N}$.

COROLLARY 54.3. *Let* $f \in C(\mathbb{Z}_p \to K)$. *Then for each* $n \in \mathbb{N} \cup \{\infty\}$ *its indefinite sum* Sf *is a* C^n*-function if and only if* f *is a* C^n*-function.*
Proof. Theorem 54.1 and Example 52.6.

THEOREM 54.4. (Characterization of analytic functions by Mahler coefficients) *Let* $f = \sum_{m=0}^{\infty} a_m \binom{x}{m} \in C(\mathbb{Z}_p \to K)$. *Then* f *is analytic if and only if* $\lim_{m\to\infty} a_m/m! = 0$.
Proof. Suppose f is analytic. There are $b_0, b_1, \ldots \in K$ with $\lim_{n\to\infty} b_n = 0$ such that $f(x) = \sum_{n=0}^{\infty} b_n x^n$ for all $x \in \mathbb{Z}_p$. According to Exercise 52.C we have $x^n = \sum_{m=0}^{\infty} a_{mn}\binom{x}{m}$ where a_{mn} is an integer divisible by $m!$ and $a_{mn} = 0$ for $m > n$. We have $\lim_{n+m\to\infty} b_n a_{mn} = 0$ so that for all $x \in \mathbb{Z}_p$

$$f(x) = \sum_{n=0}^{\infty} \sum_{m=0}^{\infty} b_n a_{mn}\binom{x}{m} = \sum_{m=0}^{\infty} \left(\sum_{n=m}^{\infty} b_n a_{mn} \right) \binom{x}{m}$$

Hence $|a_m/m!| \leqslant \sup_{n\geqslant m} |\sum_{n=m}^{\infty} b_n a_{mn}/m!| \leqslant \sup_{n\geqslant m} |b_n|$. It follows that $\lim_{m\to\infty} a_m/m! = 0$. Conversely, suppose that this limit is 0. Let $x \in \mathbb{Z}_p$. For each m we have

$$\binom{x}{m} = \frac{x(x-1)\ldots(x-m+1)}{m!} = \frac{1}{m!} \sum_{n=0}^{\infty} t_{nm} x^n$$

where the coefficients t_{nm} are integers and $t_{nm} = 0$ for $n > m$. Hence $\lim_{m+n\to\infty} t_{nm} a_m/m! = 0$ and

$$f(x) = \sum_{m=0}^{\infty} a_m \binom{x}{m} = \sum_{m=0}^{\infty} \sum_{n=0}^{\infty} \frac{a_m}{m!} t_{nm} x^n =$$

$$\sum_{n=0}^{\infty} \left(\sum_{m=n}^{\infty} \frac{a_m}{m!} t_{nm} \right) x^n$$

We see that f is analytic.

Remark. For a characterization of locally analytic functions in terms of Mahler coefficients see Y. Amice: Interpolation p-adique. *Bull. Soc. Math. France* 92 (1964), 117-80.

Exercise 54.B. Show that the condition '$\lim_{m\to\infty} a_m/m! = 0$' for analyticity of $f = \sum_{m=0}^{\infty} a_m \binom{x}{m}$ is indeed stronger than the condition

'$\lim_{m\to\infty} |a_m|\, m^n = 0$ for each n' that characterizes C^∞-functions.

Exercise 54.C. Prove the following somewhat surprising fact. The statement

if $f : \mathbb{Z}_p \to K$ is analytic then so is Sf

is *false*. (Compare Corollary 54.3 and Exercise 55.E.)

Exercise 54.D. (On Mahler series $\Sigma\, a_n \binom{x}{n}$ for $x \notin \mathbb{Z}_p$) Let $a_0, a_1, \ldots \in \mathbb{C}_p$. Show that the following conditions (α)-(γ) are equivalent.
(α) $\Sigma\, a_n \binom{x}{n}$ converges for all $x \in \mathbb{Q}_p$.
(β) $x \mapsto \Sigma_{n=0}^{\infty}\, a_n \binom{x}{n}$ is analytic on $\mathbb{C}_p$.
(γ) For all $r > 0$, $\lim_{n\to\infty} |a_n|\, r^n = 0$.
Show also that the conditions (α)'-(γ)' are equivalent (compare Theorem 54.4).
(γ)' $\Sigma\, a_n \binom{x}{n}$ converges for all $x \in \mathbb{C}_p$, $|x|_p \leqslant 1$.
(β)' $x \mapsto \Sigma_{n=0}^{\infty}\, a_n \binom{x}{n}$ is analytic on $\{x \in \mathbb{C}_p : |x|_p \leqslant 1\}$.
(γ)' $\lim_{n\to\infty} a_n/n! = 0$.
Reconsider the proof you gave for the equivalence of (α)', (β)', (γ)' and show the following. Let $\mathbb{Q}_p \subset K \subset \mathbb{C}_p$ and let the residue class field of K be a proper extension of the residue class field of $\mathbb{Q}_p$. Then if $\Sigma\, a_n \binom{x}{n}$ converges for all x in the 'closed' unit disc of K then $x \mapsto \Sigma\, a_n \binom{x}{n}$ is analytic on that disc.

55. The Volkenborn integral

In Section 30 we discussed several ways to set up an integration theory for functions on $\mathbb{Z}_p$. Here is one.

Let $f \in C^1(\mathbb{Z}_p \to K)$. Recall that the indefinite sum Sf of f is the unique continuous function g satisfying $g(x+1) - g(x) = f(x)$ for all $x \in \mathbb{Z}_p$, $g(0) = 0$. In Theorem 53.6 we have seen that Sf is also a C^1-function and that $\|Sf\|_1 \leqslant p\, \|f\|_1$. We have

$$\frac{f(0) + f(1) + \ldots + f(p^n - 1)}{p^n} = \frac{Sf(p^n) - Sf(0)}{p^n}$$

and the limit for $n \to \infty$ of the right-hand side exists and equals $(Sf)'(0)$.

DEFINITION 55.1. The *Volkenborn integral* of $f \in C^1(\mathbb{Z}_p \to K)$ is

$$\int_{\mathbb{Z}_p} f(x)dx := \lim_{n\to\infty} p^{-n} \sum_{j=0}^{p^n - 1} f(j) = (Sf)'(0)$$

PROPOSITION 55.2. $\int_{\mathbb{Z}_p}$ *is a K-linear continuous function on* $C^1(\mathbb{Z}_p \to K)$ *In fact*

$$\left| \int_{\mathbb{Z}_p} f(x)dx \right| \leqslant p \, \|f\|_1 \qquad (f \in C^1(\mathbb{Z}_p \to K))$$

In particular, if $f, f_1, f_2, \ldots \in C^1(\mathbb{Z}_p \to K)$ *and* $\lim_{n\to\infty} f_n = f$ *in the sense of the norm* $\| \;\; \|_1$ *then*

$$\lim_{n\to\infty} \int_{\mathbb{Z}_p} f_n(x)dx = \int_{\mathbb{Z}_p} f(x)dx$$

Proof. Theorem 53.6.

PROPOSITION 55.3. (The Volkenborn integral in terms of the Mahler coefficients) *Let* $f = \sum_{n=0}^{\infty} a_n \binom{*}{n} \in C^1(\mathbb{Z}_p \to K)$. *Then*

$$\int_{\mathbb{Z}_p} f(x)dx = \sum_{n=0}^{\infty} a_n \frac{(-1)^n}{n+1}$$

Proof. From Example 52.6 it follows that $S\binom{*}{n} = \binom{*}{n+1}$ for all n so that $\int_{\mathbb{Z}_p} \binom{x}{n} dx = \lim_{x\to 0} x^{-1}\binom{x}{n+1} = \lim_{x\to 0}(n+1)^{-1}\binom{x-1}{n} = (-1)^n/(n+1)$. Now $\Sigma \, a_n\binom{*}{n}$ converges in the sense of C^1. Hence

$$\int_{\mathbb{Z}_p} \left(\sum_{n=0}^{\infty} a_n \binom{x}{n} \right) dx = \sum_{n=0}^{\infty} a_n \int_{\mathbb{Z}_p} \binom{x}{n} dx = \sum_{n=0}^{\infty} a_n \frac{(-1)^n}{n+1}$$

PROPOSITION 55.4. (The Volkenborn integral of an analytic function) *Let* $f : \mathbb{Z}_p \to K$ *be an analytic function,* $f(x) = \sum_{n=0}^{\infty} a_n x^n$ $(x \in \mathbb{Z}_p)$. *Then*

$$\int_{\mathbb{Z}_p} \left(\sum_{n=0}^{\infty} a_n x^n \right) dx = \sum_{n=0}^{\infty} a_n \int_{\mathbb{Z}_p} x^n dx$$

Proof. Lemma 40.3 and Proposition 55.2.

PROPOSITION 55.5. (Shift versus integration)
(i) *For each* $f \in C^1(\mathbb{Z}_p \to K)$ *the function* $(Sf)' - Sf'$ *is constant. Its value is* $\int_{\mathbb{Z}_p} f(x)dx$.
(ii) *Let* $f \in C^1(\mathbb{Z}_p \to K)$ *and let* $s \in \mathbb{Z}_p$. *Then*

$$\int_{\mathbb{Z}_p} f(x+s)dx = (Sf)'(s)$$

$$\int_{\mathbb{Z}_p} f(x+s)dx - \int_{\mathbb{Z}_p} f(x)dx = (Sf')(s)$$

$$\int_{\mathbb{Z}_p} f(x+s+1)dx - \int_{\mathbb{Z}_p} f(x+s)dx = f'(s)$$

In particular,

$$\int_{\mathbb{Z}_p} f(x+1)dx = \int_{\mathbb{Z}_p} f(x)dx + f'(0)$$

(iii) *Let* $f \in C(\mathbb{Z}_p \to K)$ *and let* Pf *be any* C^1*-antiderivative of* f. *Then for* $s \in \mathbb{Z}_p$

$$(Sf)(s) = \int_{\mathbb{Z}_p} Pf(x+s)dx - \int_{\mathbb{Z}_p} Pf(x)dx$$

$$f(s) = \int_{\mathbb{Z}_p} Pf(x+s+1)dx - \int_{\mathbb{Z}_p} Pf(x+s)dx$$

Proof. Let $D : C^1(\mathbb{Z}_p \to K) \to C(\mathbb{Z}_p \to K)$ be the differentiation operator and, as in Section 52, let $\Delta : C(\mathbb{Z}_p \to K) \to C(\mathbb{Z}_p \to K)$ be given by $\Delta f(x) = f(x+1) - f(x)$ ($f \in C(\mathbb{Z}_p \to K)$, $x \in \mathbb{Z}_p$). Then $\Delta D = D\Delta$ on $C^1(\mathbb{Z}_p \to K)$, ΔS is the identity, $(S\Delta f)(s) = f(s) - f(0)$ ($f \in C(\mathbb{Z}_p \to K)$, $s \in \mathbb{Z}_p$). Using the formula $SD = SD\Delta S = S\Delta DS$ we arrive at $(SDf)(s) = (DSf)(s) - (DSf)(0)$ ($f \in C^1(\mathbb{Z}_p \to K)$, $s \in \mathbb{Z}_p$) implying $\int_{\mathbb{Z}_p} f(x)dx = (DSf)(0) = (Sf)'(s) - (Sf')(s)$ which proves (i). Let $L_s f(x) := f(x+s)$ ($f \in C^1(\mathbb{Z}_p \to K)$; $x, s \in \mathbb{Z}_p$). Then $\int_{\mathbb{Z}_p} f(x+s)dx = DSL_s f(0) = L_s DSf(0) = (Sf)(s)$. The remaining formulas of (ii) and (iii) are now easy to prove.

COROLLARY 55.6. (Invariant integral on $\{f \in C^1(\mathbb{Z}_p \to K) : f' = 0\}$) *Let* $N^1(\mathbb{Z}_p \to K) := \{f \in C^1(\mathbb{Z}_p \to K) : f' = 0\}$. *Then* $\int_{\mathbb{Z}_p}$ *is shift invariant on* $N^1(\mathbb{Z}_p \to K)$, i.e.

$$\int_{\mathbb{Z}_p} f(x+s)dx = \int_{\mathbb{Z}_p} f(x)dx \qquad (f \in N^1(\mathbb{Z}_p \to K), s \in \mathbb{Z}_p)$$

The function m *defined on* $\Omega := \{U \subset \mathbb{Z}_p : U \text{ open compact}\}$ *by the formula* $m(U) := \int_{\mathbb{Z}_p} \xi_U(x)dx$ *satisfies*

(i) $m(B_a(p^{-n})) = p^{-n}$ $\qquad (a \in \mathbb{Z}_p, n \in \{0, 1, 2, \ldots\})$
(ii) $m(U \cup V) = m(U) + m(V)$ $\qquad (U, V \in \Omega, U \cap V = \emptyset)$
(iii) $m(s + U) = m(U)$ $\qquad (s \in \mathbb{Z}_p, U \in \Omega)$
(iv) $m(U) \in \mathbb{Q}$ $\qquad (U \in \Omega)$

PROPOSITION 55.7. *Let* $f \in C^1(\mathbb{Z}_p \to K)$. *Then*

$$\int_{\mathbb{Z}_p} f(-x)dx = \int_{\mathbb{Z}_p} f(x+1)dx$$

If, in addition, f *is an odd function then* $\int_{\mathbb{Z}_p} f(x)dx = -\frac{1}{2}f'(0)$.

Proof. $\int_{\mathbb{Z}_p} f(-x)dx = \lim_{n \to \infty} p^{-n} \sum_{j=0}^{p^n-1} f(-j) = \lim_{n \to \infty} p^{-n} \sum_{j=1-p^n}^{0} f(j) = \lim_{n \to \infty} p^{-n} (Sf(1) - Sf(1-p^n)) = (Sf)'(1) = \int_{\mathbb{Z}_p} f(x+1)dx$. If $f(-x) = -f(x)$ for all x we have $-\int_{\mathbb{Z}_p} f(x)dx = \int_{\mathbb{Z}_p} f(-x)dx = \int_{\mathbb{Z}_p} f(x+1)dx = \int_{\mathbb{Z}_p} f(x)dx + f'(0)$. It follows that $-2\int_{\mathbb{Z}_p} f(x)dx = f'(0)$.

**Exercise* 55.A. Show that

$$\int_{\mathbb{Z}_p} a^x dx = \frac{\log_p a}{a-1} \qquad (a \in \mathbb{C}_p^+, a \neq 1)$$

$$\int_{\mathbb{Z}_p} \exp ax\, dx = \frac{a}{\exp a - 1} \qquad (a \in E, a \neq 0)$$

$$\int_{\mathbb{Z}_p} \cos a x\, dx = \frac{a \sin a}{2(1-\cos a)} \qquad (a \in E, a \neq 0)$$

$$\int_{\mathbb{Z}_p} \sin a x\, dx = -\frac{a}{2} \qquad (a \in E)$$

$$\int_{\mathbb{Z}_p} x dx = -\tfrac{1}{2}, \qquad \int_{\mathbb{Z}_p} x^2 dx = \tfrac{1}{6}, \qquad \int_{\mathbb{Z}_p} x^3 dx = 0$$

**Exercise* 55.B. Let $f \in C^1(\mathbb{Z}_p \to K)$, $m \in \mathbb{N}$. Prove that

$$\int_{\mathbb{Z}_p} f(x)dx = \frac{1}{m} \sum_{j=0}^{m-1} \int_{\mathbb{Z}_p} f(j+mx)dx$$

Exercise 55.C. Show that

$$\int_{\mathbb{Z}_p} Sf(x)dx = -\int_{\mathbb{Z}_p} (x+1) f(x)dx \qquad (f \in C^1(\mathbb{Z}_p \to K))$$

Exercise 55.D. Let $f \in N^1(\mathbb{Z}_p \to K)$ and $\epsilon > 0$. Show that there exists an $N \in \mathbb{N}$ such that for all $n \geq N$ and all choices $\xi_1, \xi_2, \ldots, \xi_{p^n}$ of representatives in $\mathbb{Z}_p$ modulo $p^n\mathbb{Z}_p$ we have

$$\left| \int_{\mathbb{Z}_p} f(x)dx - p^{-n} \sum_{j=1}^{p^n} f(\xi_j) \right| < \epsilon$$

Is this statement also true for $f \in C^1(\mathbb{Z}_p \to K)$?

Exercise 55.E. Use Proposition 55.5 to show the following. If $f : \mathbb{Z}_p \to K$ is analytic and if it possesses an analytic antiderivative $\mathbb{Z}_p \to K$ then Sf is analytic. (Compare Exercise 54.C.) If, in the above, we replace 'analytic' by 'locally analytic of order 1' then do we obtain a correct statement?

Exercise 55.F. Some authors call a (continuous) function $\mathbb{Z}_p \to K$ 'integrable' if $\lim_{n\to\infty} p^{-n}(f(0) + f(1) + \ldots + f(p^n - 1))$ exists. In this exercise we borrow this notion.

(i) Prove the existence of a continuous function $\mathbb{Z}_p \to K$ that is not integrable. (Hint. Let $f \in C(\mathbb{Z}_p \to K)$ such that $\lim_{n\to\infty} p^{-n}(f(p^n) - f(0))$ does not exist. Consider Δf.)

(ii) Find a nowhere differentiable continuous function $\mathbb{Z}_5 \to \mathbb{Z}_5$ which is integrable. (Hint. Consider $\sum a_n 5^n \mapsto \sum \sigma(a_n) 5^n$ where σ is a suitable permutation of $\{0, 1, 2, 3, 4\}$.)

Exercise 55.G. Let $f \in C^2(\mathbb{Z}_p \to K)$. Show that f' and the function $s \mapsto \int_{\mathbb{Z}_p} f(x + s)\, dx$ are in C^1 and that

$$\frac{d}{ds} \int_{\mathbb{Z}_p} f(x + s) dx = \int_{\mathbb{Z}_p} f'(x + s) dx$$

56. The Bernoulli numbers

In this section we study the *Bernoulli numbers* $B_0, B_1, \ldots$ given by the hypocritical definition

$$B_n := \int_{\mathbb{Z}_p} x^n dx \qquad (n = 0, 1, 2, \ldots)$$

The formula $\int_{\mathbb{Z}_p} f(x + 1)dx - \int_{\mathbb{Z}_p} f(x)dx = f'(0)$ yields

$$(*) \qquad \int_{\mathbb{Z}_p} (x + 1)^n dx - \int_{\mathbb{Z}_p} x^n dx = \begin{cases} 1 \text{ if } n = 1 \\ 0 \text{ if } n \in \{0, 2, 3, \ldots\} \end{cases}$$

On the other hand, from $\int_{\mathbb{Z}_p} (x+1)^n dx = \sum_{j=0}^{n} \binom{n}{j} \int_{\mathbb{Z}_p} x^j dx = \sum_{j=0}^{n} \binom{n}{j} B_j$ we get

$$(*)' \qquad \int_{\mathbb{Z}_p} (x+1)^n dx - \int_{\mathbb{Z}_p} x^n dx = \begin{cases} 0 \text{ if } n = 0 \\ \sum_{j=0}^{n-1} \binom{n}{j} B_j \text{ if } n \in \{1, 2, \ldots\} \end{cases}$$

Combining (*) and (*)' we obtain

$$(**) \qquad B_0 = 1 \text{ and } \sum_{j=0}^{n-1} \binom{n}{j} B_j = 0 \text{ for } n \geqslant 2$$

The relations (**) determine the numbers $B_0, B_1, \ldots$ So we may conclude that *the numbers $B_0, B_1, \ldots$ are rational and 'do not depend on p'* in the sense that if p and q are prime numbers then $\int_{\mathbb{Z}_p} x^n dx$ and $\int_{\mathbb{Z}_q} x^n dx$, considered as elements of $\mathbb{Q}$, are equal.

By Proposition 55.7 and (*)

$$\int_{\mathbb{Z}_p} (-x)^n dx = \int_{\mathbb{Z}_p} (x + 1)^n dx = \begin{cases} B_1 + 1 \text{ if } n = 1 \\ B_n \text{ if } n \in \{0, 2, 3, \ldots\} \end{cases}$$

so that $B_1 = -\frac{1}{2}$ and $(-1)^n B_n = B_n$ for $n \geqslant 2$. Therefore

$$B_3 = B_5 = B_7 = \ldots = 0$$

The Bernoulli numbers occur as power series coefficients of certain analytic functions. For example, from Exercise 55.A we have

$$\int_{\mathbb{Z}_p} \exp(ax)dx = \frac{a}{\exp a - 1} \qquad (a \in E, a \neq 0)$$

On the other hand, by Proposition 55.4

$$\int_{\mathbb{Z}_p} \exp(ax)dx = \int_{\mathbb{Z}_p} \sum_{n=0}^{\infty} \frac{a^n x^n}{n!}\, dx = \sum_{n=0}^{\infty} B_n \frac{a^n}{n!}$$

so that

$$\frac{a}{\exp a - 1} = \sum_{n=0}^{\infty} B_n \frac{a^n}{n!} \qquad (a \in E, a \neq 0)$$

which formula, interpreted for $a \in \mathbb{C}$, is one of the classical definitions of the Bernoulli numbers. Similarly, from

$$\int_{\mathbb{Z}_p} \cos ax\, dx = \frac{a \sin a}{2 - 2\cos a} = \frac{a}{2} \cotan \frac{a}{2} \qquad (p \neq 2)$$

we obtain

$$\frac{a}{2} \cotan \frac{a}{2} = \sum_{n=0}^{\infty} (-1)^n \frac{a^{2n} B_{2n}}{(2n)!} \qquad (a \in E, a \neq 0, p \neq 2)$$

The properties of the Bernoulli numbers treated in this section are well known. But the point is that our proofs use techniques from p-adic analysis.

THEOREM 56.1. *The denominator of B_n is free of squares.*
Proof. For each prime p we have by Proposition 55.2

$$\left| \int_{\mathbb{Z}_p} f(x)dx \right|_p \leqslant p \, \|f\|_1 \qquad (f \in C^1(\mathbb{Z}_p \to \mathbb{Q}_p))$$

so that

$$|B_n|_p = \left| \int_{\mathbb{Z}_p} x^n dx \right| \leqslant p$$

If $B_n = tm^{-1}$ where $t, m \in \mathbb{Z}$ are relatively prime then the number of factors p in m is either 0 or 1. Hence m is a product of distinct primes.

The following theorem is a little more precise. It says *what* primes occur in the denominator of B_n.

THEOREM 56.2. (von Staudt) *If n is even then*

$$B_n + \sum_{p-1 \mid n} \frac{1}{p} \in \mathbb{Z}$$

For the proof we need the following lemma which was kindly pointed out to me by L. van Hamme.

LEMMA 56.3. *The formula*

$$|B_n - \frac{1}{p}(0^n + 1^n + \ldots + (p-1)^n)|_p \leqslant 1$$

is true in each of the following cases.
(i) $p \neq 2, n \in \mathbb{N}$.
(ii) $p = 2, n \in \mathbb{N}$, *n even*.
Proof. For $k \in \{1, 2, \ldots\}, n \in \{0, 1, 2, \ldots\}$ set

$$R_n(k) := p^{-k}(0^n + 1^n + \ldots + (p^k - 1)^n)$$

We have $R_0(k) = 1$ for all k, $\lim_{k \to \infty} R_n(k) = B_n$ p-adically. It is to be shown that in the cases (i) and (ii) $|B_n - R_n(1)|_p \leqslant 1$. For $k \in \mathbb{N}$ and $n \in \{0, 1, 2, \ldots\}$ we have $R_n(k+1) = p^{-k-1} \sum_{s=0}^{p^{k+1}-1} s^n = p^{-k-1} \sum_{i=0}^{p^k-1} \sum_{j=0}^{p-1} (i + jp^k)^n = p^{-k-1} \sum_{i=0}^{p^k-1} \sum_{j=0}^{p-1} \sum_{s=0}^{n} \binom{n}{s} i^{n-s} (jp^k)^s = \sum_{s=0}^{n} \binom{n}{s} R_{n-s}(k) R_s(1) p^{ks}$. We see that for $n \geqslant 1$

$$R_n(k+1) - R_n(k) = \sum_{s=1}^{n} \binom{n}{s} R_{n-s}(k) R_s(1) p^{ks}$$

Now if $s > 1$ then

$$\binom{n}{s} R_{n-s}(k) R_s(1) p^{ks} = \binom{n}{s} p^k R_{n-s}(k) pR_s(1) p^{ks-k-1} \in \mathbb{Z}$$

For $s = 1$ and odd p we find

$$\binom{n}{1} R_{n-1}(k) R_1(1) p^k = \binom{n}{1} R_{n-1}(k) \tfrac{1}{2}(p-1) p^k \in \mathbb{Z}$$

For $s = 1, p = 2, n$ even

$$\binom{n}{1} R_{n-1}(k) R_1(1) 2^k = \tfrac{1}{2} n R_{n-1}(k) 2^k \in \mathbb{Z}$$

It follows that in either case (i) or (ii) above we have

$$|R_n(k+1) - R_n(k)|_p \leqslant 1 \qquad (k \in \mathbb{N})$$

We see that for $k > m$ ($k, m \in \mathbb{N}$)

$$|R_n(k) - R_n(m)|_p \leqslant 1$$

By taking k large and $m = 1$ we arrive at

$$|B_n - R_n(1)|_p \leqslant 1$$

which was to be shown.
Proof of Theorem 56.2. Let $\theta, \theta^2, \ldots, \theta^{p-1}$ be the $(p-1)$th roots of unity in $\mathbb{Q}_p$ (see Exercise 27.G). Both $\{0, 1, \ldots, p-1\}$ and $\{0, \theta, \theta^2, \ldots, \theta^{p-1}\}$ are full sets of representatives in $\mathbb{Z}_p$ modulo $p\mathbb{Z}_p$. Thus, for $n \in \mathbb{N}$, $0^n + 1^n$

$+\ldots+(p-1)^n \equiv \theta^n+\theta^{2n}+\ldots+\theta^{(p-1)n}$ (modulo $p\mathbb{Z}_p$).

If n is not a multiple of $p-1$ then $\theta^n \neq 1$ so that $\Sigma_{j=0}^{p-1}\theta^{nj}=0$. If n is a multiple of $p-1$ then $\theta^n=1$ so that $\Sigma_{j=0}^{p-1}\theta^{nj}=p-1$. It follows that

$$\frac{1}{p}(0^n+1^n+\ldots(p-1)^n)\in\mathbb{Z}_p \text{ if } p-1\nmid n$$
$$\frac{1}{p}(0^n+1^n+\ldots(p-1)^n)+\frac{1}{p}\in\mathbb{Z}_p \text{ if } p-1\mid n$$

After applying Lemma 56.3 we obtain

$$\left|B_n+\frac{1}{p}\right|_p \leqslant 1 \text{ if } n \text{ even}, p-1\mid n$$
$$|B_n|_p \leqslant 1 \text{ if } n \text{ even}, p-1\nmid n$$

For each prime number q we therefore have

$$\left|B_n+\sum_{p-1\mid n}\frac{1}{p}\right|_q \leqslant 1$$

from which it follows easily that $B_n+\sum_{p-1\mid n}\frac{1}{p}\in\mathbb{Z}$.

**Exercise* 56.A. The *Bernoulli polynomials* are defined by the formula

$$B_n(x) = \int_{\mathbb{Z}_p}(x+t)^n dt \qquad (n=0,1,2,\ldots; x\in\mathbb{Z}_p)$$

(i) Show by integrating $\exp a(x+t)$ in two different ways that

$$\frac{a\exp(ax)}{\exp a-1} = \sum_{n=0}^{\infty} B_n(x)\frac{a^n}{n!} \qquad (x\in\mathbb{Z}_p, a\in E, a\neq 0)$$

(ii) Let $x\in\mathbb{Z}_p$. Show that $B_n(x) = \Sigma_{j=0}^{n}\binom{n}{j}x^{n-j}B_j$ and

$$B_n(x+1)-B_n(x) = \begin{cases} 0 \text{ if } n=0 \\ nx^{n-1} \text{ if } n\in\mathbb{N}\end{cases}$$

(iii) Let $n\in\{0,1,2,\ldots\}$. Show that the indefinite sum of $x\mapsto x^n$ ($x\in\mathbb{Z}_p$) is the function

$$x \mapsto \frac{1}{n+1}(B_{n+1}(x)-B_{n+1}(0)) \qquad (x\in\mathbb{Z}_p)$$

Observe that $B_{n+1}(0) = B_{n+1}$.

57. Integration over subsets

Let U be a compact open subset of $\mathbb{Z}_p$. For $f\in C^1(\mathbb{Z}_p\to K)$ we set

$$\int_U f(x)dx := \int_{\mathbb{Z}_p} f(x)\,\xi_U(x)dx$$

For $f \in C^1(U \to K)$ we define

$$\int_U f(x)dx := \int_{\mathbb{Z}_p} g(x)dx$$

where $g(x) := f(x)$ if $x \in U$ and $g(x) := 0$ if $x \in \mathbb{Z}_p \setminus U$.

In the following proposition we collect some formulas needed later on. Recall that $T_p = \mathbb{Z}_p \setminus p\mathbb{Z}_p$, and that $x \mapsto B_m(x)$ is the Bernoulli polynomial defined in Exercise 56.A.

PROPOSITION 57.1. *Let* $f \in C^1(\mathbb{Z}_p \to K)$, $n \in \{0, 1, 2, \ldots\}$, $j \in \{0, 1, \ldots, p^n - 1\}$. *Then*

$$\int_{j + p^n \mathbb{Z}_p} f(x)dx = \int_{p^n \mathbb{Z}_p} f(j + x)dx = p^{-n} \int_{\mathbb{Z}_p} f(j + p^n x)dx$$

$$\int_{T_p} f(x)dx = p^{-1} \int_{\mathbb{Z}_p} (pf(x) - f(px))dx$$

For $m \in \{0, 1, 2, \ldots\}$

$$\int_{j + p^n \mathbb{Z}_p} x^m dx = p^{n(m-1)} B_m\left(\frac{j}{p^n}\right)$$

$$\int_{T_p} x^m dx = (1 - p^{m-1}) B_m$$

Proof. We have $\int_{j + p^n \mathbb{Z}_p} f(x)dx = \int_{p^n \mathbb{Z}_p} f(j + x)dx = \lim_{s \to \infty} p^{-s}(f(j) + f(j + p^n) + \ldots + f(j + (p^{s-n} - 1)p^n))$. With $h(x) := f(j + p^n x)$ $(x \in \mathbb{Z}_p)$ we can write this limit as $\lim_{s \to \infty} p^{-s}(h(0) + h(1) + \ldots + h(p^{s-n} - 1)) = p^{-n} \int_{\mathbb{Z}_p} h(x)dx$. The first formula follows. Next, $\int_{T_p} f(x)dx = \int_{\mathbb{Z}_p} f(x)dx - \int_{p\mathbb{Z}_p} f(x)dx = \int_{\mathbb{Z}_p} f(x)dx - p^{-1} \int_{\mathbb{Z}_p} f(px)dx = p^{-1} \int_{\mathbb{Z}_p} (pf(x) - f(px))dx$. The last two formulas are direct consequences of our definition of the Bernoulli polynomials (Exercise 56.A) and the first two formulas.

**Exercise* 57.A. Let $f : T_p \to \mathbb{Q}_p$ be a C^1-function and let $f(-x) = -f(x)$ $(x \in T_p)$. Show that $\int_{T_p} f(x)dx = 0$. Deduce that

$$\int_{T_p} x^{-1} dx = \int_{T_p} x^{-3} dx = \int_{T_p} x^{-5} dx = \ldots = 0.$$

Exercise 57.B. Let Ω denote the collection of all compact open subsets of $\mathbb{Z}_p$. Let $k \in \{0, 1, 2, \ldots\}$. Define

$$\nu_k(U) := \int_U x^k dx \qquad (U \in \Omega)$$

(i) Show that ν_k is additive (i.e. if $U, V \in \Omega$ are disjoint then $\nu_k(U \cup V)$

$= \nu_k(U) + \nu_k(V)$) but that ν_k is not bounded.
(ii) Let $a \in \mathbb{Z}_p$, $|a|_p = 1$. Define

$$\mu_k(U) := \nu_k(U) - a^{-k}\, \nu_k(aU) \qquad (U \in \Omega)$$

Show that μ_k is additive and that $|\mu_k(U)|_p \leq 1$ for all $U \in \Omega$. (Hint. It suffices to show that $|\mu_k(U)|_p \leq 1$ for $U = i + p^n\mathbb{Z}_p$ where $n \in \mathbb{N}$, $i \in \{0, 1, \ldots, p^n - 1\}$. Write $aU = j + p^n\mathbb{Z}_p$ where $j \in \{0, 1, \ldots, p^n - 1\}$, $|ai - j|_p \leq p^{-n}$. Evaluate $\mu_k(U)$ using Proposition 57.1 and estimate $|\mu_k(U)|_p$ with the help of Exercise 56.A.)
Note. The 'measures' μ_k are used in Koblitz (1978) to define the p-adic zeta functions.

PART 2: THE p-ADIC GAMMA AND ZETA FUNCTIONS

We shall use p-adic integration to derive basic properties of the p-adic gamma function Γ_p, and also to introduce a p-adic analogue of the zeta function.

58. Local analyticity of Γ_p

We shall prove that the p-adic gamma function Γ_p introduced in Section 35 is locally analytic (Theorem 58.4). In Proposition 35.3 we have seen that Γ_p satisfies the functional relation

$$\Gamma_p(x+1) = h_p(x)\,\Gamma_p(x) \qquad (x \in \mathbb{Z}_p)$$

where $h_p(x) = -x$ if $|x|_p = 1$, $h_p(x) = -1$ if $|x|_p < 1$. After taking the (Iwasawa) logarithm of both sides we get

$$\log_p \Gamma_p(x+1) - \log_p \Gamma_p(x) = \begin{cases} \log_p x & \text{if } |x|_p = 1 \\ 0 & \text{if } |x|_p < 1 \end{cases}$$

Since also $\log_p \Gamma_p(0) = 0$ we may conclude that $\log_p \Gamma_p$ is the indefinite sum of the function $g : \mathbb{Z}_p \to K$ defined by

$$g(x) := \begin{cases} \log_p x & \text{if } |x|_p = 1 \\ 0 & \text{if } |x|_p < 1 \end{cases}$$

By Proposition 55.5 (iii) we then have

$$\log_p \Gamma_p(x) = \int_{\mathbb{Z}_p} (Pg(x+u) - Pg(u))\, du$$

where Pg is any C^1-antiderivative of g, for which we choose

$$G(x) := \begin{cases} x\log_p x - x & \text{if } |x|_p = 1 \\ 0 & \text{if } |x|_p < 1 \end{cases}$$

We obtain, using the fact that $\int_{\mathbb{Z}_p} G(u)\, du = 0$ (Proposition 55.7)

$$\log_p \Gamma_p(x) = \int_{\mathbb{Z}_p} G(x+u)\,du \qquad (x \in \mathbb{Z}_p)$$

It follows that for $|x|_p < 1$

$$\log_p \Gamma_p(x) = \int_{T_p} ((x+u)\log_p(x+u) - x - u)\,du$$

which after using the power series of $\log_p(1 + xu^{-1})$ becomes

$$\log_p \Gamma_p(x) = \int_{T_p} (x \log_p u + \sum_{n=1}^{\infty} \frac{(-1)^{n+1}}{n(n+1)} x^{n+1} u^{-n})\,du$$

As $|x|_p < 1$ the terms of the series tend to zero in the sense of C^1 so we may integrate term by term. Observing also that $\int_{T_p} u^{-n} du = 0$ for odd n we may conclude the following.

LEMMA 58.1. $\log_p \Gamma_p$ *is analytic on* $p\mathbb{Z}_p$. *In fact, we have for* $x \in p\mathbb{Z}_p$

$$\log_p \Gamma_p(x) = \lambda_0 x - \sum_{n=1}^{\infty} \frac{\lambda_n}{2n(2n+1)} x^{2n+1}$$

where

$$\lambda_0 := \int_{T_p} \log_p u\,du$$

$$\lambda_n := \int_{T_p} u^{-2n}\,du \qquad (n \in \mathbb{N})$$

Next we want to prove that Γ_p itself is analytic on $p\mathbb{Z}_p$. The idea is of course to consider $\exp \log_p \Gamma_p(x)$, but we shall be a little careful. We first estimate the coefficients λ_n of the power series of $\log_p \Gamma_p$. By Exercise 36.A we have $|\lambda_0|_p = |\Gamma_p'(0)|_p \leqslant 1$. Now let $n \in \mathbb{N}$. An easy computation shows that the C^1-norm of the function

$$u \mapsto \begin{cases} u^{-2n} & \text{if } |u|_p = 1 \\ 0 & \text{if } |u|_p < 1 \end{cases}$$

equals 1 so that by Proposition 55.2

$$|\lambda_n|_p = \left| \int_{T_p} u^{-2n} du \right|_p \leqslant p$$

LEMMA 58.2. *The formula*

$$t(x) = \lambda_0 x - \sum_{n=1}^{\infty} \frac{\lambda_n}{2n(2n+1)} x^{2n+1}$$

defines an analytic function t on the open unit disc of $\mathbb{C}_p$. *If* $p \neq 2$ *then t maps* $\{x \in \mathbb{C}_p : |x|_p \leqslant p^{-1}\}$ *into itself. If* $p = 2$ *then t maps* $\{x \in \mathbb{C}_p : |x|_p \leqslant p^{-2}\}$ *into itself.*

Proof. It is easy to see that the power series converges for $x \in \mathbb{C}_p$, $|x|_p \leqslant 1$. Now let p be an odd prime and let $x \in \mathbb{C}_p$, $|x|_p \leqslant p^{-1}$. Then clearly $|\lambda_0 x|_p \leqslant |x|_p \leqslant p^{-1}$. Now let $n \in \mathbb{N}$. Using the fact that $|2n(2n+1)|_p$ is either $|2n|_p$ or $|2n+1|_p$ and that $|m|_p \geqslant m^{-1}$ for all $m \in \mathbb{N}$ we arrive at $|2n(2n+1)|_p^{-1} \leqslant 2n+1$ and we get

$$\left| \frac{\lambda_n}{2n(2n+1)} x^{2n+1} \right|_p \leqslant p(2n+1)p^{-2n-1} \leqslant p^{-1}$$

It follows that $|t(x)|_p \leqslant p^{-1}$. For the case $p = 2$, let $x \in \mathbb{C}_2$, $|x|_2 \leqslant 2^{-2}$. Then again $|\lambda_0 x|_2 \leqslant 2^{-2}$ and for $n \geqslant 1$ we have

$$\left| \frac{\lambda_n}{2n(2n+1)} x^{2n+1} \right|_2 \leqslant 4n\, 2^{-4n-2} = n 2^{-4n} \leqslant 2^{-2}$$

and we get $|t(x)|_2 \leqslant 2^{-2}$.

LEMMA 58.3. *Let $p \neq 2$. Then Γ_p is analytic on $\{x \in \mathbb{Q}_p : |x|_p < 1\}$. Γ_2 is analytic on $\{x \in \mathbb{Q}_2 : |x|_2 < \frac{1}{4}\}$. Even stronger, the function f given by*

$$f(x) := \begin{cases} \Gamma_2(x) & \text{if } |x|_2 < \frac{1}{4} \\ -\Gamma_2(x) & \text{if } |x|_2 = \frac{1}{4} \end{cases}$$

is analytic on $\{x \in \mathbb{Q}_2 : |x|_2 \leqslant \frac{1}{4}\}$.

Proof. First let $p \neq 2$. The function t of the previous lemma maps $\{x \in \mathbb{C}_p : |x|_p \leqslant 1/p\}$ into itself, hence by Theorem 42.4 the function $\exp \circ t$ is analytic on $\{x \in \mathbb{C}_p : |x|_p \leqslant 1/p\}$. For $x \in p\mathbb{Z}_p$ we have $\exp \log_p \Gamma_p(x) = \exp t(x)$ so that $\exp \log_p \Gamma_p$ is analytic on $p\mathbb{Z}_p$. By Proposition 35.3 (ii), $|\Gamma_p(x) - \Gamma_p(0)|_p \leqslant |x|_p < 1$ if $x \in p\mathbb{Z}_p$. It follows that $\Gamma_p(x) \in 1 + E$ for $x \in p\mathbb{Z}_p$ so that $\Gamma_p(x) = \exp \log_p \Gamma_p(x)$ and we may conclude that Γ_p is analytic on $p\mathbb{Z}_p$.

To treat the case $p = 2$ observe that $\log_2 f(x) = \log_2 \Gamma_2(x)$ for all $|x|_2 \leqslant \frac{1}{4}$. By a reasoning similar to the one before (t maps $\{x \in \mathbb{C}_2 : |x|_2 \leqslant \frac{1}{4}\}$ into itself) we arrive at the analyticity of $\exp t = \exp \log_2 f$ on $\{x \in \mathbb{Q}_2 : |x|_2 \leqslant \frac{1}{4}\}$. By Exercise 35.A(iv) we have for $x \in \mathbb{Q}_2$, $|x|_2 < \frac{1}{4}$ that $|\Gamma_2(x) - 1| \leqslant \frac{1}{4}$ so that $\Gamma_2(x) \in 1 + E$. It is an easy matter to show that if $|x|_2 = \frac{1}{4}$ then $|-\Gamma_2(x) - 1|_2 \leqslant \frac{1}{4}$ (for example, $|\Gamma_2(4(2n-1)) + 1|_2 \leqslant \frac{1}{4}$ by induction on n). Hence we have $f(x) \in 1 + E$ for $|x|_2 \leqslant \frac{1}{4}$, and so $f = \exp \log_2 f$ is analytic on $\{x : |x|_2 \leqslant \frac{1}{4}\}$.

THEOREM 58.4. *Let $p \neq 2$. Then Γ_p is locally analytic of order 1 on $\mathbb{Z}_p$. Γ_2 is locally analytic of order 3. Γ_p is not analytic on $\mathbb{Z}_p$, Γ_2 is not locally analytic of order 2.*

Proof. The formula

$$\Gamma_p(x+1) = \Gamma_p(x)\, h_p(x) \qquad (x \in \mathbb{Z}_p)$$

where

$$h_p(x) = \begin{cases} -x & \text{if } |x|_p = 1 \\ -1 & \text{if } |x|_p < 1 \end{cases}$$

shows that in the case $p \neq 2$ the function $x \mapsto \Gamma_p(x+1)$ is analytic on $p\mathbb{Z}_p$. In general, for $j \in \{1, 2, \ldots, p-1\}$ we have

$$\Gamma_p(x+j) = h_p(x) h_p(1+x) \ldots h_p(j-1+x)\, \Gamma_p(x) \qquad (x \in \mathbb{Z}_p)$$

which yields the analyticity of $x \mapsto \Gamma_p(x+j)$ on $p\mathbb{Z}_p$. It follows that Γ_p is analytic on $1 + p\mathbb{Z}_p, \ldots, p-1+p\mathbb{Z}_p$, hence locally analytic of order 1. In a similar manner one proves that Γ_2 is locally analytic of order 3. That Γ_p is not analytic on the whole of $\mathbb{Z}_p$ follows, for example, from the fact that

$$\Gamma_p(x)\, \Gamma_p(1-x) = (-1)^{l(x)} \qquad (x \in \mathbb{Z}_p)$$

(Proposition 37.2). If Γ_p were analytic then so would be $x \mapsto (-1)^{l(x)}$. But the latter is locally constant, not constant and hence not analytic on $\mathbb{Z}_p$. If Γ_2 were analytic of order 2 then $\Gamma_2 - f$ where

$$f(x) := \begin{cases} \Gamma_2(x) & \text{if } |x|_2 < \frac{1}{4} \\ -\Gamma_2(x) & \text{if } |x|_2 = \frac{1}{4} \end{cases}$$

(see Lemma 58.3) would be analytic on $\{x : |x|_2 \leqslant \frac{1}{4}\}$. But this is impossible since $\Gamma_2 - f = 0$ on $\{x : |x|_2 < \frac{1}{4}\}$.

Exercise 58.A. Show that $\Gamma_p''(0) = \gamma_p^2$ where γ_p is the p-adic Euler constant of Section 36.

59. A formula for $\log_p 2$

The formula

$$\log_p(1+x) = \sum_{n=1}^{\infty} \frac{(-1)^{n+1}}{n} x^n$$

holds for $x \in \mathbb{C}_p$, $|x|_p < 1$. If nevertheless we substitute $x = 1$ we get the formula

$$\log_p 2 = \sum_{n=1}^{\infty} \frac{(-1)^{n+1}}{n}$$

which is meaningless since the series diverges. Yet, we have

THEOREM 59.1. *Let $p \neq 2$. Then*

$$\log_p 2 = \frac{p}{2(p-1)} \lim_{n\to\infty} \sum_{j=0}^{p^n-1}{}' \frac{(-1)^{j+1}}{j}$$

Proof. For $x \in \mathbb{Z}_p$ we have

$$(*) \qquad \log_p \Gamma_p(x+1) - \log_p \Gamma_p(x) = \begin{cases} \log_p x & \text{if } |x|_p = 1 \\ 0 & \text{if } |x|_p < 1 \end{cases}$$

Differentiation yields

$$\frac{\Gamma_p'(x+1)}{\Gamma_p(x+1)} - \frac{\Gamma_p'(x)}{\Gamma_p(x)} = \begin{cases} \frac{1}{x} & \text{if } |x|_p = 1 \\ 0 & \text{if } |x|_p < 1 \end{cases}$$

It follows that $x \mapsto \frac{\Gamma_p'(x)}{\Gamma_p(x)} - \frac{\Gamma_p'(0)}{\Gamma_p(0)}$ is the indefinite sum of the function

$$x \mapsto \begin{cases} \frac{1}{x} & \text{if } |x|_p = 1 \\ 0 & \text{if } |x|_p < 1 \end{cases}$$

so that

$$(**) \qquad \frac{\Gamma_p'(\frac{1}{2})}{\Gamma_p(\frac{1}{2})} = \frac{\Gamma_p'(0)}{\Gamma_p(0)} + \lim_{n\to\infty} \sum_{j=0}^{\frac{1}{2}(p^n-1)}{}' \frac{1}{j}$$

On the other hand, after substituting $\frac{1}{2}x$ for x in (*) we get

$$f(x+2) - f(x) = \begin{cases} \log_p x - \log_p 2 & \text{if } |x|_p = 1 \\ 0 & \text{if } |x|_p < 1 \end{cases}$$

where

$$f(x) := \log_p \Gamma_p\left(\frac{x}{2}\right) \qquad (x \in \mathbb{Z}_p)$$

Integration yields

$$f'(1) + f'(0) = \int_{\mathbb{T}_p} (\log_p x - \log_p 2)\, dx$$

or

$$\frac{1}{2}\frac{\Gamma_p'(\frac{1}{2})}{\Gamma_p(\frac{1}{2})} + \frac{1}{2}\frac{\Gamma_p'(0)}{\Gamma_p(0)} = \frac{\Gamma_p'(0)}{\Gamma_p(0)} - \frac{p-1}{p} \log_p 2$$

whence

$$\frac{\Gamma_p'(\frac{1}{2})}{\Gamma_p(\frac{1}{2})} - \frac{\Gamma_p'(0)}{\Gamma_p(0)} = -\frac{2(p-1)}{p}\log_p 2$$

Together with (**) this gives us

$$\log_p 2 = -\frac{p}{2(p-1)} \lim_{n\to\infty} \sum_{j=0}^{\frac{1}{2}(p^n-1)}{}' \frac{1}{j}$$

Now observe that $\lim_{|n|_p \to 0} \Sigma'^{\,n-1}_{j=0}\ 1/j = 0$, so that

$$\lim_{n\to\infty}\left(\sum_{\substack{j=0\\ j \text{ even}}}^{p^n-1}{}' \frac{1}{j} + \sum_{\substack{j=0\\ j \text{ odd}}}^{p^n-1}{}' \frac{1}{j}\right) = 0$$

Further,

$$\lim_{n\to\infty} \sum_{j=0}^{\frac{1}{2}(p^n-1)}{}' \frac{1}{j} = 2\lim_{n\to\infty} \sum_{\substack{j=0\\ j \text{ even}}}^{p^n-1}{}' \frac{1}{j}$$

$$= \lim_{n\to\infty}\left(\sum_{\substack{j=0\\ j \text{ even}}}^{p^n-1}{}' \frac{1}{j} - \sum_{\substack{j=0\\ j \text{ odd}}}^{p^n-1}{}' \frac{1}{j}\right) = \lim_{n\to\infty} \sum_{j=0}^{p^n-1}{}' \frac{(-1)^j}{j}$$

We get

$$\log_p 2 = \frac{p}{2(p-1)} \lim_{n\to\infty} \sum_{j=0}^{p^n-1}{}' \frac{(-1)^{j+1}}{j}.$$

Note. For a different proof of the formula for $\log_p 2$ see Koblitz (1980), p. 38.

Exercise 59.A. Let h be the indefinite sum of

$$x \mapsto \begin{cases} 1/x & \text{if } |x|_p = 1 \\ 0 & \text{if } |x|_p < 1 \end{cases}$$

Show that

$$\frac{1}{m}\sum_{k=0}^{m-1} h\left(\frac{x+k}{m}\right) = h(x) - \frac{p-1}{p}\log_p m \quad (x \in \mathbb{Z}_p,\, m \in \mathbb{N},\, |m|_p = 1).$$

60. Diamond's log gamma function

In this section we study Diamond's function. Although it is not equal to $\log_p \Gamma_p$ it deserves the name 'log gamma function'.

Recall that $\log_p \Gamma_p$ is a solution of the difference equation

$$f(x+1) - f(x) = \begin{cases} \log_p x & \text{if } |x|_p = 1 \\ 0 & \text{if } |x|_p < 1 \end{cases} \qquad (x \in \mathbb{Z}_p)$$

and that

$$\Gamma_p(x+1) = h_p(x)\,\Gamma_p(x) \qquad (x \in \mathbb{Z}_p)$$

where

$$h_p(x) = \begin{cases} -x & \text{if } |x|_p = 1 \\ -1 & \text{if } |x|_p < 1 \end{cases}$$

In Section 35 we have seen that there is no continuous function φ on $\mathbb{Z}_p$ for which

$$\varphi(x+1) = x\,\varphi(x) \qquad (x \in \mathbb{Z}_p)$$
$$\varphi(1) = 1$$

There is also no continuous function ψ on $\mathbb{Z}_p \backslash \{0\}$ for which

$$\psi(x+1) - \psi(x) = \log_p x \qquad (x \in \mathbb{Z}_p,\, x \neq 0)$$

(If there is such a ψ then $\lim_{x \to -1} \psi(x+1) = \psi(-1) + \log_p(-1) = \psi(-1)$, hence $\lim_{x \to 0} \psi(x)$ exists. But then $\lim_{x \to 0} (\psi(x+1) - \psi(x))$ exists whereas $\lim_{x \to 0} \log_p x$ does not.) It is possible, however, to find a continuous function G_p satisfying

$$G_p(x+1) - G_p(x) = \log_p x$$

as long as we stay out of $\mathbb{Z}_p$.

DEFINITION 60.1. For $x \in \mathbb{C}_p \backslash \mathbb{Z}_p$ define (compare the integral formula for $\log_p \Gamma_p$ in Section 58)

$$G_p(x) := \int_{\mathbb{Z}_p} ((x+u)\log_p(x+u) - (x+u))\,du$$

G_p is the *log gamma function of Diamond*.

Observe that the integrand is a C^1-function of u, so that G_p is well defined. Property (v) of the next theorem resembles the classical asymptotic formula of $\log(\Gamma(x)/\sqrt{2\pi})$.

THEOREM 60.2. (Properties of G_p)
(i) $G_p(x+1) - G_p(x) = \log_p x \quad (x \in \mathbb{C}_p \backslash \mathbb{Z}_p)$.
(ii) $G_p(x) + G_p(1-x) = 0 \quad (x \in \mathbb{C}_p \backslash \mathbb{Z}_p)$.

(iii) *For each* $m \in \mathbb{N}$

$$G_p(x) = (x - \tfrac{1}{2}) \log_p m + \sum_{j=0}^{m-1} G_p\left(\frac{x+j}{m}\right) \qquad (x \in \mathbb{C}_p \setminus \mathbb{Z}_p)$$

In particular, for $n \in \mathbb{N}$

$$G_p(x) = \sum_{j=0}^{p^n - 1} G_p\left(\frac{x+j}{p^n}\right) \qquad (x \in \mathbb{C}_p \setminus \mathbb{Z}_p)$$

(iv) G_p *is locally analytic on* $\mathbb{C}_p \setminus \mathbb{Z}_p$. *In fact, if* $a \in \mathbb{C}_p \setminus \mathbb{Z}_p$ *and* $\rho = d(a, \mathbb{Z}_p)$ $:= \inf\{|a-x|_p : x \in \mathbb{Z}_p\}$, *then* G_p *is the sum of a power series in* $x - a$ *on* $\{x : |x-a|_p < \rho\}$.

(v)

$$G_p(x) = (x - \tfrac{1}{2})\log_p x - x + \sum_{j=1}^{\infty} \frac{B_{j+1}}{j(j+1)} x^{-j} \qquad (x \in \mathbb{C}_p, |x|_p > 1)$$

(vi) (Connection with Morita's function Γ_p)

$$\log_p \Gamma_p(x) = \sum_{\substack{j=0 \\ |x+j|_p = 1}}^{p-1} G_p\left(\frac{x+j}{p}\right) \qquad (x \in \mathbb{Z}_p)$$

Proof. Let $f(x, u) := (x+u)\log_p(x+u) - (x+u)$ $(x \in \mathbb{C}_p \setminus \mathbb{Z}_p, u \in \mathbb{Z}_p)$. Then $f(x+1, u) - f(x, u) = f(x, u+1) - f(x, u)$, hence $G_p(x+1) - G_p(x) = (\frac{\partial f}{\partial u})(x, 0) = \log_p x$, which establishes (i). To prove (ii), observe that since $\log_p(-1) = 0$ we have $f(-x, u) = -f(x, -u)$. By Proposition 55.7 we get

$$G_p(-x) = -\int_{\mathbb{Z}_p} f(x, -u)du = -\int_{\mathbb{Z}_p} f(x, u+1)du = -\int_{\mathbb{Z}_p} f(x+1, u)du = -G_p(1+x)$$

so that $G_p(-x) + G_p(1+x) = 0$, which is (ii). We proceed to prove (iii). According to Exercise 55.B we have

$$G_p(x) = \int_{\mathbb{Z}_p} f(x, u)du = \frac{1}{m} \sum_{j=0}^{m-1} \int_{\mathbb{Z}_p} f(x, j + mu)du$$

For every $y \in \mathbb{C}_p \setminus \mathbb{Z}_p$ we have

$$\begin{aligned} f(y, mu) &= (y + mu)\log_p(y + mu) - (y + mu) \\ &= m(\tfrac{y}{m} + u)(\log_p(\tfrac{y}{m} + u) + \log_p m) - m(\tfrac{y}{m} + u) \\ &= (y + mu)\log_p m + mf(\tfrac{y}{m}, u) \end{aligned}$$

Then $f(x, j+mu) = f(x+j, mu) = (x+j+mu)\log_p m + mf(\frac{x+j}{m}, u)$. Hence,

$$\int_{\mathbb{Z}_p} f(x, j + mu)du = (x + j - \tfrac{1}{2}m) \log_p (m) + m\, G_p\left(\frac{x+j}{m}\right)$$

Now

$$\frac{1}{m} \sum_{j=0}^{m-1} (x + j - \tfrac{1}{2}m) \log_p m = (x - \tfrac{1}{2}) \log_p m$$

It follows that

$$G_p(x) = (x - \tfrac{1}{2}) \log_p m + \sum_{j=0}^{m-1} G_p\left(\frac{x+j}{m}\right)$$

Next, we prove (iv). Let $x \in \mathbb{C}_p$, $|x|_p < \rho = d(a, \mathbb{Z}_p)$. Then $G_p(a + x) = \int_{\mathbb{Z}_p} f(a + x, u)du$. Now

$$\begin{aligned} f(a + x, u) &= (a + u + x) \log_p (a + u + x) - (a + u + x) \\ &= (a + u + x) \left(\log_p \left(1 + \frac{x}{a+u}\right) + \log_p (a + u)\right) - (a + u + x) \\ &= (a + u + x)\left(\sum_{n=1}^{\infty} \frac{(-1)^{n+1}}{n} \frac{x^n}{(a+u)^n}\right) + (a + u + x)(\log_p (a + u) - 1) \\ &= x + \sum_{n=1}^{\infty} \frac{(-1)^{n+1}}{n(n+1)} \frac{x^{n+1}}{(a+u)^n} + (a + u + x)(\log_p (a + u) - 1) \end{aligned}$$

The series converges (as a function of u) in the sense of C^1. Integration yields

$$G_p(a + x) = G_p(a) + \tau_1 x + \sum_{n=1}^{\infty} \frac{(-1)^{n+1}}{n(n+1)} \tau_{n+1} x^{n+1} \qquad (|x|_p < \rho)$$

where

$$\tau_1 = \int_{\mathbb{Z}_p} \log(a + u)du$$

$$\tau_{n+1} = \int_{\mathbb{Z}_p} (a + u)^{-n} du \qquad (n \in \mathbb{N})$$

To prove (v), let $|x|_p > 1$. For $u \in \mathbb{Z}_p$ we have $\left|\frac{u}{x}\right|_p < 1$ so that

$$(x + u) \log_p (x + u) - (x + u) =$$

$$(x + u) (\log_p (1 + \tfrac{u}{x}) + \log_p x) - (x + u) =$$

$$u + x \sum_{n=1}^{\infty} \frac{(-1)^{n+1}}{n(n+1)} \frac{u^{n+1}}{x^{n+1}} + (x + u) \log_p x - (x + u)$$

Hence,

$$G_p(x) = (x - \tfrac{1}{2}) \log_p x - x + \sum_{n=1}^{\infty} \frac{(-1)^{n+1}}{n(n+1)} B_{n+1} x^{-n}$$

If n is even then $B_{n+1} = 0$, so that we may omit the symbol $(-1)^{n+1}$ in the above formula. To prove the last formula, let $x \in \mathbb{Z}_p$. In Section 58 we have seen that

$$\log_p \Gamma_p(x) = \int_{\mathbb{Z}_p} G(x+u)du$$

where

$$G(y) = \begin{cases} y \log_p y - y & \text{if } |y|_p = 1 \\ 0 & \text{if } |y|_p < 1 \end{cases}$$

We have

$$\int_{\mathbb{Z}_p} G(x+u)du = \sum_{j=0}^{p-1} \int_{j+p\mathbb{Z}_p} G(x+u)du = (\text{by Proposition 57.1})$$

$$= \sum_{j=0}^{p-1} p^{-1} \int_{\mathbb{Z}_p} G(x+j+pu)du = \sum_{\substack{j=0 \\ |j+x|_p = 1}}^{p-1} p^{-1} \int_{\mathbb{Z}_p} f(x+j, pu)du =$$

$$= \sum_{\substack{j=0 \\ |j+x|_p = 1}}^{p-1} \int_{\mathbb{Z}_p} f(\frac{x+j}{p}, u)du = \sum_{\substack{j=0 \\ |j+x|_p = 1}}^{p-1} G_p\left(\frac{x+j}{p}\right)$$

61. The p-adic zeta functions

For each $x \in T_p$ we have $|x^{p-1} - 1|_p \leqslant p^{-1}$ so that, according to Definition 47.9 and Theorem 47.10, $(x^{p-1})^s$ is defined for each $s \in \mathbb{C}_p$ for which $|s|_p < p^{(p-2)/(p-1)}$ and

$$(x^{p-1})^s = \exp(s(p-1)\log_p x) = \sum_{n=0}^{\infty} \binom{s}{n} (x^{p-1}-1)^n$$

We define the functions $f_0, f_1, \ldots, f_{p-2}$ as follows. For $j \in \{0, 1, \ldots, p-2\}$ let

$$f_j(s) := \int_{T_p} x^j (x^{p-1})^s dx \qquad (|s|_p < p^{(p-2)/(p-1)})$$

Except when $p = 2$ the $f_0, f_1, \ldots, f_{p-2}$ are defined on a disc strictly containing the closed unit of $\mathbb{C}_p$.

PROPOSITION 61.1 *The functions $f_0, f_1, \ldots, f_{p-2}$ are analytic.*
Proof. Let $j \in \{0, 1, \ldots, p-2\}$ and let s be in the domain of f_j. By the above formulas we have

$$(*) \qquad f_j(s) = \int_{T_p} x^j \left(\sum_{n=0}^{\infty} (s^n (p-1)^n \log_p^n x)/n! \right) dx$$

By using the inequalities

$$|\log_p x|_p \leqslant p^{-1}$$

$$\left|\frac{\log_p^n x - \log_p^n y}{x-y}\right|_p \leqslant \max(|\log_p x|_p, |\log_p y|_p)^{n-1}$$

for $x, y \in T_p$ $(x \neq y)$ and $n \in \mathbb{N}$ one proves in a standard way that $\lim_{n\to\infty} \|g_n\|_1 = 0$ where

$$g_n(x) := \begin{cases} (s^n(p-1)^n \log_p^n x)/n! \text{ if } x \in T_p \\ 0 \text{ if } x \in p\,\mathbb{Z}_p \end{cases}$$

It follows that the series $\Sigma\, g_n$ converges in the C^1-sense so that we may integrate term by term in (*) obtaining

$$f_j(s) = \sum_{n=0}^{\infty} a_n s^n$$

where, for each $n \in \{0, 1, 2, \ldots\}$

$$a_n = \frac{(p-1)^n}{n!} \int_{T_p} x^j \log_p^n x\, dx$$

We see that f_j is analytic.

Exercise 61.A. Show that $f_j = 0$ for odd j.

Exercise 61.B. Let $p \neq 2$ and $j \in \{0, 1, \ldots, p-2\}$. Show that the restriction of f_j to $\mathbb{Z}_p$ has the Mahler coefficients $n \mapsto \int_{T_p} x^j (x^{p-1} - 1)^n\, dx$.

DEFINITION 61.2. The *p-adic zeta functions* $\zeta_{p,0}, \zeta_{p,1}, \ldots, \zeta_{p,p-2}$ are given by the formula

$$\zeta_{p,j}(s) = \frac{1}{j+(p-1)s} \int_{T_p} x^j (x^{p-1})^s dx \quad (|s|_p < p^{(p-2)/(p-1)},\ s \neq -j/(p-1))$$

For the reason why these functions are called zeta functions, see below.

THEOREM 61.3 *The functions* $s \mapsto \zeta_{p,0} - (1/ps)$ *and* $\zeta_{p,1}, \ldots, \zeta_{p,p-2}$ *extend to analytic functions on* $\{s \in \mathbb{C}_p : |s|_p < p^{(p-2)/(p-1)}\}$.

Proof. $\zeta_{p,0}(s) = (p-1)^{-1} s^{-1} f_0(s) = (p-1)^{-1} s^{-1} \sum_{n=0}^{\infty} a_n s^n$. Now $a_0 = \int_{T_p} dx = (p-1)/p$. We see that $\zeta_{p,0}(s) = (1/ps) +$ an analytic function of s. Now let $j \in \{1, \ldots, p-2\}$. We develop f_j in a power series in $j + (p-1)s$ as follows. According to Exercise 45.A (vii) we have $\exp \log_p x^j = x^j \omega_p(x^{-j})$ where ω_p is the Teichmüller character (observe that $p \neq 2$). We get

$$x^j (x^{p-1})^s = \omega_p(x^j) \exp\{(j + (p-1)s) \log_p x\}$$

so that (justification of the interchange of summation and integration is similar to the one in the proof of Proposition 61.1)

$$f_j(s) = \int_{T_p} x^j (x^{p-1})^s \, dx = \sum_{n=0}^{\infty} b_n (j+(p-1)s)^n \quad (|s|_p < p^{(p-2)/(p-1)})$$

where

$$b_n = \frac{1}{n!} \int_{T_p} \omega_p(x^j) \log_p^n x \, dx \qquad (n \in \{0, 1, 2, \ldots\})$$

We compute b_0. $\omega_p(x^j)$ takes the values $1, \theta^j, \theta^{2j}, \ldots, \theta^{(p-2)j}$ where θ is a primitive $(p-1)$th root of unity. ω_p is constant on each one of the sets $1 + p\mathbb{Z}_p, 2 + p\mathbb{Z}_p, \ldots, (p-1) + p\mathbb{Z}_p$ so we get

$$b_0 = \int_{T_p} \omega_p(x^j) dx = p^{-1} (1 + \theta^j + \ldots + \theta^{(p-2)j}) = 0$$

We see that $f_j(s) = b_1(j+(p-1)s) + b_2(j+(p-1)s)^2 + \ldots$ Hence, $\zeta_{p,j}(s) = b_1 + b_2(j+(p-1)s) + \ldots$ is an analytic function of s.

To understand the connection between the p-adic and the classical zeta functions we first determine the values of $\zeta_{p,j}$ in $0, 1, 2, \ldots$ Using the last formula of Proposition 57.1 we find for $p \neq 2$ and $j \in \{1, 2, \ldots, p-1\}$

$$\zeta_{p,0}(n) = (1 - p^{(p-1)n-1}) \frac{B_{(p-1)n}}{(p-1)n} \qquad (n \in \{1, 2, \ldots\})$$

$$\zeta_{p,j}(n) = (1 - p^{j-1+(p-1)n}) \frac{B_{j+(p-1)n}}{j+(p-1)n} \qquad (n \in \{0, 1, \ldots\})$$

whereas

$$\zeta_{2,0}(n) = (1 - p^{n-1}) \frac{B_n}{n} \qquad (n \in \{2, 4, 6, \ldots\})$$

The classical zeta function ζ defined by

$$\zeta(s) = \sum_{n=1}^{\infty} \frac{1}{n^s} \qquad (\mathrm{Re}\, s > 1)$$

can be written in the form

$$\zeta(s) = \prod_q (1 - q^{-s})^{-1}$$

where the product is taken over all primes q. Let us remove the factor $(1-p^{-s})^{-1}$, i.e. define

$$\zeta^*(s) = \prod_{q \neq p} (1 - q^{-s})^{-1} = (1 - p^{-s}) \zeta(s)$$

It is known that ζ can be extended to a meromorphic function (again denoted ζ) on the complex plane with a simple pole at 1. It can be proved that

$$\zeta(1-n) = -\frac{B_n}{n} \qquad (n \in \{1,2,\ldots\})$$

Our function ζ^* then obviously extends to a meromorphic function and

$$\zeta^*(1-n) = -(1-p^{n-1})\frac{B_n}{n}$$

Therefore we have the following equalities of rational numbers. For $p \neq 2$ and $j \in \{1, 2, \ldots, p-1\}$

$$\zeta_{p,0}(n) = -\zeta^*(1-(p-1)n) \qquad (n \in \{1,2,\ldots\})$$
$$\zeta_{p,j}(n) = -\zeta^*(1-(j+(p-1)n)) \qquad (n \in \{0,1,2,\ldots\})$$

and

$$\zeta_{2,0}(n) = -\zeta^*(1-n) \qquad (n \in \{2,4,\ldots\})$$

In other words, we can construct the p-adic zeta functions out of the classical zeta function by p-adic interpolation of sequences of values of ζ^* at certain negative integers.

Remark. For the p-adic analogues of the L-functions of Dirichlet see Iwasawa (1972) or Koblitz (1980).

Exercise 61.C. (Kummer congruence) Let $p \neq 2$ and $j \in \{1, 2, \ldots, p-2\}$. Show that for all $s, t \in \mathbb{Z}_p$

$$|\zeta_{p,j}(s) - \zeta_{p,j}(t)|_p \leqslant p^{-1}|s-t|_p$$

Interpret this result in a number theoretic way as follows. If $s, t \in \mathbb{N}$ and $s \equiv t \pmod{(p-1)p^n}$ and $s \not\equiv 0 \pmod{(p-1)}$ then

$$(1-p^{s-1})\frac{B_s}{s} \equiv (1-p^{t-1})\frac{B_t}{t} \pmod{p^{n+1}}$$

Exercise 61.D. Show by considering $\zeta_{p,0}((1-p)^{-1})$ that in $\mathbb{Q}_p$ we have

$$\int_{T_p} \omega_p(x)x^{-1}\,dx = \lim_{n\to\infty} B_{p^n-1} \qquad (p \neq 2)$$

Exercise 61.E. Show that for each $s \in \mathbb{Z}_p$, $s \neq 0$

$$|\zeta_{p,0}(s)|_p = p|s|_p^{-1}$$

Exercise 61.F. Show that

$$\Gamma_p'(0) = \int_{T_p} \log_p x\,dx = \lim_{N\to\infty}\left\{-\frac{1}{p}\sum_{j=1}^{N}\frac{1}{j} + \sum_{j=1}^{N}(-1)^{j+1}\binom{N}{j}\zeta_{p,0}(j)\right\}$$

PART 3: VAN DER PUT'S BASE AND ANTIDERIVATION

In this part we shall consider antiderivation and a few examples of differential equations with C^1-solutions. To this end we shall introduce an orthonormal base of $C(\mathbb{Z}_p \to K)$ (van der Put's base) consisting of locally constant functions. We have seen in Part 1 of this chapter that the Mahler basis is suited to characterize C^n-functions, analytic functions, Lip_1-functions and that the indefinite sum operator written with respect to this base gets a simple form. The base of van der Put shall enable us to characterize Lip_α-functions, C^n-functions with zero derivative (in Section 69 we shall even characterize differentiable functions with zero derivative) and to define an antiderivation operator $P : C(\mathbb{Z}_p \to K) \to C^1(\mathbb{Z}_p \to K)$. van der Put's base is not a base of $C^1(\mathbb{Z}_p \to K)$; in Section 68 we shall extend it to an orthogonal base of $C^1(\mathbb{Z}_p \to K)$.

62. van der Put's base of $C(\mathbb{Z}_p \to K)$

We start with some terminology. For $x = \sum_{j=-\infty}^{\infty} a_j p^j \in \mathbb{Q}_p$ define its *p-adic entire part* $[x]_p$ by

$$[x]_p := \sum_{j=-\infty}^{-1} a_j p^j$$

and set

$$x_n := p^n [p^{-n} x]_p = \sum_{j=-\infty}^{n-1} a_j p^j \qquad (n \in \{0, 1, \ldots\})$$

This way we have assigned to each $x \in \mathbb{Q}_p$ a *standard sequence* $x_0, x_1, \ldots$ converging to x. The standard sequence of an element x of $\mathbb{Z}_p$ consists of nonnegative integers; it is eventually constant if $x \in \{0, 1, \ldots\}$. We write

$$m \lhd x \qquad (m \in \mathbb{N} \cup \{0\}, x \in \mathbb{Z}_p)$$

if m is one of the numbers $x_0, x_1, \ldots$ We sometimes refer to the relation $\lhd$ between m and x as 'x starts with m' or 'm is an initial part of x'. If $n \in \mathbb{N}$ then

$$\{m : m \lhd n, m \neq n\}$$

is finite and has a largest element (with respect to $\lhd$) which is what we have called n_- in Definition 47.3.

In the following we list some elementary facts concerning these notions.

PROPOSITION 62.1.
(i) *Let* $x \in \mathbb{Z}_p$, $n \in \{0, 1, 2, \ldots\}$. *Then* $|x - x_n|_p \leqslant p^{-n}$ *and* $x_n \in \{0, 1, \ldots, p^n - 1\}$. *Conversely,* $y \in \{0, 1, \ldots, p^n - 1\}$, $|x - y|_p \leqslant p^{-n}$ *implies* $y = x_n$.
(ii) *Let* $x, y \in \mathbb{Z}_p$, $n \in \{0, 1, 2, \ldots\}$. *Then*

$$|x - y|_p \leqslant p^{-n} \text{ if and only if } x_n = y_n$$
$$|x - y|_p = p^{-n} \text{ if and only if } x_n = y_n \text{ and } x_{n+1} \neq y_{n+1}$$

(iii) $x \mapsto x_n$ $(x \in \mathbb{Z}_p)$ *is constant on the cosets of* $p^n\mathbb{Z}_p$ $(n \in \{0, 1, 2, \ldots\})$.
(iv) *Let* $x, y \in \mathbb{Z}_p$, $n \in \{0, 1, 2, \ldots\}$. *Then*

$$|x_n - y_n|_p = \begin{cases} 0 & \text{if } |x - y|_p \leqslant p^{-n} \\ |x - y|_p & \text{if } |x - y|_p > p^{-n} \end{cases}$$

(v) $(x_n)_m = x_{\min(n,m)}$ $(x \in \mathbb{Z}_p,\ n, m \in \{0, 1, 2, \ldots\})$.
(vi) *Let* $x \in \mathbb{Z}_p$, $m \in \mathbb{N}$. *Then*

$$m \lhd x \text{ if and only if } |x - m|_p < \frac{1}{m}$$

(vii)
$$|m - m_-|_p = p^{-s(m)} \qquad (m \in \mathbb{N})$$

where $s(m) := [\log m / \log p]$ *is also determined by*

$$m = a_0 + a_1 p + \ldots + a_{s(m)} p^{s(m)}, a_{s(m)} \neq 0$$

Proof. We only prove (vi) and (vii) leaving the inspection of (i)-(v) to the reader. If $m \lhd x$ then $m = a_0 + a_1 p + \ldots + a_s p^s$ for some s, $x = \sum_{j=0}^{\infty} a_j p^j$. So $|x - m|_p \leqslant p^{-(s+1)} < 1/m$. Conversely, if $|x - m|_p < 1/m$ and $m = a_0 + a_1 p + \ldots + a_s p^s$ $(a_s \neq 0)$ then $1/m \leqslant p^{-s}$ so that $|x - m|_p < p^{-s}$, which means that the first $s + 1$ coefficients in the p-adic expansion of x must be equal to those of m. That is, $m \lhd x$. To prove (vii), let $m = a_0 + a_1 p + \ldots + a_s p^s$ $(a_s \neq 0)$. Then $p^s \leqslant m < p^{s+1}$, hence $s = \left[\frac{\log m}{\log p}\right]$ and $|m - m_-|_p = p^{-s}$.

THEOREM 62.2. *The functions* $e_0, e_1, \ldots$ *defined by*

$$e_n(x) := \begin{cases} 1 & \text{if } n \lhd x \\ 0 & \text{otherwise} \end{cases} \qquad (x \in \mathbb{Z}_p,\ n \in \{0, 1, 2, \ldots\})$$

form an orthonormal base (the van der Put base) of $C(\mathbb{Z}_p \to K)$. *If* $f \in C(\mathbb{Z}_p \to K)$ *has the expansion*

$$f(x) = \sum_{n=0}^{\infty} a_n e_n(x) \qquad (x \in \mathbb{Z}_p)$$

then $a_0 = f(0)$ *and* $a_n = f(n) - f(n_-)$ *for* $n \in \mathbb{N}$.
Proof. Let $f \in C(\mathbb{Z}_p \to K)$ and suppose there are $a_0, a_1, \ldots \in K$ such that

$$f(x) = \sum_{n=0}^{\infty} a_n e_n(x) \qquad (x \in \mathbb{Z}_p)$$

Then $f(0) = 0$ and for $m \in \mathbb{N}$

$$f(m) = \sum_{n \lhd m} a_n, \quad f(m_-) = \sum_{n \lhd m_-} a_n$$

so that

$$f(m) - f(m_-) = a_m$$

Now let $f \in C(\mathbb{Z}_p \to K)$ be arbitrary and consider the series

$$g(x) := f(0)\, e_0 + \sum_{n=1}^{\infty} (f(n) - f(n_-))\, e_n(x) \qquad (x \in \mathbb{Z}_p)$$

Since f is (uniformly) continuous, $\lim_{n\to\infty} (f(n) - f(n_-)) = 0$, and the series converges uniformly, hence g is a well-defined element of $C(\mathbb{Z}_p \to K)$. By the above we have $f(0) = g(0), f(n) - f(n_-) = g(n) - g(n_-)$ for all $n \in \mathbb{N}$. We conclude that $f(n) = g(n)$ for all $n \in \mathbb{N} \cup \{0\}$. By continuity, $f = g$. At this stage we know that for each $f \in C(\mathbb{Z}_p \to K)$, $x \in \mathbb{Z}_p$

$$f(x) = f(0)\, e_0 + \sum_{n=1}^{\infty} (f(n) - f(n_-))\, e_n(x) = \sum_{n=0}^{\infty} a_n e_n(x)$$

That $\|f\|_\infty \leqslant \sup |a_n|$ is obvious. On the other hand, $|f(0)| \leqslant \|f\|_\infty$; $|f(n) - f(n_-)| \leqslant \|f\|_\infty$, so

$$\|f\|_\infty = \sup_n |a_n| = \max_n |a_n|$$

This fact, together with the observation that $\lim_{n\to\infty} a_n = 0$, implies that $e_0, e_1, \ldots$ is an orthonormal base of $C(\mathbb{Z}_p \to K)$, which finishes the proof.

In the following exercises (and also in Sections 63 and 68) we shall ask questions similar to the ones of Part 1 of this chapter and compare the behaviour of the bases of Mahler and van der Put. (Of course there is a bijective linear isometry $c_0 \to c_0$ such that for each $f \in C(\mathbb{Z}_p \to K)$ the sequence of its Mahler coefficients is mapped into the sequence of its coefficients with respect to the van der Put base, see Mahler (1980).)

IN THE REST OF THIS CHAPTER $e_0, e_1, \ldots$ IS THE VAN DER PUT BASE OF THEOREM 62.2.

Exercise 62.A. (Pointwise representation of a noncontinuous function, compare Exercise 51.B) Let $f(x) := 1$ if $x \in \mathbb{Z}_p \backslash \{0\}$, $f(0) := 0$. Show that there exist $a_0, a_1, \ldots \in \mathbb{Q}_p$ such that

$$f(x) = \sum_{n=0}^{\infty} a_n e_n(x) \qquad (x \in \mathbb{Z}_p)$$

but that the series does not converge uniformly and that $a_0, a_1, \ldots$ is not a null sequence.

Exercise 62.B. (Sequel to the previous exercise) Let $f: \mathbb{Z}_p \to \mathbb{Q}_p$. Show that the following conditions (α) and (β) are equivalent.
(α) There are $a_0, a_1, \ldots \in \mathbb{Q}_p$ such that $f = \sum_{n=0}^{\infty} a_n e_n$ (pointwise).
(β) For each $x \in \mathbb{Z}_p$ (and with x_n as in the beginning of this section)

$$f(x) = \lim_{n \to \infty} f(x_n)$$

Show that if f is continuous at all points of $\mathbb{Z}_p \setminus \{0, 1, 2, \ldots\}$ then condition (β) is satisfied but that the converse is not true.

**Exercise* 62.C. (Compare Exercise 51.D and 52.A) Let $f = \sum_{n=0}^{\infty} a_n e_n \in C(\mathbb{Z}_p \to K)$. For each $n \in \{0, 1, 2, \ldots\}$ let V_n be the space of all functions $\mathbb{Z}_p \to K$ that are constant on cosets of $p^n \mathbb{Z}_p$. Prove the following.
(i) $V_n = [\![e_0, e_1, \ldots, e_{p^n - 1}]\!]$. f is locally constant if and only if $a_n = 0$ for large n.
(ii) Set $f_n := \sum_{j=0}^{p^n - 1} a_j e_j$. Then f_n is the unique element g of V_n for which $g(0) = f(0), g(1) = f(1), \ldots, g(p^n - 1) = f(p^n - 1)$.
(iii) f_n is a best approximation of f in V_n.
(iv) For each $n \in \{0, 1, 2, \ldots\}$

$$\max(|f(0)|, |f(1)|, \ldots, |f(n)|) = \max(|a_0|, |a_1|, \ldots, |a_n|)$$

Exercise 62.D. (van der Put base for $C(\mathbb{Z}_p \times \mathbb{Z}_p \to K)$, compare Exercise 52.D) Prove that the functions

$$(x, y) \mapsto e_m(x) e_n(y) \qquad (m, n \in \{0, 1, 2, \ldots\})$$

form an orthonormal base of $C(\mathbb{Z}_p \times \mathbb{Z}_p \to K)$. Let $f \in C(\mathbb{Z}_p \times \mathbb{Z}_p \to K)$. Find a formula expressing the coefficients of f in terms of values of f.

Exercise 62.E. (Volkenborn integral versus van der Put base)
(i) Show that $\int_{\mathbb{Z}_p} e_0(x)dx = 1$ and that for $n \in \mathbb{N}$

$$\int_{\mathbb{Z}_p} e_n(x)dx = p^{-s(n)-1}$$

where $s(n) = [\log n / \log p]$.
(ii) Let $f = \sum_{n=0}^{\infty} a_n e_n \in C(\mathbb{Z}_p \to K)$. Prove that for each $m \in \mathbb{N}$

$$p^{-m}\sum_{j=0}^{p^m-1} f(j) = a_0 + \sum_{n=1}^{p^m-1} a_n p^{-s(n)-1}$$

and draw the conclusion

$$\int_{\mathbb{Z}_p} f(x)dx = a_0 + \lim_{m\to\infty} \sum_{n=1}^{p^m-1} a_n p^{-s(n)-1}$$

for $f \in C^1(\mathbb{Z}_p \to K)$.

Exercise 62.F. (Shift versus van der Put base) In Section 52 we have shown that, for $f = \sum_{n=0}^{\infty} a_n \binom{x}{n} \in C(\mathbb{Z}_p \to K)$, the Mahler expansion of $\Delta f(x) = f(x+1) - f(x)$ is simply $\sum_{n=0}^{\infty} a_{n+1}\binom{x}{n}$. The connection between the 'van der Put expansions' of Δf and f is a little more complicated. In fact, prove that if $f = \sum_{n=0}^{\infty} a_n e_n$ then $\Delta f = \sum_{n=0}^{\infty} b_n e_n$ where

$$b_n = \begin{cases} a_1 & \text{if } n = 0 \\ a_{n+1} - a_n - a_{p^s} & \text{if } n = ap^s - 1 \ (s \in \{0, 1, 2, \ldots\}, a \in \{2, 3, \ldots, p\}) \\ a_{n+1} - a_n & \text{in all other cases} \end{cases}$$

Exercise 62.G. ($\mathbb{Q}_q$-valued Haar integral on $C(\mathbb{Z}_p \to \mathbb{Q}_q)$) Let q be a prime different from p. Show that there exists a unique translation invariant continuous $\mathbb{Q}_q$-linear function μ: $C(\mathbb{Z}_p \to \mathbb{Q}_q) \to \mathbb{Q}_q$ such that $\mu(\xi_{\mathbb{Z}_p}) = 1$. Define a 'van der Put base' $\epsilon_0, \epsilon_1, \ldots$ of $C(\mathbb{Z}_p \to \mathbb{Q}_q)$ and show that for $f = \sum_{n=0}^{\infty} a_n \epsilon_n \in C(\mathbb{Z}_p \to \mathbb{Q}_q)$ we have $\mu(f) = a_0 + \sum_{n=1}^{\infty} a_n p^{-s(n)-1}$ (See Exercise 62.E.)

63. Characterizations by means of coefficients

Our key lemma is the following.

LEMMA 63.1. *Let $f \in C(\mathbb{Z}_p \to K)$, let B be a ball in $\mathbb{Z}_p$, let S be a ball in K. Suppose that*

$$\Phi_1 f(n, n_-) = \frac{f(n) - f(n_-)}{n - n_-} \in S \qquad (n \in \mathbb{N}, n, n_- \in B)$$

Then

$$\Phi_1 f(x, y) = \frac{f(x) - f(y)}{x - y} \in S \qquad (x, y \in B, x \neq y)$$

Proof. First we make two remarks. If suffices to prove the statement only for $x, y \in B \cap \mathbb{N}$ ($\mathbb{N}$ is dense in $\mathbb{Z}_p$, f is continuous and S is closed in K).

Further, S is 'convex' in the following sense. If $x_1, \ldots, x_n \in S$ and $\lambda_1, \ldots, \lambda_n \in K$, $|\lambda_i| \leqslant 1$ for all i, $\Sigma \lambda_i = 1$ then $\Sigma \lambda_i x_i \in S$ (we shall not offend the reader by giving a proof). Now let $x, y \in \mathbb{N} \cap B$, $x \neq y$. Let z be the common initial part of x and y (in other words, if $x = a_0 + a_1 p + \ldots, y = b_0 + b_1 p + \ldots, |x-y|_p = p^{-n} < 1$, then $a_0 = b_0, \ldots, a_{n-1} = b_{n-1}, a_n \neq b_n$, and $z = a_0 + a_1 p + \ldots + a_{n-1} p^{n-1}$; if $|x-y|_p = 1$, then $z = 0$). Then $z \in B$, $z \lhd x$, $z \lhd y$ and $\max(|z-x|_p, |z-y|_p) = |x-y|_p$. Now

$$\Phi_1 f(x, y) = \Phi_1 f(x, z) \frac{x-z}{x-y} + \Phi_1 f(z, y) \frac{z-y}{x-y}$$

which shows that $\Phi_1 f(x, y) \in S$ as soon as $\Phi_1 f(x, z)$ and $\Phi_1 f(z, y)$ are in S. This means that to prove $\Phi_1 f(x, y) \in S$ we may add the assumption $y \lhd x$. Then there exist distinct $t_1 \lhd t_2 \lhd t_3 \ldots \lhd t_n$ where $t_1 = y$, $t_n = x$ and such that $(t_j)_- = t_{j-1}$ $(j = 2, \ldots n)$. Clearly $t_j \in B$ for all j. Write

$$\Phi_1 f(x, y) = \sum_{j=2}^{n} \lambda_j \Phi_1 f(t_j, t_{j-1})$$

where $\lambda_j = (x-y)^{-1}(t_j - t_{j-1})$ $(j = 2, \ldots, n)$. It is easily verified that all $|\lambda_j|$ are $\leqslant 1$ and that $\Sigma_{j=2}^{n} \lambda_j = 1$. Since $\Phi_1 f(t_j, t_{j-1}) \in S$ for each j, the lemma follows.

As a corollary we obtain the following characterization (compare Theorem 53.4).

THEOREM 63.2. (Characterization of Lipschitz functions by van der Put coefficients) *Let* $f = \Sigma_{n=0}^{\infty} a_n e_n \in C(\mathbb{Z}_p \to K)$. *Then* $f \in \mathrm{Lip}_1(\mathbb{Z}_p \to K)$ *if and only if* $\sup_n |a_n| n < \infty$. *More precisely, for* $f = \Sigma_{n=0}^{\infty} a_n e_n \in C(\mathbb{Z}_p \to K)$ *we have the following.*

(i) $\|\Phi_1 f\|_\infty = \sup\{|a_n|\, |\gamma_n|_p^{-1} : n \in \mathbb{N}\}$.

(ii) $\|\Phi_1 f\|_\infty \leqslant \sup\{|a_n| n : n \in \mathbb{N}\} \leqslant p \|\Phi_1 f\|_\infty$.

(iii) *If* $f \in \mathrm{Lip}_1(\mathbb{Z}_p \to K)$ *then* $\|f\|_1 = \sup\{|a_n|\, |\gamma_n|_p^{-1} : n \in \{0, 1, 2, \ldots\}\}$.

Proof. The preceding lemma (for $B = \mathbb{Z}_p$) yields $\|\Phi_1 f\|_\infty = \sup\{|\Phi_1 f(n, n_-)| : n \in \mathbb{N}\} = \sup\{|n - n_-|_p^{-1}\, |f(n) - f(n_-)| : n \in \mathbb{N}\} = \sup\{|\gamma_n|_p^{-1}\, |a_n| : n \in \mathbb{N}\}$. The rest of the proof is obvious.

It is quite remarkable that the conditions on the coefficients of f in order that $f \in \mathrm{Lip}_1(\mathbb{Z}_p \to K)$ with respect to either Mahler's base or van der Put's base are exactly the same. This fact may become more striking if we look at the next characterization. Recall that $N^1(\mathbb{Z}_p \to K) = \{f \in C^1(\mathbb{Z}_p \to K) : f' = 0\}$.

THEOREM 63.3. (Characterization of C^1-functions with zero derivative. Compare also Theorem 69.7) *Let* $f = \sum_{n=0}^{\infty} a_n e_n \in C(\mathbb{Z}_p \to K)$. *Then*

$$f \in N^1(\mathbb{Z}_p \to K) \Leftrightarrow \lim_{n \to \infty} |a_n|\, n = 0 \Leftrightarrow \lim_{n \to \infty} a_n (n - n_-)^{-1} = 0$$

Proof. Let $f \in N^1(\mathbb{Z}_p \to K)$. Then $\overline{\Phi}_1 f$ (the unique continuous extension of $\Phi_1 f$) is a uniformly continuous function on $\mathbb{Z}_p \times \mathbb{Z}_p$ that vanishes on the diagonal. Therefore, $\lim_{n \to \infty} \Phi_1 f(n, n_-) = \lim_{n \to \infty} (\Phi_1 f(n, n_-) - \overline{\Phi}_1 f(n, n)) = 0$, because $\lim_{n \to \infty} (n - n_-) = 0$. Now $|\Phi_1 f(n, n_-)| = |n - n_-|_p^{-1} \cdot |a_n| \geqslant p^{-1} n |a_n|$ (Lemma 53.3). Hence $\lim_{n \to \infty} |a_n|\, n = 0$. If conversely $\lim_{n \to \infty} |a_n|\, n = 0$ then also $\lim_{n \to \infty} a_n \gamma_n^{-1} = 0$, i.e. $\lim_{n \to \infty} \Phi_1 f(n, n_-) = 0$. Let $a \in \mathbb{Z}_p$. If $n \in \mathbb{N}$ $(n \neq a)$ approaches a p-adically then n tends to infinity in the ordinary sense. It follows that

$$\lim_{n \to a} \Phi_1 f(n, n_-) = 0 \qquad (a \in \mathbb{Z}_p)$$

Let $a \in \mathbb{Z}_p$, $\epsilon > 0$. There is $\delta > 0$ such that $n \in \mathbb{N}$, $0 < |n - a|_p < \delta$ implies $|\Phi_1 f(n, n_-)| < \epsilon$. If a happens to be itself an element of $\mathbb{N}$, assume also $\delta < |a - a_-|_p$. Then the condition of Lemma 63.1 is fulfilled with $B = \{x \in \mathbb{Z}_p : |x - a|_p < \delta\}$ and $S = \{x \in K : |x| < \epsilon\}$. We conclude that $|\Phi_1 f(x, y)| < \epsilon$ for all $x, y \in B$, $x \neq y$, i.e. f is C^1 at a and $f'(a) = 0$.

We see that it is one and the same condition on the coefficients that characterizes C^1-functions in the case of the Mahler base and N^1-functions in the case of the van der Put base.

**Exercise* 63.A. ('Characterization' of C^1-functions by means of van der Put coefficients) Let $f = \sum_{n=0}^{\infty} a_n e_n \in C(\mathbb{Z}_p \to K)$. Show that $f \in C^1(\mathbb{Z}_p \to K)$ if and only if $\lim_{n \to a} a_n \gamma_n^{-1}$ exists for each $a \in \mathbb{Z}_p$.

Exercise 63.B. (Characterization of Lip_α-functions by means of van der Put coefficients) Let $f = \sum_{n=0}^{\infty} a_n e_n \in C(\mathbb{Z}_p \to K)$, let $\alpha > 0$. Prove that $f \in \mathrm{Lip}_\alpha(\mathbb{Z}_p \to K)$ if and only if $\sup \{|a_n|\, n^\alpha : n \in \mathbb{N}\} < \infty$. (Your solution applied to the case $\alpha = 1$ may very well yield a new proof of Theorem 63.2.).

Exercise 63.C. (Sequel to the previous exercise) Prove that

$$\bigcup_{\alpha > 1} \mathrm{Lip}_\alpha (\mathbb{Z}_p \to K) \subset N^1 (\mathbb{Z}_p \to K)$$

and that the inclusion is strict. (Hint. Find a p-adic sequence $a_0, a_1, \ldots$ for which $|a_n|\, n \to 0$ but $|a_n|\, n^\alpha$ is unbounded for each $\alpha > 1$.)

Exercise 63.D. (Characterization of C^n-functions with zero derivative by means of van der Put coefficients) Let $f = \sum_{m=0}^{\infty} a_m e_m \in C(\mathbb{Z}_p \to K)$.

Use condition (β) of Theorem 29.12 to show that for each $n \in \mathbb{N}$

$$f \in C^n(\mathbb{Z}_p \to K), f' = 0 \Leftrightarrow \lim_{m \to \infty} |a_m| \, m^n = 0$$

Deduce that

$$f \in C^\infty(\mathbb{Z}_p \to K), f' = 0 \Leftrightarrow \lim_{m \to \infty} |a_m| \, m^a = 0 \text{ for each } a > 0$$

Exercise 63.E. (Characterization of monotone functions by means of van der Put coefficients) Let $f = \sum_{n=0}^{\infty} a_n e_n \in C(\mathbb{Z}_p \to K)$. Show that f is increasing (Definition 24.6) if and only if $a_n \gamma_n^{-1}$ is positive for each $n \in \mathbb{N}$. Show also that f is monotone of type α if and only if $a_n \gamma_n^{-1} \in \alpha$ for each $n \in \mathbb{N}$.

Exercise 63.F. Let $f \in C^1(\mathbb{Z}_p \to K)$. Show that

$$\int_{\mathbb{Z}_p} f(x)dx = f(0) + p^{-1} \lim_{m \to \infty} \sum_{n=1}^{p^m - 1} \Phi_1 f(n, n_-) \, \tau(n)$$

where

$$\tau(n) := a_s \qquad (n = a_0 + a_1 p + \ldots + a_s p^s : a_s \neq 0)$$

(Use Exercises 63.A and 62.E.)

64. Antiderivation

In Section 30 we have seen that there are many antiderivation maps $C(\mathbb{Z}_p \to K) \to C^1(\mathbb{Z}_p \to K)$. In this section we single out a particular one by imposing extra conditions. (Compare also Exercise 30.C.)

For a given $f \in C(\mathbb{Z}_p \to K)$, a continuous function $g : \mathbb{Z}_p \to K$ for which $g(0) = 0$ and $\Phi_1 g(n, n_-) = f(n_-)$ for all $n \in \mathbb{N}$ is uniquely determined. Indeed, the conditions on g prescribe its van der Put coefficients so g necessarily has the form $g = \sum_{n=1}^{\infty} f(n_-)(n - n_-) e_n$. Conversely, the latter formula defines a continuous function g (since $\lim_{n \to \infty} f(n_-)(n - n_-) = 0$) and $g(n) - g(n_-) = f(n_-)(n - n_-)$ for all $n \in \mathbb{N}$, $g(0) = 0$. Therefore the following definition is meaningful.

DEFINITION 64.1. For $f \in C(\mathbb{Z}_p \to K)$ we denote by Pf the unique continuous function satisfying $Pf(0) = 0$ and $Pf(n) - Pf(n_-) = (n - n_-) f(n_-)$ for all $n \in \mathbb{N}$.

THEOREM 64.2. (Properties of P) *P is a linear isometry of $C(\mathbb{Z}_p \to K)$ into*

$C^1(\mathbb{Z}_p \to K)$. For each $f \in C(\mathbb{Z}_p \to K)$, Pf is an antiderivative of f with the following mean value property.

$$\left|\frac{Pf(x) - Pf(y)}{x - y}\right| \leqslant \max \{|f(z)| : z \in [x, y]\} \quad (x, y \in \mathbb{Z}_p, x \neq y)$$

(where $[x, y]$ denotes the smallest disc containing x and y).

Proof. From the definition of P it follows that for each $a \in \mathbb{Z}_p$ we have $\lim_{n \to a} (Pf(n) - Pf(n_-))/(n - n_-) = \lim_{n \to a} f(n_-) = f(a)$. (If $n \in \mathbb{N}$, $n \neq a$, is p-adically close to a then n is large in the ordinary sense so that $|n - n_-|_p$ is small. Hence, n_- is also p-adically close to a.) A simple application of Lemma 63.1 where S is a disc in K with centre $f(a)$ and B is a sufficiently small disc in $\mathbb{Z}_p$ with centre a leads to $\lim_{x, y \to a} (Pf(x) - Pf(y))/(x - y) = f(a)$, i.e. Pf is a C^1-antiderivative of f. Likewise one can arrive at the mean value property of Pf (choose $B := [x, y]$, $S := \{t \in K : |t| \leqslant \max \{|f(z)| : z \in B\}\}$). The linearity of P is easy. Let $f \in C(\mathbb{Z}_p \to K)$. By Exercise 53.A and the mean value property we have $\|Pf\|_1 = |Pf(0)| \vee \|\Phi_1 Pf\|_\infty = \|\Phi_1 Pf\|_\infty \leqslant \|f\|_\infty$. On the other hand, $\|f\|_\infty = \|(Pf)'\|_\infty \leqslant \|Pf\|_1$. Hence, P is an isometry.

THEOREM 64.3. (Formulas for Pf) *Let x_n be as in Proposition 62.1. We have for $f \in C(\mathbb{Z}_p \to K)$*

$$Pf(x) = \sum_{n=0}^{\infty} f(x_n)(x_{n+1} - x_n) \qquad (x \in \mathbb{Z}_p)$$

More concretely, if $x = \sum_{n=0}^{\infty} b_n p^n \in \mathbb{Z}_p$ then

$$Pf(x) = f(0)b_0 + \sum_{n=1}^{\infty} f\left(\sum_{j=0}^{n-1} b_j p^j\right) b_n p^n$$

If $f = \sum_{n=0}^{\infty} a_n e_n$ then

$$Pf(x) = \sum_{n=0}^{\infty} a_n (x - n) e_n(x) \qquad (x \in \mathbb{Z}_p)$$

Finally, we have

$$Pf = \sum_{n=1}^{\infty} f(n_-)(n - n_-)e_n$$

Proof. To prove the first formula, let $h(x) := \sum_{n=0}^{\infty} f(x_n)(x_{n+1} - x_n)$ for all $x \in \mathbb{Z}_p$. h is a well-defined continuous function, $h(0) = 0$. If $m \in \mathbb{N}$, $m = b_0 + b_1 p + \ldots + b_s p^s$ $(b_s \neq 0)$ we have $h(m) = f(0)b_0 + f(b_0)b_1 p +$

$f(b_0 + b_1 p)b_2 p^2 + \ldots + f(b_0 + b_1 p + \ldots + b_{s-1} p^{s-1})\, b_s p^s$. By considering a similar formula for $h(m_-)$ and subtracting we get $h(m) - h(m_-) = f(m_-) b_s p^s = f(m_-)\,(m - m_-)$. Thus, h satisfies the conditions of Definition 64.1, i.e. $h = Pf$. We prove the third formula in a similar way. Let $t(x) := \sum_{n=0}^{\infty} a_n (x - n)\, e_n(x)$ $(x \in \mathbb{Z}_p)$. Then t is continuous, $t(0) = 0$ and for $m \in \mathbb{N}$ we have $t(m) = \sum_{n \lhd m} a_n (m - n)$, $t(m_-) = \sum_{n \lhd m_-} a_n (m_- - n)$ so that $t(m) - t(m_-) = \sum_{n \lhd m_-} a_n (m - m_-) = (m - m_-) \sum_{n \lhd m_-} a_n = (m - m_-) f(m_-)$. Hence, $t = Pf$. For the remaining two formulas no proof is needed.

Exercise 64.A. Show that $P(\xi_{\mathbb{Z}_p}) = \ast$ but that $P(\ast) \neq \frac{1}{2}\ast^2$ (compare Exercise 30.C.). In fact, prove that $P\ast = \frac{1}{2}\ast^2 - \frac{1}{2} \sum_{n=1}^{\infty} (n - n_-)^2 e_n$.

**Exercise* 64.B. Let $f \in C(\mathbb{Z}_p \to K)$ be locally constant. Show that Pf is locally linear (i.e. each point of $\mathbb{Z}_p$ has a neighbourhood U on which f has the form $x \mapsto ax + b$ for some $a, b \in K$ depending on U).

**Exercise* 64.C. ('Compactness' of the antiderivation operator) For $m \in \mathbb{N}$, $f \in C(\mathbb{Z}_p \to K)$ set

$$Q_m f(x) := \sum_{n=m}^{\infty} f(x_n)(x_{n+1} - x_n) \qquad (x \in \mathbb{Z}_p)$$

(i) Show that $Q_m f$ is a C^1-antiderivative of f. (Consider $P - Q_m$.) Deduce that *a continuous function on $\mathbb{Z}_p$ has arbitrarily small antiderivatives.*
(ii) Prove that $\lim_{m \to \infty} \|Q_m\| = 0$ (see Proposition 13.5) so that $\lim_{m \to \infty} (P - Q_m) = P$, and also that the range of $P - Q_m$ is finite dimensional. (Thus, P is compact in the sense that it is the limit of a sequence of operators with finite dimensional range.)

Exercise 64.D. (Other constructions of antiderivation operators) The formula $Pf(x) = \sum_{n=0}^{\infty} a_n (x - n) e_n(x)$ of Theorem 64.3 can be read as

$$Pf = \sum_{n=0}^{\infty} a_n P e_n$$

This leads to the following idea. Let $u_0, u_1, \ldots$ be *any* orthonormal base of $C(\mathbb{Z}_p \to K)$. For each n we select an antiderivative $U_n \in C^1(\mathbb{Z}_p \to K)$. Can we do it in such a way that for any null sequence $a_0, a_1, \ldots$ in K we have $\sum_{n=0}^{\infty} a_n U_n \in C^1(\mathbb{Z}_p \to K)$ and $(\sum_{n=0}^{\infty} a_n U_n)' = \sum_{n=0}^{\infty} a_n u_n$? To answer this question, prove the following.
(i) (Concrete part) Let $g \in C(\mathbb{Z}_p \to K)$, $\|g\|_\infty = 1$. Then g has a C^1-antiderivative G for which $\|G\|_1 = 1$.

(ii) (Abstract part) Let $u_0, u_1, \dots$ be an orthonormal base of $C(\mathbb{Z}_p \to K)$, and let $U_0, U_1, \dots \in C^1(\mathbb{Z}_p \to K)$ be such that $\|U_n\|_1 = 1$ and $U_n' = u_n$ for each n. Then the formula

$$\sum_{n=0}^{\infty} a_n u_n \mapsto \sum_{n=0}^{\infty} a_n U_n \qquad ((a_0, a_1, \dots) \in c_0)$$

defines an isometrical antiderivation operator $C(\mathbb{Z}_p \to K) \to C^1(\mathbb{Z}_p \to K)$.

65. The differential equation $y' = F(x, y)$

By way of example we shall treat the p-adic counterpart of the well-known classical differential equation $y' = F(x, y)$ where F is a continuous function of two variables satisfying the Lipschitz condition $|F(x, y) - F(x, z)| \leqslant M|y - z|$. We put the question in a slightly more general way as follows.

PROBLEM. Let A be a (not necessarily linear) map of $C(\mathbb{Z}_p \to K)$ into itself satisfying

$$\|Af - Ag\|_\infty \leqslant M\|f - g\|_\infty \qquad (f, g \in C(\mathbb{Z}_p \to K))$$

where M is a fixed positive real number. Describe the solution set of the differential equation

$$(*) \qquad f' = Af \qquad (f \in C^1(\mathbb{Z}_p \to K))$$

(It is easy to see that the equation $y' = F(x, y)$ is a particular case of (*): choose $Af(x) := F(x, f(x))$ $(f \in C(\mathbb{Z}_p \to K),\ x \in \mathbb{Z}_p)$, but (*) covers also other (weird) types of differential equations such as $f'(x) = f(x^2)$ $(x \in \mathbb{Z}_p)$.)

Suppose f is a solution of (*). Let $Q : C(\mathbb{Z}_p \to K) \to C^1(\mathbb{Z}_p \to K)$ be any antiderivation map. Then $f = f - Qf' + Qf' = h + QAf$ where $h' = 0$, in other words $h \in N^1(\mathbb{Z}_p \to K)$. Thus, if f is a solution of (*) then there is an $h \in N^1(\mathbb{Z}_p \to K)$ such that $f = h + QAf$. Conversely, if $f = h + QAf$ for some $h \in N^1(\mathbb{Z}_p \to K)$ then $f \in C^1(\mathbb{Z}_p \to K)$ and $f' = (QAf)' = Af$ so that f is a solution of (*). So we have 'integrated' (*) in the following sense. We obtain all solutions of (*) by solving for each $h \in N^1(\mathbb{Z}_p \to K)$

$$(**) \qquad f = h + QAf \qquad (f \in C(\mathbb{Z}_p \to K))$$

(Solutions of (**) are automatically C^1.) We now prove that, with a proper choice of Q, (**) has a unique solution.

Define $T_h : C(\mathbb{Z}_p \to K) \to C(\mathbb{Z}_p \to K)$ by the formula

$$T_h(f) = h + QAf$$

We have

$$\|T_h f - T_h g\|_\infty = \|QAf - QAg\|_\infty \qquad (f, g \in C(\mathbb{Z}_p \to K))$$

If we choose

$$Qf(x) = \sum_{n=m}^{\infty} f(x_n)(x_{n+1} - x_n) \qquad (x \in \mathbb{Z}_p, f \in C(\mathbb{Z}_p \to K))$$

where m is sufficiently large we can arrange that $\|Qf\|_\infty \leq (2M)^{-1} \|f\|_\infty$ for all $f \in C(\mathbb{Z}_p \to K)$ (see Exercise 64.C). It follows that

$$\|T_h f - T_h g\|_\infty \leq (2M)^{-1} \|Af - Ag\|_\infty \leq \tfrac{1}{2} \|f - g\|_\infty \quad (f, g \in C(\mathbb{Z}_p \to K))$$

By Banach's contraction theorem (Appendix A.1) T_h has a fixed point t_h which is the unique solution of (**). So we see that *the collection* $\{t_h : h \in N^1(\mathbb{Z}_p \to K)\}$ *is a complete set of solutions of* (*). We claim that the map $h \mapsto t_h$ $(h \in N^1(\mathbb{Z}_p \to K))$ is an isometry with respect to $\| \ \|_\infty$. In fact, if $h_1, h_2 \in N^1(\mathbb{Z}_p \to K)$ and $h_1 \neq h_2$ then also $t_{h_1} \neq t_{h_2}$ and $t_{h_1} - t_{h_2} = h_1 - h_2 + QAt_{h_1} - QAt_{h_2}$, where $\|QAt_{h_1} - QAt_{h_2}\| \leq \frac{1}{2} \|t_{h_1} - t_{h_2}\|_\infty$ so that by the strong triangle inequality $\|t_{h_1} - t_{h_2}\|_\infty = \|h_1 - h_2\|_\infty$. We have proved the following.

THEOREM 65.1. *Let* $A : C(\mathbb{Z}_p \to K) \to C(\mathbb{Z}_p \to K)$ *satisfy a Lipschitz condition* $\|Af - Ag\|_\infty \leq M \|f - g\|_\infty$ $(f, g \in C(\mathbb{Z}_p \to K))$. *Then there is an isometrical map (with respect to* $\| \ \|_\infty$*) of* $N^1(\mathbb{Z}_p \to K)$ *onto the solution set of the differential equation* $f' = Af$ $(f \in C^1(\mathbb{Z}_p \to K))$.

Exercise 65.A. Does the correspondence $h \mapsto t_h$ behave well too with respect to the norm $\| \ \|_1$?

66. C^1-solutions of a meromorphic differential equation

From the theory of p-adic differential equations with *analytic* solutions (which goes beyond the scope of this book) we select one particular problem for which the C^1-theory can be of some help. Let $\lambda \in \mathbb{Z}_p$. Consider the differential equation

$$(*) \qquad xf'(x) - \lambda f(x) = (1 - x)^{-1} \qquad (x \in D)$$

PROBLEM. Does there exist a neighbourhood $D \subset B_0(1^-) \subset \mathbb{Z}_p$ of 0 and an analytic function $f : D \to \mathbb{C}_p$ such that (*) is satisfied?

To solve the problem we set $f(x) = \sum_{n=0}^{\infty} b_n x^n$, $(1 - x)^{-1} = \sum_{n=0}^{\infty} x^n$. Substitution of these power series in (*) yields

$$(n - \lambda) b_n = 1 \qquad (n \in \{0, 1, 2, \dots\})$$

To avoid needless complications, assume $\lambda \notin \{0, 1, 2, \dots\}$. The crucial question is whether the radius of convergence of $\Sigma\, b_n x^n = \Sigma\,(n-\lambda)^{-1} x^n$ is positive, i.e. whether $\underline{\lim}_{n\to\infty} \sqrt[n]{|n-\lambda|_p} > 0$. This leads to

DEFINITION 66.1. For each $\lambda \in \mathbb{Z}_p$ set $\nu(\lambda) := \underline{\lim}_{n\to\infty} \sqrt[n]{|n-\lambda|_p}$. λ is a *p-adic Liouville number* if $\nu(\lambda) = 0$.

**Exercise* 66.A. Show that $\nu(\lambda) = 1$ for $\lambda \in \{0, 1, 2, \ldots\}$.

We may conclude the following. *If $\lambda \in \mathbb{Z}_p$ is a Liouville number then* (*) *has an analytic solution for no neighbourhood D of* 0.

In the following two exercises we shall solve a more general form of (*). Existence and further properties of p-adic Liouville numbers shall be discussed in Section 67.

Exercise 66.B. Let $\lambda \in \mathbb{Z}_p \backslash \{0, 1, 2, \ldots\}$. Show that the following conditions are equivalent.
(α) For every neighbourhood U of 0 and for every analytic function g: $U \to K$ there is an analytic function f defined on some neighbourhood V of 0 ($V \subset U$) such that $xf'(x) - \lambda f(x) = g(x)$ for all $x \in V$.
(β) $\nu(\lambda) > 0$.

Exercise 66.C. Let $\lambda \in \{0, 1, 2, \ldots\}$ and let g be an analytic function, defined on some neighbourhood U of 0, with power series $f(x) = \Sigma_{n=0}^{\infty} a_n x^n$. Show that the following conditions are equivalent. (Compare Exercise 66.A.)
(α) There is a neighbourhood V of 0 ($V \subset U$) and an analytic function f on V such that $xf'(x) - \lambda f(x) = g(x)$ for all $x \in V$.
(β) $a_\lambda = 0$.

The fact that for a Liouville number λ and certain analytic functions g the differential equation $xf'(x) - \lambda f(x) = g(x)$ has no analytic solutions has led to the following problem stated by van der Put (Meromorphic differential equations over valued fields. *Indag. Math.* 42, Fasc. 3 (1980)).

PROBLEM. Let $g : \mathbb{Z}_p \to K$ be a continuous function and let $\lambda \in \mathbb{Z}_p$. Does there exist a C^1-function $f : \mathbb{Z}_p \to K$ such that $xf'(x) - \lambda f(x) = g(x)$ for all $x \in \mathbb{Z}_p$?

We shall prove that the answer, with trivial restrictions, is 'yes'. Indeed, there are a few conditions we are forced to impose on g. First, if f is a C^1-solution then $(g(x) - g(0))/x = (xf'(x) - \lambda(f(x) - f(0)))/x$ ($x \in \mathbb{Z}_p, x \neq 0$). Apparently we have to require that g is differentiable at 0. Further, from $g'(0) = (1-\lambda)f'(0)$ we see that $g'(0) = 0$ if $\lambda = 1$. Finally, $g(0) = 0$ if $\lambda = 0$.

These conditions turn out to be sufficient according to the following theorem.

THEOREM 66.2. *Let $g : \mathbb{Z}_p \to K$ be a continuous function that is differentiable at 0, and let $\lambda \in \mathbb{Z}_p$. If $\lambda = 0$, assume $g(0) = 0$, if $\lambda = 1$ assume $g'(0) = 0$. Then there is a C^1-function $f : \mathbb{Z}_p \to K$ such that $xf'(x) - \lambda f(x) = g(x)$ for all $x \in \mathbb{Z}_p$.*

Proof. First suppose that $g(0) = g'(0) = 0$. Let P be the antiderivation map of Definition 64.1, restricted to $C_0 := \{f \in C(\mathbb{Z}_p \to K) : f(0) = 0\}$. We shall find a $u \in C_0$ such that $f := Pu$ is a solution of the differential equation. The latter is the case if and only if u satisfies the equation $xu(x) - \lambda Pu(x) = g(x)$ $(x \in \mathbb{Z}_p)$. One easily verifies that the map Q defined by

$$Qu(x) := \begin{cases} x^{-1}(\lambda Pu(x) + g(x)) & \text{if } x \in \mathbb{Z}_p \setminus \{0\} \\ 0 & \text{if } x = 0 \end{cases}$$

maps C_0 into C_0. To prove that Q has a fixed point we estimate $Pu(x)$ for $x \in \mathbb{Z}_p$, $|x|_p = p^{-m}$. We have $x_0 = x_1 = \ldots = x_m = 0$, $x_{m+1} \neq 0$ (terminology of Section 62) so that (Theorem 64.3) $Pu(x) = \sum_{n \geqslant m+1} u(x_n)(x_{n+1} - x_n)$. Hence $|Pu(x)| \leqslant \|u\|_\infty \max_{n \geqslant m+1} |x_{n+1} - x_n|_p = \|u\|_\infty |x_{m+1} - x_{m+2}|_p \leqslant \|u\|_\infty p^{-m-1} = p^{-1}|x|_p \|u\|_\infty$. For $u, v \in C_0$ we therefore have $\|Qu - Qv\|_\infty = \sup_{x \neq 0} |x|_p^{-1} |\lambda| |Pu(x) - Pv(x)| \leqslant \sup_{x \neq 0} |x|_p^{-1} \cdot |P(u - v)(x)| \leqslant \sup_{x \neq 0} |x|_p^{-1} p^{-1} |x|_p \|u - v\|_\infty \leqslant p^{-1}\|u - v\|_\infty$. We see that Q is a contraction and has a fixed point u. Then $f = Pu$ is the required solution. To solve the general case we apply the first part of this proof to $x \mapsto g(x) - g'(0)x - g(0)$ in place of g yielding a C^1-function f_1 for which $xf_1'(x) - \lambda f_1(x) = g(x) - g'(0)x - g(0)$ for all $x \in \mathbb{Z}_p$. It is easily verified that $f_2(x)$ defined for $x \in \mathbb{Z}_p$ by the formula

$$f_2(x) := \begin{cases} (1-\lambda)^{-1} g'(0)x - \lambda^{-1} g(0) & \text{if } \lambda \in \mathbb{Z}_p \setminus \{0, 1\} \\ g'(0)x & \text{if } \lambda = 0 \\ -g(0) & \text{if } \lambda = 1 \end{cases}$$

satisfies $xf_2'(x) - \lambda f_2(x) = g'(0)x + g(0)$. It follows that $f := f_1 + f_2$ is the required solution.

Remarks.

1. It is quite striking that in Theorem 66.2 a condition of *ordinary* differentiability (rather than a C^1-condition) appears in a natural context.
2. I do not know (and it seems to be an open problem) whether the differential equation $xy' - \lambda y = g$ has C^∞-solutions for every C^∞-function g with $g(0) = g'(0) = 0$.

Exercise 66.D. (Nonuniqueness of solutions of $xy' - \lambda y = g$) Let λ, g satisfy the conditions of Theorem 66.2. Show that $\{f \in C^1(\mathbb{Z}_p \to K) : xf'(x) - \lambda f(x) = g(x)$ for all $x \in \mathbb{Z}_p\}$ is an additive coset of $H := \{f \in C^1(\mathbb{Z}_p \to K) : xf'(x) - \lambda f(x) = 0\}$. Show that H is infinite dimensional.

67. *p*-adic Liouville numbers

In the previous section we have met the p-adic Liouville numbers, i.e. numbers $\lambda \in \mathbb{Z}_p$ for which $\underline{\lim}_{n\to\infty} \sqrt[n]{|n-\lambda|_p} = 0$. The properties of these numbers will show a striking analogy with the real Liouville numbers. See, for example, J. C. Oxtoby's book (*Measure and Category*, Springer-Verlag, New York (1971), Section 2).

A *gap* in the p-adic expansion $x = \sum_{j=0}^{\infty} a_j p^j$ of an element x of $\mathbb{Z}_p$ is a pair of numbers $s < t$ such that $a_s \neq 0, a_{s+1} = a_{s+2} = \ldots = a_{t-1} = 0, a_t \neq 0$. The *length* of such a gap is the number $[t/p^s]$. Condition (β) of the following theorem makes it more visible which p-adic integers are Liouville numbers and also yields a method to construct them.

THEOREM 67.1. *Let $\lambda \in \mathbb{Z}_p$. The following conditions are equivalent.*
(α) *λ is a Liouville number.*
(β) *The expansion of λ has arbitrarily long gaps.*

Proof. (α) $\Rightarrow$ (β). Let λ be a Liouville number and let $k \in \mathbb{N}$. We shall determine a gap in the expansion of λ with length $\geqslant k$. There is an $n \in \mathbb{N}$ such that $\sqrt[n]{|n-\lambda|_p} < p^{-k}$, in other words $|n-\lambda|_p < p^{-nk}$, so the coefficients with index $\leqslant nk$ of the expansions of n and λ coincide. Let $n = a_0 + a_1 p + \ldots + a_s p^s$ ($a_s \neq 0$), then $\lambda = \sum_{j=0}^{\infty} a_j p^j$ where $a_{s+1} = a_{s+2} = \ldots = a_{nk} = 0$. By Exercise 66.A we have $\lambda \notin \{0, 1, 2, \ldots\}$ so that $a_j \neq 0$ for some $j > nk$. We see that the expansion of λ has a gap with length $\geqslant [nk/p^s] \geqslant [p^s k/p^s] = k$.
(β) $\Rightarrow$ (α). Let $s < t$ be a gap in the p-adic expansion of $\lambda = \sum_{j=0}^{\infty} a_j p^j$ of length m. Choose $n = \sum_{j=0}^{s} a_j p^j$. Then $|n-\lambda|_p = |a_t p^t|_p = p^{-t}$ so that $\sqrt[n]{|n-\lambda|_p} \leqslant p^{-t/n}$. Now $t/n \geqslant t/p^{s+1} \geqslant p^{-1}\,[t/p^s] = p^{-1} m$. Let $\epsilon > 0$, $N \in \mathbb{N}$. By choosing m large enough we can arrange that $n \geqslant N$ and $\sqrt[n]{|n-\lambda|_p} < \epsilon$. It follows that $\underline{\lim}_{n\to\infty} \sqrt[n]{|n-\lambda|_p} = 0$.

Exercise 67.A. Find $a_0, a_1, \ldots \in \{0, 1, \ldots, p-1\}$ such that $\sum_{j=0}^{\infty} a_j p^j$ is a p-adic Liouville number.

THEOREM 67.2. *A p-adic Liouville number is not algebraic over $\mathbb{Q}$.*

Proof. Let $a \in \mathbb{Z}_p$ be algebraic over $\mathbb{Q}$: we prove that $\nu(a) = \underline{\lim}_{n\to\infty} \sqrt[n]{|n-\lambda|_p}$

$= 1$. There is a nonzero polynomial function $f: \mathbb{Z}_p \to \mathbb{Q}_p$ with coefficients in $\mathbb{Z}$ such that $f(a) = 0$. Let

$$f(x) = a_0 + a_1 x + \ldots + a_d x^d \qquad (x \in \mathbb{Z}_p, a_d \neq 0)$$

Since $f \in C^1 (\mathbb{Z}_p \to \mathbb{Q}_p)$ it satisfies a Lipschitz condition

$$|f(x) - f(y)|_p \leqslant c|x - y|_p \qquad (x, y \in \mathbb{Z}_p)$$

so that for each $n \in \mathbb{N}$

$$(*) \qquad |f(n)|_p = |f(n) - f(a)|_p \leqslant c|n - a|_p$$

f has at most d zeros in $\mathbb{Q}$ so $f(n) \neq 0$ for sufficiently large n. Further, $f(n) \in \mathbb{Z}$ and we may apply the product formula of Exercise 10.B obtaining $|f(n)|_p \geqslant |f(n)|_\infty^{-1}$ (here $|\ \ |_\infty$ is the ordinary absolute value function on $\mathbb{Q}$). We have

$$(**) \qquad |f(n)|_p^{-1} \leqslant |f(n)|_\infty = |a_0 + a_1 n + \ldots + a_d n^d|_\infty \leqslant c' n^d$$

where, for example, $c' = |a_0|_\infty + |a_1|_\infty + \ldots + |a_d|_\infty$. Combining (*) and (**) we find for large n

$$|n - a|_p \geqslant c'' n^{-d}$$

where c'' is a positive constant. It follows that $\nu(a) = 1$.

We now show that the set of the p-adic Liouville numbers is 'big' in one sense but 'small' in another. To this end we introduce the following notions. A subset of $\mathbb{Z}_p$ is a *G_δ-set* if it is a countable intersection of open sets. A subset of $\mathbb{Z}_p$ is a *null set* if for every $\epsilon > 0$ it can be covered by countably many discs $B_1, B_2, \ldots$ such that $\Sigma_j d(B_j) < \epsilon$. (Recall that $d(B_j)$ is the diameter of B_j.)

THEOREM 67.3. *The p-adic Liouville numbers form a dense G_δ-subset of $\mathbb{Z}_p$.*

Proof. Let $n \in \mathbb{N}$. By a proper choice of $n_1 < n_2 < \ldots$ we can arrange that $n + p^{n_1} + p^{n_2} + \ldots$ is close to n and has arbitrarily long gaps. So the closure of the set of Liouville numbers contains $\mathbb{N}$, hence equals $\mathbb{Z}_p$. To show that the Liouville numbers form a G_δ-set observe that a is a Liouville number if and only if for all $k \in \mathbb{N}$ and $n \in \mathbb{N}$ there is $m \geqslant n$ such that $|m - a|_p < k^{-m}$. Thus, a is a Liouville number if and only if

$$a \in \bigcap_{k \in \mathbb{N}} \bigcap_{n \in \mathbb{N}} \bigcup_{\substack{m \geqslant n \\ m \in \mathbb{N}}} U_{mk}$$

where $U_{mk} := \{\beta \in \mathbb{Z}_p \mid |m - \beta|_p < k^{-m}\}$, and we are done.

(The Baire category theorem (Appendix A.1) implies that the non-Liouville numbers in $\mathbb{Z}_p$ are rare in the sense that they form a meagre F_σ -set.)

THEOREM 67.4. *The p-adic Liouville numbers form a null set in* $\mathbb{Z}_p$.
Proof. Let U_{mk} be as in the proof of the preceding theorem. For each $k, n \in \mathbb{N}$ the set of Liouville numbers is contained in $\bigcup_{m \geq n} U_{mk}$. The diameter $d(U_{mk})$ is of course less than k^{-m} so that (if $k \geq 2$) $\sum_{m \geq n} d(U_{mk}) \leq \sum_{m \geq n} k^{-m} \leq k^{-n+1}$. By choosing k or n large we can make k^{-n+1} smaller than a prescribed $\epsilon > 0$. The theorem follows.

Exercise 67.B. Let $\lambda \in \mathbb{Z}_p$ be a Liouville number and $n \in \mathbb{N}$. Are $n + \lambda$ and $n\lambda$ Liouville numbers?

Exercise 67.C. Let $\lambda \in \mathbb{Z}_p$ be a Liouville number. Show that $\lim_{n\to\infty} \sqrt[n]{|\binom{\lambda}{n}|_p} = 0$. (Compare Exercise 47.D.)

68. van der Put's base of C^1 ($\mathbb{Z}_p \to K$)

To end this chapter we return to the subject of the van der Put base $e_0, e_1, \ldots$ of $C(\mathbb{Z}_p \to K)$. Does the latter act also as a base of C^1 ($\mathbb{Z}_p \to K$)? (Compare the behaviour of Mahler's base in this respect, see Theorem 53.5 (iii).) It is easily seen that the answer is 'no'. In fact, the elements $e_0, e_1, \ldots$ are locally constant so their K-linear span is in N^1 ($\mathbb{Z}_p \to K$) and the latter is a closed proper subspace of C^1 ($\mathbb{Z}_p \to K$).

THEOREM 68.1. *Let* $\gamma_0, \gamma_1, \ldots$ *be as in Definition* 53.2. *Let P be the antiderivation map of Definition* 64.1. *Then the functions* $\gamma_0 e_0, \gamma_1 e_1, \ldots, Pe_0, Pe_1, \ldots$ *form an orthonormal base of* C^1 ($\mathbb{Z}_p \to K$); $\gamma_0 e_0, \gamma_1 e_1, \ldots$ *is an orthonormal base of* N^1 ($\mathbb{Z}_p \to K$).
Proof. The functions $e_0, e_1, \ldots$ form an orthonormal set in $C(\mathbb{Z}_p \to K)$. Since $P : C(\mathbb{Z}_p \to K) \to C^1(\mathbb{Z}_p \to K)$ is an isometry (Theorem 64.2) we have that $Pe_0, Pe_1, \ldots$ is orthonormal in C^1 ($\mathbb{Z}_p \to K$). From Theorem 63.2 (iii) applied to a finite linear combination of $e_0, e_1, \ldots$ it follows that $\gamma_0 e_0, \gamma_1 e_1, \ldots$ is orthonormal. For $f \in N^1$ ($\mathbb{Z}_p \to K$) and $g \in C(\mathbb{Z}_p \to K)$ we have $\|f + Pg\|_1 \geq \|(f + Pg)'\|_\infty = \|(Pg)'\|_\infty = \|g\|_\infty = \|Pg\|_1$. By Lemma 13.4 we also have $\|f + Pg\|_1 \geq \|f\|_1$ so $N^1(\mathbb{Z}_p \to K) \perp \operatorname{Im} P$, in particular $[\![e_0, e_1, \ldots]\!] \perp [\![Pe_0, Pe_1, \ldots]\!]$. We see that $\{\gamma_0 e_0, \gamma_1 e_1, \ldots, Pe_0, Pe_1, \ldots\}$ is an orthonormal set in C^1 ($\mathbb{Z}_p \to K$). To show that it is in fact a base, let $f \in C^1$ ($\mathbb{Z}_p \to K$). We have $f = f - Pf' + Pf'$ and $f' = f'(0)\, e_0 + \sum_{n=1}^{\infty}$

$(f'(n)-f'(n_-))\, e_n$ in the sense of $\| \quad \|_\infty$. P is a linear isometry so we have

$$Pf' = f'(0)\, Pe_0 + \sum_{n=1}^{\infty} (f'(n)-f'(n_-))\, Pe_n$$

in the sense of $\| \quad \|_1$; $g := f - Pf'$ is in $N^1(\mathbb{Z}_p \to K)$, let

$$(*) \qquad g = g(0)\, e_0 + \sum_{n=1}^{\infty} (g(n)-g(n_-))\, e_n$$

be its expansion in $C(\mathbb{Z}_p \to K)$. According to Theorem 63.3, $\lim_{n\to\infty} \Phi_1 g(n, n_-) = 0$ so that $\|(g(n)-g(n_-))\ e_n\|_1 = \|\Phi_1 g(n, n_-)\ \gamma_n e_n\|_1 = |\Phi_1 g(n, n_-)|$ tends to zero for $n \to \infty$. The series of (*) therefore converges in the sense of $\| \quad \|_1$ and (*) is true as an identity in $C^1(\mathbb{Z}_p \to K)$. We see that $f = f - Pf' + Pf'$ can be written as a convergent linear combination of $e_0, e_1, \ldots, Pe_0, Pe_1, \ldots$ and that if $f \in N^1(\mathbb{Z}_p \to K)$ then $f = f - Pf'$ is a combination of $e_0, e_1, \ldots$ The theorem follows.

COROLLARY 68.2. (Coefficients with respect to $e_0, e_1, \ldots, Pe_0, Pe_1, \ldots$) *Let $f \in C^1(\mathbb{Z}_p \to K)$ have the expansion*

$$f = \sum_{n=0}^{\infty} a_n e_n + \sum_{n=0}^{\infty} b_n Pe_n$$

Then

$$a_n = \begin{cases} f(0) \text{ if } n = 0 \\ f(n) - f(n_-) - (n - n_-)\, f'(n_-) \text{ if } n \in \mathbb{N} \end{cases}$$

$$b_n = \begin{cases} f'(0) \text{ if } n = 0 \\ f'(n) - f'(n_-) \text{ if } n \in \mathbb{N} \end{cases}$$

Proof. From the proof of Theorem 68.1 it follows that (with $g = f - Pf'$) $f = g(0)\, e_0 + \Sigma_{n=1}^{\infty} (g(n) - g(n_-))\, e_n + f'(0)\, Pe_0 + \Sigma_{n=1}^{\infty} (f'(n) - f'(n_-)) Pe_n$. The corollary follows after observing that $a_0 = g(0) = f(0) - Pf'(0) = f(0)$ and $g(n) - g(n_-) = f(n) - f(n_-) - Pf'(n) + Pf'(n_-) = f(n) - f(n_-) - (n - n_-) f'(n_-)$ by Definition 64.1.

COROLLARY 68.3. *The locally constant functions form a dense subset of $N^1(\mathbb{Z}_p \to K)$. The locally linear functions form a dense subset of $C^1(\mathbb{Z}_p \to K)$.*

Proof. A finite linear combination of $e_0, e_1, \ldots$ is locally constant. A finite linear combination of $Pe_0, Pe_1, \ldots$ is locally linear (Exercise 64.B). Now apply Theorem 68.1.

Exercise 68.A. (The Volkenborn integral versus the base of $C^1(\mathbb{Z}_p \to K)$) Compute $\int_{\mathbb{Z}_p} Pe_n(x)\,dx$ for each n (yes, the answers are rather surprising). Use Exercise 62.E (i) to express $\int_{\mathbb{Z}_p} f(x)\,dx$ for $f = \sum_{n=0}^{\infty} a_n e_n + \sum_{n=0}^{\infty} b_n Pe_n \in C^1(\mathbb{Z}_p \to K)$ in terms of $a_0, a_1, \ldots, b_0, b_1, \ldots$

Exercise 68.B. (Extensions of the Volkenborn integral on $N^1(\mathbb{Z}_p \to K)$) Let $\mu : N^1(\mathbb{Z}_p \to K) \to K$ be the Volkenborn integral, restricted to $N^1(\mathbb{Z}_p \to K)$. In Corollary 55.6 we saw that μ is shift invariant.

(i) Prove that if $\nu \in N^1(\mathbb{Z}_p \to K)'$ is shift invariant then ν is a scalar multiple of μ.

(ii) Use Theorem 68.1 to show that μ can be extended in many ways to an element of $C^1(\mathbb{Z}_p \to K)'$ but that none of these extensions is shift invariant.

4

More General Theory of Functions

THROUGHOUT CHAPTER 4 X IS A NONEMPTY SUBSET OF K

In Chapter 4 we shall study continuity, differentiability, monotonicity, . . . of functions $X \to K$. However this does not mean that we are aiming at mere generalizations of the results of Chapter 3. On the contrary, it will turn out that the case $X = \mathbb{Z}_p$, $K = \mathbb{C}_p$ often yields new and non-trivial statements whose proofs just happen to be valid in a more general setting.

PART 1: CONTINUITY AND DIFFERENTIABILITY

In this part we shall have a closer look at differentiability, derivative functions and compare the notions 'differentiability' and 'continuous differentiability'. We shall study differentiable homeomorphisms and isometries and prove several theorems about extensions of functions $X \to K$ of a certain kind to functions $K \to K$ of the same kind.

69. Convergent sequences of differentiable functions

IN THIS SECTION X HAS NO ISOLATED POINTS

For later use we shall introduce a natural notion of convergence in the K-linear space $BD(X \to K) := \{f : X \to K, f \text{ differentiable}, \|f\|_\infty < \infty\}$, avoiding the use of locally convex spaces.

For $f : X \to K$, $a \in X$ we define (admitting the value ∞)

$$\|f\|^a := \|f\|_\infty \vee \sup\{|\Phi_1 f(x, a)| : x \neq a\}$$

(Recall that $\Phi_1 f(x, a) = (f(x) - f(a))/(x - a)$.) It is easy to see that $\|f\|^a < \infty$ for each $a \in X$, $f \in BD(X \to K)$ and that $\| \ \|^a$ is a norm on $BD(X \to K)$ for each $a \in X$. Our interest lies in convergence of a sequence of functions with respect to *all* norms $\| \ \|^a$.

DEFINITION 69.1. Let $f, f_1, f_2, \ldots : X \to K$. The sequence $f_1, f_2, \ldots$ is *d-convergent to* f (notation $d\text{-}\lim_{n\to\infty} f_n = f$) if $\lim_{n\to\infty} \|f-f_n\|^a = 0$ for each $a \in X$. The sequence $f_1, f_2, \ldots$ is a *d-Cauchy sequence* if $\lim_{n,m\to\infty} \|f_n-f_m\|^a = 0$ for each $a \in X$.

PROPOSITION 69.2. *A sequence $f_1, f_2, \ldots$ of functions $X \to K$ is d-Cauchy if and only if it is d-convergent.*

Proof. Let $f_1, f_2, \ldots$ be d-Cauchy; we prove it to be d-convergent. As $\| \ \|^a \geqslant \| \ \|_\infty$ for each $a \in X$ the sequence $f_1, f_2, \ldots$ converges uniformly to some $f: X \to K$. Let $a \in X$. For $x \in X$ $(x \neq a)$, $n, m \in \mathbb{N}$ we have

$$|\Phi_1 f(x,a) - \Phi_1 f_n(x,a)| \leqslant |\Phi_1 f(x,a) - \Phi_1 f_m(x,a)| \vee \|f_m - f_n\|^a$$

By taking m large we see that

$$|\Phi_1 f(x,a) - \Phi_1 f_n(x,a)| \leqslant \overline{\lim_{m\to\infty}} \ \|f_m - f_n\|^a$$

It follows that $\lim_{m\to\infty} \|f-f_n\|^a = 0$ for each a, i.e. $d\text{-}\lim_{n\to\infty} f_n = f$.

PROPOSITION 69.3. *Let $f_1, f_2, \ldots \in BD(X\to K)$, $f = d\text{-}\lim_{n\to\infty} f_n$. Then $f \in BD(X \to K)$ and $f' = \lim_{n\to\infty} f_n'$.*

Proof. For $a \in X$ we have $|f'_n(a) - f'_m(a)| \leqslant \|f_n - f_m\|^a$ $(n, m \in \mathbb{N})$. Thus, $g := \lim_{n\to\infty} f'_n$ exists (pointwise). We prove that $f' = g$. In fact, we have for $a, x \in X$ $(x \neq a)$ and $n \in \mathbb{N}$

$$|\Phi_1 f(x,a) - g(a)| \leqslant \|f - f_n\|^a \vee |\Phi_1 f_n(x,a) - f'_n(a)| \vee |f'_n(a) - g(a)|$$

By choosing n large enough the first and the third term can be made small. Next, the middle term gets small if we choose x close to a. Hence, f is differentiable and $f' = g$. The boundedness of f follows from the fact that f is the uniform limit of the sequence $f_1, f_2, \ldots$ which consists of bounded functions.

From the above propositions we may conclude that $BD(X\to K)$ is 'complete with respect to d-convergence of sequences', i.e. each d-Cauchy sequence in $BD(X \to K)$ is d-convergent to a function in $BD(X \to K)$. To express d-convergence in a more direct way we shall use the notion of equidifferentiability.

DEFINITION 69.4. A set S of differentiable functions $X \to K$ is *equidifferentiable* if for each $a \in X$ and $\epsilon > 0$ there exists a $\delta > 0$ such that for all $f \in S$, $0 < |x-a| < \delta$ implies $|\Phi_1 f(x,a) - f'(a)| < \epsilon$.

PROPOSITION 69.5. *Let $f_1, f_2, \ldots \in BD(X\to K)$. The following conditions are equivalent.*

(α) *$f_1, f_2, \ldots$ is a d-Cauchy sequence.*

(β) *$\lim_{n\to\infty} f_n$ exists uniformly, $\lim_{n\to\infty} f'_n$ exists pointwise, $\{f_1, f_2, \ldots\}$ is equidifferentiable.*

Proof. For the implication (α) ⇒ (β) we only have to prove equidifferentiability of $\{f_1, f_2, \ldots\}$. Let $\epsilon > 0$, $a \in X$. There is $m \in \mathbb{N}$ such that $\|f_n - f_m\|^a < \epsilon$ for all $n \geqslant m$. There is $\delta > 0$ such that $0 < |x-a| < \delta$ implies $|\Phi_1 f_j(x, a) - f'_j(a)| < \epsilon$ for $j \in \{1, 2, \ldots, m\}$. For such x and $n \geqslant m$ we then have

$$|\Phi_1 f_n(x, a) - f'_n(a)|$$
$$\leqslant |\Phi_1 f_n(x, a) - \Phi_1 f_m(x, a)| \vee |\Phi_1 f_m(x, a) - f'_m(a)| \vee |f'_m(a) - f'_n(a)|$$
$$\leqslant \|f_n - f_m\|^a \vee |\Phi_1 f_m(x, a) - f'_m(a)| < \epsilon$$

which proves the equidifferentiability. To prove (β) ⇒ (α) let $a \in X$, $\epsilon > 0$. There is a δ $(0 < \delta < 1)$ such that if $0 < |x-a| < \delta$ then $|\Phi_1 f_n(x, a) - f'_n(a)| < \epsilon$ for all $n \in \mathbb{N}$. Further, there is an N such that for all $n, m \geqslant N$

$$|f'_n(a) - f'_m(a)| < \epsilon$$
$$\|f_n - f_m\|_\infty < \epsilon\delta$$

Now let $n, m \geqslant N$, $x \in X$, $x \neq a$. We shall prove that

$$(*) \qquad |\Phi_1 f_n(x, a) - \Phi_1 f_m(x, a)| < \epsilon$$

(from which one easily obtains that $\|f_n - f_m\|^a \leqslant \epsilon$, which is what we wanted to prove). If $|x-a| < \delta$ then (*) follows from

$$|\Phi_1 f_n(x, a) - \Phi_1 f_m(x, a)| \leqslant |\Phi_1 f_n(x, a) - f'_n(a)| \vee |f'_n(a) - f'_m(a)|$$
$$\vee |f'_m(a) - \Phi_1 f_m(x, a)|$$

whereas for $|x-a| \geqslant \delta$ we have

$$|\Phi_1 f_n(x, a) - \Phi_1 f_m(x, a)| \leqslant \delta^{-1} \|f_n - f_m\|_\infty < \epsilon.$$

As a first application we show that the locally constant functions $X \to K$ are 'dense' in $\{f : X \to K : f' = 0\}$. (Compare the first statement of Corollary 68.3, for the translation of the second statement, see Corollary 70.7.)

THEOREM 69.6. *Let $f : X \to K$. The following conditions are equivalent.*
(α) f is differentiable and $f' = 0$.
(β) There is an equidifferentiable sequence $f_1, f_2, \ldots$ of locally constant functions such that $\lim_{n \to \infty} f_n = f$ uniformly.
(γ) There is a sequence $f_1, f_2, \ldots$ of locally constant functions such that $f = d\text{-}\lim_{n \to \infty} f_n$.

Proof. We may assume that f is bounded. (Each one of the conditions (α), (β), (γ) implies continuity of f. By Theorem 26.2 there is a locally constant function g such that $f - g$ is bounded.) Suppose (β). By Propositions 69.5 and 69.2 the sequence $f_1, f_2, \ldots$ is d-Cauchy, hence d-convergent (to f) so we have (γ). Suppose (γ). By Proposition 69.3 $f \in BD(X \to K)$ and $f' = \lim_{n \to \infty} f'_n = 0$ and we have (α). Finally, we prove (α) ⇒ (β). Let $n \in \mathbb{N}$. Cover K with disjoint discs of the form $B_a(1/n)$, choose a centre of each disc and define a map $\sigma_n : K \to K$ assigning to each $x \in K$ the centre of the disc to which x belongs. Then σ_n is locally constant, $|\sigma_n(x)-x| < 1/n$ for all $x \in K$, $|\sigma_n(x)-\sigma_n(y)| \leqslant |x-y|$ for all $x, y \in K$. Define

$$f_n := \sigma_n \circ f \quad (n \in \mathbb{N})$$

Clearly each f_n is locally constant, $\lim_{n \to \infty} f_n = f$ uniformly. The equidifferentiability at $a \in X$ follows from $f'(a) = 0$ and

$$\left| \frac{f_n(x)-f_n(a)}{x-a} \right| \leqslant \left| \frac{f(x)-f(a)}{x-a} \right| \qquad (x \in X, x \neq a, n \in \mathbb{N})$$

We now present a characterization of $f \in C(\mathbb{Z}_p \to K)$: f is differentiable and $f' = 0$ by means of the coefficients of f with respect to the base of van der Put (compare Theroem 63.3).

THEOREM 69.7. *Let $K \supset \mathbb{Q}_p$ and let $e_0, e_1, \ldots$ be van der Put's base of $C(\mathbb{Z}_p \to K)$. Let $f = \sum_{n=0}^{\infty} a_n e_n \in C(\mathbb{Z}_p \to K)$. The following conditions are equivalent.*
(α) f is differentiable and $f' = 0$.
(β) $\lim_{n \to \infty} |a_n| (n \wedge |a-n|_p^{-1}) = 0$ for each $a \in \mathbb{Z}_p$.

This theorem is an easy consequence of the following.

PROPOSITION 69.8. *Let $K, e_0, e_1, \ldots, f$ be as in Theorem 69.7. Then for each $a \in \mathbb{Z}_p$*

$$\|f\|^a = \sup_{n \geqslant 0} |a_n| \, \|e_n\|^a$$

and

$$\|e_n\|^a = \begin{cases} 1 \text{ if } n = 0 \\ (|n-n\_|_p \vee |a-n|_p)^{-1} \text{ if } n \in \mathbb{N} \end{cases}$$

Proof. We compute $\|e_n\|^a$ for $n \geqslant 1$. If $n \lhd a$ then $e_n(a) = 1$ so $\sup \{|\Phi_1 e_n(x, a)| : x \neq a\} = \max\{|x-a|_p^{-1} : |x-a|_p \geqslant 1/n\} = |n-n\_|_p^{-1} = (|n-n\_|_p \vee |a-n|_p)^{-1}$. If not $n \lhd a$ then $e_n(a) = 0$ so $\sup\{|\Phi_1 e_n(x, a)| : x \neq a\} = \max\{|x-a|_p^{-1} : |x-n|_p < 1/n\} = |a-n|_p^{-1} = (|n-n\_|_p \vee |a-n|_p)^{-1}$. The proposed formulas for $\|e_n\|^a$ follow easily. From the inequalities $|\Phi_1 f(x, a)| \leqslant \max_n |a_n| \, |x-a|_p^{-1} \, |e_n(x)-e_n(a)| \leqslant \sup_n |a_n| \, \|e_n\|^a$ and $|f(x)| \leqslant \max_n |a_n| \leqslant \sup_n |a_n| \, \|e_n\|^a$ we obtain

$$\|f\|^a \leqslant \sup_{n \geqslant 0} |a_n| \, \|e_n\|^a \qquad (a \in \mathbb{Z}_p)$$

To prove the opposite inequality, let $n \in \mathbb{N}$. Then by Theorem 62.2 $|a_n| = |f(n)-f(n\_)| \leqslant |f(n)-f(a)| \vee |f(n\_)-f(a)| \leqslant \|f\|^a(|n-a|_p \vee |n\_-a|_p) = \|f\|^a(\|e_n\|^a)^{-1}$ which finishes the proof.

Proof of Theorem 69.7. First observe that (by Lemma 53.3(ii)) condition (β) is equivalent to $\lim_{n\to\infty} |a_n| \, \|e_n\|^a = 0$. Suppose ($\beta$) and set $f_n := \sum_{j=0}^n a_j e_j$. Then $\lim_{n\to\infty} \|f-f_n\|^a = 0$, i.e. $d\text{-}\lim_{n\to\infty} f_n = f$. By the implication ($\gamma$) $\Rightarrow$ (α) of Theorem 69.6 we have $f' = 0$. Conversely, suppose (α). Again by Theorem 69.6 there is for each $\epsilon > 0$ a locally constant function g for which $\|f-g\|^a < \epsilon$. By Exercise 62.C(i), g is a finite linear combination of the e_n, $g = \sum_{j=1}^m b_j e_j$, say. Now $\|f-g\|^a < \epsilon$ implies $\sup_{n>m} \|a_n e_n\|^a < \epsilon$ and (β) follows.

Exercise 69.A. Obtain a second proof of Theorem 63.2 by taking in the formula $\|f\|^a = \sup_{n\geq 0} |a_n| \, \|e_n\|^a$ the supremum over all $a \in \mathbb{Z}_p$.

70. A function of the first class has an antiderivative

IN THIS SECTION X HAS NO ISOLATED POINTS

A function $f : X \to K$ is of the *first class of Baire* if there are continuous functions $f_1, f_2, \ldots : X \to K$ such that $f = \lim_{n\to\infty} f_n$ (pointwise). Our aim is to show the following.

THEOREM 70.1. *Let $f : X \to K$. Then f has an antiderivative if and only if f is of the first class of Baire.*

This theorem has no analogue in the theory of real functions. It is true that every f: $\mathbb{R} \to \mathbb{R}$ having an antiderivative is of the first class of Baire, but the converse does not hold. In fact, even a simple characterization of $\{f : \mathbb{R} \to \mathbb{R} : f \text{ has an antiderivative}\}$ does not seem to be known.

The proof of Theorem 70.1 runs in several steps. One half is simple.

PROPOSITION 70.2. *Let $f : X \to K$ be differentiable. Then f is of the first class of Baire.*

Proof. For each $n \in \mathbb{N}$ we construct a continuous map $\sigma_n : X \to X$ such that $0 < |\sigma_n(x) - x| < n^{-1}$ for all $x \in X$ as follows. Let B be a nonempty (relatively) clopen subset of X of diameter $< n^{-1}$. Since X has no isolated points we can write B as a disjoint union of nonempty (relatively) clopen sets B_1 and B_2. Choose $a_1 \in B_1$, $a_2 \in B_2$ and let $\sigma_n(x) := a_1$ if $x \in B_2$ and $\sigma_n(x) = a_2$ if $x \in B_1$. Then σ_n is continuous on B and $0 < |\sigma_n(x) - x| < n^{-1}$ for $x \in B$. We obtain the desired map by carrying out this construction for every B belonging to a disjoint covering of X by means of clopen subsets of diameter $< n^{-1}$. Now set for $n \in \mathbb{N}$

$$f_n(x) := \Phi_1 f(\sigma_n(x), x) \qquad (x \in X)$$

Each f_n is continuous, $\lim_{n\to\infty} f_n(x) = f'(x)$ for all $x \in X$.

For the proof of the other half of Theorem 70.1 we shall write a function f of the first class of Baire as an infinite sum of locally constant functions f_n (Lemma 70.3), then select locally linear antiderivatives F_n of f_n (Lemmas 70.4 and 70.5) in such a way that ΣF_n is an antiderivative of f.

LEMMA 70.3. *Let $f : X \to K$ be a function of the first class of Baire. Then there exist locally constant functions $f_1, f_2, \ldots : X \to K$ such that*

$$f(x) = \sum_{n=1}^{\infty} f_n(x) \qquad (x \in X)$$

If f is bounded then f_n can be chosen such that $\|f_n\|_\infty \leqslant \|f\|_\infty$ for all n.
Proof. There are continuous functions $g_1, g_2, \ldots$ such that $\lim_{n\to\infty} g_n = f$. By Theorem 26.2 there are locally constant functions $h_1, h_2, \ldots$ such that $\|g_n - h_n\|_\infty < 1/n$ for each n. Then also $\lim_{n\to\infty} h_n = f$. Define $j_1, j_2, \ldots : X \to K$ as follows.

$$j_n(x) := \begin{cases} h_n(x) \text{ if } |h_n(x)| \leqslant \|f\|_\infty \\ 0 \quad \text{if } |h_n(x)| > \|f\|_\infty \end{cases}$$

Then j_n is locally constant, $\lim_{n\to\infty} j_n = f$, $\|j_n\|_\infty \leqslant \|f\|_\infty$ for all n. Finally, let

$$f_1 := j_1, \quad f_n := j_n - j_{n-1} \quad (n > 1)$$

Then $f = \Sigma_{n=1}^{\infty} f_n$ pointwise anf f_n is locally constant, $\|f_n\|_\infty \leqslant \|f\|_\infty$ for each n.

LEMMA 70.4 (Preparation for Lemma 70.5) *Let B be a 'closed' ball in X and let $\epsilon > 0$. There exists a locally linear $F : X \to K$ such that $F' = \xi_B$, $\|F\|_\infty \leqslant \epsilon$, $F(x) = 0$ if $x \in X \backslash B$ and for all $x, y \in X$*

$$|F(x) - F(y)| \leqslant |x - y| \quad \textit{if } x, y \in B$$
$$|F(x) - F(y)| \leqslant \epsilon |x - y| \quad \textit{otherwise}$$

Proof. B has the form $\{x \in X : |x - \rho| \leqslant \rho\}$ for some $a \in X$, $\rho \in \mathbb{R}^+$. We may assume $\epsilon < 1$. Let $r := \min(\epsilon\rho, \epsilon)$. The relation $|x - y| \leqslant r$ decomposes B into a disjoint union of balls B_i of the form

$$B_i = \{x \in X : |x - a_i| \leqslant r\}$$

Define $F : X \to K$ as follows.

$$F(x) := \begin{cases} x - a_i \text{ if } x \in B_i \\ 0 \quad \text{if } x \in X \backslash B \end{cases}$$

Obviously, $F' = \xi_B$, $\|F\|_\infty \leqslant r \leqslant \epsilon$. To prove the crucial property, let $x, y \in X$. If none of these are in B then $|F(x) - F(y)| = 0 \leqslant \epsilon\, |x - y|$. If $x \in B_i$

for some i and $y \notin B$ then $|x - y| \geqslant d(y, B) > \rho$, hence $|F(x) - F(y)| = |F(x)| = |x - a_i| \leqslant r \leqslant \epsilon\rho < \epsilon|x - y|$. If $x \in B_i$ and $y \in B_j$ and $i \neq j$ then $|F(x) - F(y)| = |x - a_i - y + a_j| \leqslant \max(|x - a_i|, |y - a_j|) \leqslant r \leqslant |x - y|$. Finally, if $x, y \in B_i$, then $F(x) - F(y) = x - a_i - (y - a_i) = x - y$.

LEMMA 70.5. *Let $f : X \to K$ be a locally constant function and let $\epsilon > 0$. Then f has a locally linear antiderivative $F : X \to K$ for which $\|F\|_\infty \leqslant \epsilon$ and such that for all $x, y \in X$*

$$|F(x) - F(y)| \leqslant \max(|f(x)|, \epsilon)\ |x - y|$$

Proof. X is a disjoint union of 'closed' balls S_i such that for each i, f is constant on S_i. Let c_i be the value of f on S_i. For each i, let $\epsilon_i := \epsilon(1 + c_i)^{-1}$. By Lemma 70.4 applied to S_i there exist locally linear $F_i : X \to K$ such that $F_i' = \xi_{S_i}$, $\|F_i\|_\infty \leqslant \epsilon_i$, $F_i(x) = 0$ if $x \in X\backslash S_i$ and

$$|F_i(x) - F_i(y)| \leqslant |x - y| \text{ if } x, y \in S_i$$

$$|F_i(x) - F_i(y)| \leqslant \epsilon_i |x - y| \text{ otherwise}$$

Define $F : X \to K$ by the formula

$$F(x) = c_i F_i(x) \qquad (x \in S_i)$$

Then F is locally linear, $F' = f$. For $x \in X$ we have $x \in S_i$ for some i so that $|F(x)| = |c_i|\ |F_i(x)| \leqslant |c_i|\ \epsilon_i \leqslant \epsilon$. It follows that $\|F\|_\infty \leqslant \epsilon$. Now let $x, y \in X$. Then $x \in S_i$, $y \in S_j$ for some i, j. If $i = j$ then $|F(x) - F(y)| = |c_i|\ |F_i(x) - F_i(y)| \leqslant |c_i|\ |x - y| = |f(x)|\ |x - y|$. If $i \neq j$ then $F_i(y) = F_j(x) = 0$ so that $|F(x)| = |c_i F_i(x)| = |c_i|\ |F_i(x) - F_i(y)| \leqslant |c_i|\ |\epsilon_i|\ |x - y| \leqslant \epsilon |x - y|$. Similarly, $|F(y)| \leqslant \epsilon |x - y|$. Thus $|F(x) - F(y)| \leqslant \epsilon\ |x - y|$.

We are now ready to prove the following detailed form of Theorem 70.1.

THEOREM 70.6. *Let $f : X \to K$ be of the first class of Baire, let $\epsilon > 0$. Then f has an antiderivative F with the following properties.*

(i) *$F = d\text{-}\lim_{n\to\infty} g_n$ for certain locally linear functions $g_1, g_2, \ldots : X \to K$.*

(ii) $\|F\|_\infty \leqslant \epsilon$.

(iii) *If f is bounded then $|F(x) - F(y)| \leqslant \|f\|_\infty\ |x - y|\ (x, y \in X)$.*

Proof. We may assume that $f \neq 0$ and $\epsilon < \|f\|_\infty$. By Lemma 70.3 we have $f = \sum_{n=1}^{\infty} f_n$ where each f_n is locally constant and $\|f_n\|_\infty \leqslant \|f\|_\infty$ for all n in the case f is bounded. By Lemma 70.5, for each n there is a locally linear antiderivative F_n of f_n for which $\|F_n\|_\infty \leqslant \epsilon\, n^{-1}$, $|F_n(x) - F_n(y)| \leqslant \max(|f_n(x)|, \epsilon\, n^{-1})\ |x - y|\ (x, y \in X)$. We see that for each $a \in X$

$$|F_n(x) - F_n(a)| \leqslant \max(|f_n(a)|, \epsilon n^{-1})\ |x - a| \quad (x \in X)$$

Using the fact that $\lim_{n\to\infty} f_n(a) = 0$ and $\lim_{n\to\infty} \|F_n\|_\infty = 0$, we may conclude that

$$\lim_{n\to\infty} \|F_n\|^a = 0$$

Hence the partial sums $g_n : n \mapsto \sum_{j=1}^{n} F_j$ form a d-Cauchy sequence in the sense of Definition 69.1. By Propositions 69.2 and 69.3 the sum F defined by

$$F(x) := \sum_{n=1}^{\infty} F_n(x) \qquad (x \in X)$$

is differentiable and $F' = \sum_{n=1}^{\infty} F_n' = \sum_{n=1}^{\infty} f_n = f$. This establishes (i). For $x, y \in X$ we have if f is bounded $|F(x) - F(y)| \leqslant \max |F_n(x) - F_n(y)| \leqslant \max(\|f\|_\infty, \epsilon) |x - y| \leqslant \|f\|_\infty |x - y|$, which is (iii). Finally, $\|F\|_\infty \leqslant \sup_n \|F_n\|_\infty \leqslant \epsilon$.

As a by-product we obtain the following corollary stating that the locally linear functions are 'dense' in the set of differentiable functions. Compare Theorem 69.6 and Corollary 68.3.

COROLLARY 70.7. *Let $f : X \to K$. Then the following are equivalent.*
(α) *f is differentiable.*
(β) *There is an equidifferentiable sequence $f_1, f_2, \ldots$ of locally linear functions such that* $\lim_{n\to\infty} f_n = f$ *uniformly and* $\lim_{n\to\infty} f_n'$ *exists pointwise.*
(γ) *There is a sequence $f_1, f_2, \ldots$ of locally linear functions such that $f = d\text{-}\lim_{n\to\infty} f_n$.*

Proof. The implications (β) $\Leftrightarrow$ (γ) $\Rightarrow$ (α) follow from Propositions 69.3 and 69.5. We prove (α) $\Rightarrow$ (γ). By Theorem 70.1, f' is of the first class of Baire, so it has an antiderivative g for which there exist locally linear functions $g_1, g_2, \ldots$ such that $g = d\text{-}\lim_{n\to\infty} g_n$ (Theorem 70.6). Now $(f-g)' = 0$ so by Theorem 69.6 there is a sequence $h_1, h_2, \ldots$ of locally constant functions such that $f - g = d\text{-}\lim_{n\to\infty} h_n$. We see that $f = d\text{-}\lim_{n\to\infty} (g_n + h_n)$ and $g_n + h_n$ is locally linear for each n.

Remark. For more information on the space of all functions that are of the first class of Baire we refer to Appendix A.2. After combining it with the theory of this section we can conclude that *the set of all derivative functions is uniformly closed.*

71. Points at which a differentiable function is C^1

A derivative function $K \to K$ is of the first class of Baire and therefore has (Appendix A.2) 'many' points of continuity. One may view this as a direct translation of the corresponding statement for real functions; the proofs are practically identical. However in the ultrametric theory it is more natural to consider the (smaller) set of points at which a differentiable function is C^1 rather than the continuity points of its derivative. (See Definition 27.1.)

THEOREM 71.1. *Let $f : K \to K$ be differentiable. Then $\{a \in K : f$ is C^1 at $a\}$ is a dense G_δ -subset of K.*

This theorem is a special case of the following.

THEOREM 71.2. *Let X be a G_δ-subset of K without isolated points and let $f : X \to K$ be differentiable. Then $C := \{a \in X : f$ is C^1 at $a\}$ is a G_δ -subset of K and C is dense in X.*

Proof. Let $Y := \{a \in X : f'$ is continuous at $a\}$. Then $C \subset Y \subset X$, Y is a dense G_δ-subset of X, Y is a G_δ-subset of K (Appendixes A.2 and A.1). For $m, n \in \mathbb{N}$ set

$T_{mn} := \{a \in Y :$ there is $x \in Y$ such that $0 < |x - a| < n^{-1}$ and $|\Phi_1 f(x, a) - f'(a)| \geqslant m^{-1}\}$

We shall prove successively the following facts which together imply the theorem.

(i) $Y \backslash C = \bigcup_{m=1}^{\infty} \bigcap_{n=1}^{\infty} \overline{T}_{mn} \cap Y$. (Thus C is a G_δ -subset of Y, hence of K.)

(ii) T_{mn} is open in Y for all $m, n \in \mathbb{N}$. (Thus T_{mn} is a G_δ -subset of X.)

(iii) $\bigcap_{n=1}^{\infty} T_{mn} = \emptyset$ for each $m \in \mathbb{N}$.

(iv) $\bigcap_{n=1}^{\infty} \overline{T}_{mn} \cap X$ is nowhere dense in X for each $m \in \mathbb{N}$. (Thus, $X \backslash C = (X \backslash Y) \cup (Y \backslash C)$ is meagre in X so that C is dense in X (Appendix A.2).)

Proof of (i). Let $a \in Y \backslash C$. Since f is not C^1 at a we can find $\epsilon > 0$ and sequences $x_1, x_2, \ldots, y_1, y_2, \ldots$ with $x_j \neq y_j$ for all j such that $\lim_{j \to \infty} x_j = \lim_{j \to \infty} y_j = a$, but $|\Phi_1 f(x_j, y_j) - f'(a)| \geqslant \epsilon$ $(j \in \mathbb{N})$. Choose $m \in \mathbb{N}$ such that $m^{-1} < \epsilon$ and let $n \in \mathbb{N}$ be arbitrary. Since f' is continuous at a we have for large j

$$|f'(a) - f'(x_j)| < \epsilon \leqslant |\Phi_1 f(x_j, y_j) - f'(a)|$$

Hence,

$$|\Phi_1 f(x_j, y_j) - f'(x_j)| \geqslant \epsilon > m^{-1}, 0 < |x_j - y_j| < n^{-1}$$

for large j. Therefore $x_j \in T_{mn}$ for large j so that $a \in \overline{T}_{mn}$. Since n was arbitrary we may conclude that $a \in \bigcap_{n=1}^{\infty} \overline{T}_{mn} \cap Y$ so that

$$Y \backslash C \subset \bigcup_{m=1}^{\infty} \bigcap_{n=1}^{\infty} \overline{T}_{mn} \cap Y$$

Conversely, let $a \in Y$, $a \in \bigcap_{n=1}^{\infty} \overline{T}_{mn}$ for some m. We prove that $a \notin C$. For each n, choose $b_n \in T_{mn}$ such that $|a - b_n| < n^{-1}$. By the definition of T_{mn} there is $c_n \in Y$ such that $0 < |c_n - b_n| < n^{-1}$ and $|\Phi_1 f(c_n, b_n) - f'(b_n)| \geqslant m^{-1}$. For large n we have by the continuity of f' at a that $|f'(b_n) - f'(a)| < m^{-1}$. So we obtain $|\Phi_1 f(c_n, b_n) - f'(a)| \geqslant m^{-1}$ for large n implying that f is not C^1 at a, i.e. $a \notin C$.

Proof of (ii). Let $a \in T_{mn}$. There is $b \in Y$ such that $0 < |b - a| < n^{-1}$ and $|\Phi_1 f(b, a) - f'(a)| \geqslant m^{-1}$. There is $\delta < |b - a|$ such that $0 < |x - a| < \delta$ implies

$$|f'(x) - f'(a)| < m^{-1}, \quad |\Phi_1 f(x, a) - f'(a)| < m^{-1}$$

We now claim that $0 < |x - a| < \delta$, $x \in Y$ implies $x \in T_{mn}$. Indeed, we show that $0 < |b - x| < n^{-1}$ and $|\Phi_1 f(b, x) - f'(x)| \geqslant m^{-1}$. First, since $|x - a| < \delta < |b - a|$ we have $|b - x| = |b - a|$ so $0 < |b - x| < n^{-1}$. Further we write

$$\Phi_1 f(b, x) - f'(x) = \frac{b-a}{b-x} (\Phi_1 f(b, a) - f'(a)) + \frac{a-x}{b-x} (\Phi_1 f(a, x) - f'(a)) + f'(a) - f'(x)$$

Now $|b - x| = |b - a|$ so that

$$\left| \frac{b-a}{b-x} \right| |\Phi_1 f(b, a) - f'(a)| \geqslant m^{-1}$$

Since $|a - x| < \delta < |b - x|$ we have

$$\left| \frac{a-x}{b-x} \right| |\Phi_1 f(a, x) - f'(a)| < m^{-1}$$

Also

$$|f'(a) - f'(x)| < m^{-1}$$

It follows that

$$|\Phi_1 f(b, x) - f'(x)| \geqslant m^{-1}$$

Proof of (iii). If $a \in \bigcap_{n=1}^{\infty} T_{mn}$ for some m we could find a sequence $x_1, x_2, \ldots$ converging to a such that $x_n \neq a$ for all n and $|\Phi_1 f(x_n, a) - f'(a)| \geqslant m^{-1}$, contradicting the differentiability of f at a.

Proof of (iv). Suppose that, for some m, $\bigcap_{n=1}^{\infty} \overline{T}_{mn} \cap X$ is not nowhere dense in X. There is a ball B (relative to the metric space X) such that $B \subset \overline{T}_{mn}$ for all n. Then $B \cap T_{mn}$ is dense in B for each n. By (ii), $T_{mn} \cap B$ is a G_δ -subset of B. By the Baire category theorem (Appendix A. 1) $(\bigcap_{n=1}^{\infty} T_{mn}) \cap B$ is also dense in B, but this is obviously not true, as $\bigcap_{n=1}^{\infty} T_{mn} = \emptyset$ by (iii).

Exercise 71.A. Give an example of a differentiable function $f : \mathbb{Z}_p \to \mathbb{Q}_p$ for which $\{a \in \mathbb{Z}_p : f \text{ is } C^1 \text{ at } a\}$ is strictly contained in $\{a \in \mathbb{Z}_p : f \text{ is continuous at } a\}$.

Exercise 71.B. Show that the following statement is *false*. If $f : \mathbb{Z}_p \to \mathbb{Q}_p$ is C^1 then f is C^2 at some point of $\mathbb{Z}_p$.

Exercise 71.C. (Failure of Theorem 71.1 if X is not a G_δ -set)
(i) Show that $\mathbb{N}$ is not a G_δ -subset of $\mathbb{Z}_p$.
(ii) Let $g : \mathbb{Z}_p \to \mathbb{Q}_p$ be a function such that $g' = 0$, g is C^1 at every point except 0, $\|\Phi_1 g\|_\infty \leqslant 1$. (For example, let g be as in 26.6.) Set

$$h(x) := \sum_{n=1}^{\infty} p^n g(x-n)$$

Show that h is well defined and that $h' = 0$. Further, prove that h is C^1 at every point of $\mathbb{Z}_p \backslash \mathbb{N}$, that h is C^1 at no point of $\mathbb{N}$.
(iii) Let $f := h|\mathbb{N}$. Show that f *is differentiable,* that f' *is continuous,* but that f *is nowhere* C^1. (Compare Theorem 71.1.)

72. Local behaviour of differentiable functions

We consider the following question. Let U be a nonempty open subset of K, let $f : U \to K$ be differentiable and let $f'(a) \neq 0$ for some $a \in U$. What can be said about the local behaviour of f at a?
(1) If f is also C^1 at a we have a satisfactory answer. Proposition 27.3 and Theorem 27.5 guarantee the existence of a $\delta > 0$ such that
(i) f maps $B_a(\delta)$ onto $B_{f(a)}(|f'(a)|\,\delta)$.
(ii) For all $x, y \in B_a(\delta)$

$$|f(x)-f(y)| = |f'(a)|\;|x-y|$$

(iii) The local inverse $B_{f(a)}(|f'(a)|\,\delta) \to B_a(\delta)$ is C^1 at $f(a)$.
(2) If it is not given that f is C^1 at a it is true that $|f(x)-f(a)| = |f'(a)|\,|x-a|$ if x is sufficiently close to a, but this fact (even when $f' \neq 0$ everywhere on U) is not enough to prove either (i) or (ii). In fact, let f be as in Example 26.6. We have seen there that f is not locally injective at 0 so that (ii) fails dramatically. Further, for each $n \in \mathbb{N}$, $f(B_0(p^{-m}))$ is not a ball (not even a neighbourhood of $0 = f(0)$) as it does not meet $\bigcup_n B_n$. Thus (i) is false for $a = 0$ and each δ. Yet, it is possible to save (i) by a subtle argument if it is given *a priori* that f is a homeomorphism (Theorem 72.1). Surprisingly, even in this case (ii) is false in general (Example 72.2).

THEOREM 72.1. *Let U, V be open subsets of K and let $f: U \to V$ be a (surjective) homeomorphism. Suppose that f is differentiable at $a \in U$ and $f'(a) \neq 0$. Then, for sufficiently small $\delta > 0$, the function f maps $B_a(\delta)$ onto $B_{f(a)}(|f'(a)|\ \delta)$.*

Proof. There is a $\delta_1 > 0$ such that $|x-a| \leqslant \delta_1$ implies $x \in U$ and $|f(x)-f(a)| = |f'(a)|\,|x-a|$. Let $B := B_a(\delta_1)$. Then $f(B)$ is a neighbourhood of $f(a)$, so $f(B) \supset B_{f(a)}(\epsilon)$ for some $\epsilon > 0$. There is a $\delta_2 < \delta_1$, δ_2 in the value group of K, such that for all positive $\delta \leqslant \delta_2$

$$f(B_a(\delta)) \subset B_{f(a)}(\epsilon)$$

Observe that $|f'(a)|\ \delta_2 \leqslant \epsilon$. Indeed, if $|x-a| = \delta_2$ then $|f(x)-f(a)| \leqslant \epsilon$, but also $x \in B$ so that $|f(x)-f(a)| = |x-a|\ |f'(a)| = \delta_2\ |f'(a)|$. We claim that for $0 < \delta \leqslant \delta_2$ we have

$$f(B_a(\delta)) = B_{f(a)}(|f'(a)|\ \delta)$$

In fact, let $x \in B_a(\delta)$. Then since $x \in B$

$$|f(x)-f(a)| = |f'(a)|\,|x-a| \leqslant |f'(a)|\,\delta$$

i.e. $f(x) \in B_{f(a)}(|f'(a)|\,\delta)$.

Conversely, let $y \in B_{f(a)}(|f'(a)|\,\delta)$. Then $|y-f(a)| \leqslant |f'(a)|\,\delta \leqslant \epsilon$, so that $y \in B_{f(a)}(\epsilon) \subset f(B)$. It follows that $x := f^{-1}(y) \in B$ and, in consequence,

$$|f(x)-f(a)| = |f'(a)|\ \ |x-a|$$

This, combined with

$$|f(x)-f(a)| = |y-f(a)| \leqslant |f'(a)|\,\delta$$

yields $|x-a| \leqslant \delta$, i.e. $x \in B_a(\delta)$. Then $y = f(x) \in f(B_a(\delta))$ and the theorem is proved.

Exercise 72.A. Let g be the inverse of the function f of Theorem 72.1. Show that g is differentiable at $f(a)$.

Remark. Theorem 72.1 shall be used in Section 74 in order to determine for which pairs of compact open subsets of $\mathbb{Q}_p$ there exists a diffeomorphism between them.

EXAMPLE 72.2. *There is a differentiable bijection $f: \mathbb{Z}_p \to \mathbb{Z}_p$ such that $f' = 1$ but which is not locally an isometry at* 0.

Proof. Let $x = \sum_{n=0}^{\infty} a_n p^n \in \mathbb{Z}_p$. If $|x|_p = p^{-m}$ for some $m \in \{0, 1, 2, \ldots\}$, define $f(x)$ by interchanging a_{2m} and a_{2m+1}. In other words, let

$$f(x) := \sum_{n=0}^{\infty} b_n p^n$$

where $b_n = a_n$ for $n \notin \{2m, 2m+1\}$, $b_{2m} = a_{2m+1}$, $b_{2m+1} = a_{2m}$. Further, set

$$f(0) := 0$$

On each 'annulus' $\{x : |x|_p = p^{-m}\}$ we have $f(x) - f(y) = x - y$ if x, y are sufficiently close, hence $f' = 1$. That $f'(0) = 1$ follows from

$$|x^{-1}(f(x) - f(0)) - 1|_p = |x^{-1}|_p |f(x) - x|_p \leqslant p^m p^{-2m}$$

for $|x|_p = p^{-m}$. Further, $f \circ f$ is the identity so that f is a bijection. For each $m \in \mathbb{N}$ the 'triangle' p^m, $p^m + p^{2m}$, $p^m + p^{2m+1}$ is mapped (in that order) onto p^m, $p^m + p^{2m+1}$, $p^m + p^{2m}$. Thus, in every neighbourhood of 0, $|\Phi_1 f|$ takes the values $p^{-1}, 1, p$. In particular, f is not a local isometry at 0.

Exercise 72.B. (A more dramatic example) Show that the above f is in Lip_1 $(\mathbb{Z}_p \to \mathbb{Q}_p)$. By modifying the construction in a suitable way obtain a differentiable bijection $g : \mathbb{Z}_p \to \mathbb{Z}_p$ for which $g' = 1$ but such that g is in Lip_a $(\mathbb{Z}_p \to \mathbb{Q}_p)$ for no positive a.

Exercise 72.C. Find a non-isometrical C^∞ -bijection of $\mathbb{Z}_p$ whose derivative is 1.

Exercise 72.D. Use Theorem 71.1 to prove the following. Let $f : K \to K$ be differentiable and let $f'(x) = 1$ for all $x \in K$. Then there exists an open dense subset U of K such that (i) f maps open subsets of U onto open subsets of K, (ii) $f | U$ is locally an isometry.

Exercise 72.E. Let K be not locally compact and let $f : K \to K$ be differentiable. Suppose that $f(B)$ is compact for each disc B in K. Prove that $f' = 0$.

Exercise 72.F. Let L be a closed proper subfield of K and let $f : K \to K$ be differentiable, $f(K) \subset L$. Show that $f' = 0$.

Exercise 72.G. Conclude from either one of the two preceding exercises that a differentiable $f : \mathbb{C}_p \to \mathbb{Q}_p$ has zero derivative. Find such an f that is not locally constant. (Hint. Let $g : \mathbb{C}_p \to \mathbb{Q}_p$ be a nonzero $\mathbb{Q}_p$ -linear continuous function (Exercise 50.H), let $j : \mathbb{Q}_p \to \mathbb{Q}_p$ be a non-locally constant function with zero derivative. Set $f = j \circ g$.)

73. Lusin-type theorems

In this section we shall be dealing with a p-adic translation of the following two classical theorems from the theory of real functions.
(Lusin's theorem) *Let* $f : \mathbb{R} \to \mathbb{R}$ *be differentiable. Then* $\{f(x) : f'(x) = 0\}$ *is a null set* (in the sense of Lebesgue, i.e. for each $\epsilon > 0$ it can be covered by countably many intervals whose total length is less than ϵ).
(Property (N)) *Let* $f : \mathbb{R} \to \mathbb{R}$ *be differentiable. Then* f *maps null sets into null sets.*

In Section 67 we touched upon the notion of a p-adic null set in passing.

DEFINITION 73.1. A subset of $\mathbb{Q}_p$ is a *null set* if for each $\epsilon > 0$ it can be covered by countably many discs $B_1, B_2, \ldots$ such that $\Sigma_{j=1}^{\infty} d(B_j) < \epsilon$.

THEOREM 73.2. *Let* $f : \mathbb{Z}_p \to \mathbb{Q}_p$ *be differentiable.*
(i) $\{f(x) : f'(x) = 0\}$ *is a null set.*
(ii) f *maps null sets into null sets.*

To keep the proof as elementary as possible we shall avoid an explicit use of the (real valued) Haar measure on $\mathbb{Q}_p$.

**Exercise* 73.A. Let $A, B, A_1, A_2, \ldots$ be subsets of $\mathbb{Q}_p$. Prove the following.
(i) If $A \subset B$ and B is a null set then so is A.
(ii) If $A_1, A_2, \ldots$ are null sets then so is $\bigcup_j A_j$.
(iii) If A is a null set and $a \in \mathbb{Q}_p$ then $a + A$ is a null set.
(iv) $\mathbb{Z}_p$ is not a null set. (Hint. The next lemma may be helpful.)

LEMMA 73.3.
(i) *Let* $B_1, \ldots, B_n$ *be a partition of* $\mathbb{Z}_p$ *into discs. Then* $\Sigma_{j=1}^{n} d(B_j) = 1$.
(ii) *Let* $B_1, B_2, \ldots$ *be a disjoint collection of subdiscs of* $\mathbb{Z}_p$. *Then* $\Sigma_{j=1}^{\infty} d(B_j) \leqslant 1$.

Proof. (i) Let $\delta := \min_j d(B_j)$ and decompose each B_j into discs of diameter δ. Now use the fact that if D is a disc in $\mathbb{Z}_p$ and D is a disjoint union of discs $D_1, \ldots, D_n$ all having equal diameter then $d(D) = \Sigma_{j=1}^{n} d(D_j)$.
(ii) Let $n \in \mathbb{N}$. The complement in $\mathbb{Z}_p$ of $\bigcup_{j=1}^{n} B_j$ is clopen, hence a disjoint union of discs $D_1, \ldots, D_m$. By (i) we have $1 = \Sigma_{j=1}^{n} d(B_j) + \Sigma_{j=1}^{m} d(D_j) \geqslant \Sigma_{j=1}^{n} d(B_j)$.

Proof of Theorem 73.2.
(i) Let $A := \{x \in \mathbb{Z}_p : f'(x) = 0\}$; we prove that $f(A)$ is a null set. Let $n \in \mathbb{N}$. For each $a \in A$ there is a $\delta_a \in |\mathbb{Q}_p^\times|$ $(0 < \delta_a < 1)$ such that $|x - a|_p \leqslant \delta_a$ implies $|f(x) - f(a)|_p \leqslant p^{-n} |x - a|_p$. The discs $B_a(\delta_a)$, where a runs through

A, cover A. By removing those discs that are strictly contained in other discs of this collection we obtain a (necessarily countable) disjoint subcovering $B_{a_1}(\delta_1), B_{a_2}(\delta_2), \ldots$ By Lemma 73.3(ii) we have $\sum_{j=1}^{\infty} \delta_j \leqslant 1$. We have $f(B_{a_j}(\delta_j)) \subset B_{f(a_j)}(p^{-n}\delta_j)$ for each j so that

$$f(A) \subset \bigcup_{j=1}^{\infty} B_{f(a_j)}(p^{-n}\delta_j)$$

and $\sum_{j=1}^{\infty} d(B_{f(a_j)}(p^{-n}\delta_j)) = \sum_{j=1}^{\infty} p^{-n}\delta_j \leqslant p^{-n}$. It follows that $f(A)$ is a null set.

(ii) Let $Y \subset \mathbb{Z}_p$ be a null set. For each $n \in \mathbb{N}$, let $Y_n := \{x \in Y : |f'(x)|_p \leqslant p^n\}$. Then $Y = \bigcup_n Y_n$ and it suffices to show that each $f(Y_n)$ is a null set. Let $\epsilon > 0$. We can cover Y_n with discs $B_1, B_2, \ldots$ such that $\sum_{j=1}^{\infty} d(B_j) < \epsilon p^{-n}$. By a reasoning similar to the one in (i) we can cover $B_1 \cap Y_n$ by disjoint discs $B_{a_1}(\delta_1), B_{a_2}(\delta_2), \ldots$ such that for each j

$$a_j \in B_1 \cap Y_n, \delta_j \in |\mathbb{Q}_p^\times|, f(B_{a_j}(\delta_j)) \subset B_{f(a_j)}(p^n\delta_j)$$

By disjointness and Lemma 73.3(ii) we have $\sum_{j=1}^{\infty} \delta_j \leqslant d(B_1)$. Also

$$f(B_1 \cap Y_n) \subset \bigcup_{j=1}^{\infty} B_{f(a_j)}(p^n\delta_j)$$

and $\sum_{j=1}^{\infty} d(B_{f(a_j)}(p^n\delta_j)) = p^n \sum_{j=1}^{\infty} \delta_j \leqslant p^n d(B_1)$. After repeating this reasoning for $B_2 \cap Y_n, B_3 \cap Y_n, \ldots$ we may conclude that $f(Y_n)$ can be covered by discs of which the sum of their diameters is less than $p^n \sum_{j=1}^{\infty} d(B_j) < p^n p^{-n} \epsilon = \epsilon$.

In the next exercise this result is extended to arbitrary locally compact fields.

Exercise 73.B. Let K be locally compact, let q be the number of elements of its residue class field. Normalize the valuation in such a way that $\max |K^\times| \cap (0, 1) = q^{-1}$. Show that, with this choice, Lemma 73.3 holds after replacing $\mathbb{Z}_p$ by the 'closed' unit disc of K. Define 'null set' in an obvious way and prove the statements (i) and (ii) of Theorem 73.2 for a differentiable function $f: B_0(1) \to K$.

What about non-locally compact fields (e.g. $\mathbb{C}_p$)? We shall leave the problem of defining a suitable notion of 'null set' as an open question but instead read Lusin's theorem as 'the image of $\{x : f'(x) = 0\}$ is a small set'. The following example shows that there is no such theorem for $\mathbb{C}_p$.

EXAMPLE 73.4. *Let K be not locally compact. Then there exists a C^∞-*

homeomorphism f of the 'closed' unit disc such that $f' = 0$.

Proof. Let $\pi \in K$, $0 < |\pi| < 1$. By Theorem 12.3 there is a set $R \subset B_0(1)$ with the properties $0 \in R$, if $x, y \in R$, $x \neq y$, then $|x - y| > |\pi|$, such that each $x \in B_0(1)$ can uniquely be written as a series $x = \sum_{n=0}^{\infty} a_n \pi^n$ ($a_n \in R$). Since K is not locally compact R is infinite. For each $n \in \mathbb{N}$ choose a bijection $\sigma_n : R \to R^n$, and for $1 \leqslant j \leqslant n$ denote the jth coordinate of $\sigma_n(r)$ by $[\sigma_n(r)]_j$. Define f by the formula

$$\begin{aligned} f\left(\sum_{n=0}^{\infty} a_n \pi^n\right) = {} & a_0 + \sigma_1(a_1)\pi \\ & + [\sigma_2(a_2)]_1 \pi^2 + [\sigma_2(a_2)]_2 \pi^3 \\ & + [\sigma_3(a_3)]_1 \pi^4 + [\sigma_3(a_3)]_2 \pi^5 + [\sigma_3(a_3)]_3 \pi^6 \\ & + [\sigma_4(a_4)]_1 \pi^7 + \ldots . \\ & + \ldots \end{aligned}$$

Clearly, f is a homeomorphism $B_0(1) \to B_0(1)$. To prove that f is C^∞ let $x, y \in B_0(1), x \neq y, x = \sum_{n=0}^{\infty} a_n \pi^n, y = \sum_{n=0}^{\infty} b_n \pi^n$. Suppose $a_0 = b_0, \ldots, a_{j-1} = b_{j-1}, a_j \neq b_j$ for some j. Then in the expansions of $f(x)$ and $f(y)$ the coefficients of π^0 up to $\pi^{j(j-1)/2}$ coincide. It follows easily that f satisfies Lipschitz conditions of all positive orders. Theorem 29.12 (or Exercise 29.E) now says that f is a C^∞-function and $f' = 0$.

Exercise 73.C. Let $f : \mathbb{Z}_p \to \mathbb{Q}_p$ be a C^1-function. Prove that $f' = 0$ if and only if $f(\mathbb{Z}_p)$ is a null set.

Exercise 73.D. Prove the following generalization of the p-adic Lusin theorem. Let $X \subset \mathbb{Q}_p$ have no isolated points and let $f : X \to \mathbb{Q}_p$ be differentiable. Then $\{ f(x) : f'(x) = 0 \}$ is a null set.

Exercise 73.E. Let $f \in \mathrm{Lip}_1(\mathbb{Z}_p \to \mathbb{Q}_p)$. Show that f maps null sets into null sets.

Exercise 73.F. (A continuous function $\mathbb{Z}_p \to \mathbb{Z}_p$ not having property (N)) We construct a continuous function $g : \mathbb{Z}_p \to \mathbb{Z}_p$ that maps some null set onto $\mathbb{Z}_p$. (By Theorem 73.2(ii) g is bound to be nondifferentiable at some point of $\mathbb{Z}_p$.) Let $X := \{ \sum_{n=0}^{\infty} a_n p^n \in \mathbb{Z}_p : a_1 = a_3 = a_5 = \ldots = 0 \}$. For each $x = \sum_{n=0}^{\infty} a_n p^n \in \mathbb{Z}_p$ let $t(x) := \min \{ n : n \text{ odd}, a_n \neq 0 \}$ if $x \notin X$, $t(x) := \infty$ if $x \in X$. Define $g : \mathbb{Z}_p \to \mathbb{Z}_p$ by

$$g(x) := \sum_{\substack{j < t(x) \\ j \text{ even}}} a_j p^{j/2} \qquad \left(x = \sum_{j=0}^{\infty} a_j p^j \in \mathbb{Z}_p\right)$$

and prove the following.
(i) g is continuous.
(ii) g maps the closed null set X onto $\mathbb{Z}_p$.
(iii) g is differentiable with derivative 0 at all points of $\mathbb{Z}_p \backslash X$.
Note. For a famous example of this kind in the theory of real functions, see, for example, E. Hewitt & K. Stromberg, *Real and abstract analysis*, Springer-Verlag, Berlin-Heidelberg-New York (1965), p. 113.

Exercise 73.G. Let $f : \mathbb{Z}_p \to \mathbb{Z}_p$ be a differentiable surjection for which $|f'(x)|_p \leqslant 1$ for all $x \in \mathbb{Z}_p$. Prove that actually $|f'(x)|_p = 1$ for all x, that f is a homeomorphism, but that f need not be an isometry. (Hint. Prove (i) and (ii) below.
(i) If $U_1, \ldots, U_n$ are open compact subsets of $\mathbb{Z}_p$ and $\Sigma_{j=1}^{n} d(U_j) < 1$ then they do not cover $\mathbb{Z}_p$.
(ii) If $B_1, \ldots, B_n$ are discs in $\mathbb{Z}_p$ that cover $\mathbb{Z}_p$ and $\Sigma_{j=1}^{n} d(B_j) = 1$ then $B_1, \ldots, B_n$ are disjoint.)

74. Differentiable homeomorphisms

A *homeomorphism* of a metric space Y onto a metric space Z is a bijection $\sigma : Y \to Z$ such that σ and σ^{-1} are continuous. If there is a homeomorphism between Y and Z then Y and Z are *homeomorphic*. A *diffeomorphism* of a clopen subset U of K onto a clopen subset V of K is a homeomorphism $\sigma : U \to V$ such that σ and σ^{-1} are differentiable. If there is a diffeomorphism between U and V then U and V are *diffeomorphic*. In this section we consider the following question (mainly for $K = \mathbb{Q}_p$ or $K = \mathbb{C}_p$).

QUESTION. Let U, V be clopen subsets of K. When are they homeomorphic? When are they diffeomorphic?

The case $K = \mathbb{C}_p$ is quite simple.

PROPOSITION 74.1. *Each two nonempty clopen subsets of $\mathbb{C}_p$ are diffeomorphic.*

Proof. Let U, V be nonempty clopen subsets of $\mathbb{C}_p$. $\mathbb{C}_p$ is not locally compact but separable so we can write U and V as a disjoint union of infinitely yet countably many 'closed' discs $B_1, B_2, \ldots$ and $B_1', B_2', \ldots$ respectively. For each n the discs B_n and B_n are diffeomorphic by means of a linear map g_n. By gluing together these maps g_n we obtain a diffeomorphism between U and V.

Remark. The above proof works for every non-locally compact separable field.

We move to the local compact case. The following exercise is easy.

**Exercise* 74.A. Let K be locally compact, let U, V be nonempty clopen subsets of K. Prove the following.
(i) If U is unbounded and U, V are homeomorphic then V is unbounded.
(ii) If U, V are unbounded then U and V are diffeomorphic.

We see that our question (at least for separable K) reduces to the case where U, V are compact open subsets of a locally compact field K.

PROPOSITION 74.2. *Let Y, Z be infinite compact ultrametric spaces without isolated points. Then Y and Z are homeomorphic.*
Proof. Decompose Y and Z into finitely many balls of diameter $\leqslant 1$. Since Y, Z do not have isolated points we can decompose each of these balls into a prescribed number of nonempty clopen sets. Thus, there are a number $n_1 \in \mathbb{N}$ and nonempty clopen sets $Y_1, \ldots, Y_{n_1}, Z_1, \ldots, Z_{n_1}$ of diameter $\leqslant 1$ such that $Y_1, \ldots, Y_{n_1}$ is a partition of Y and $Z_1, \ldots, Z_{n_1}$ is a partition of Z. A similar procedure applied to each pair Y_i, Z_i in place of Y, Z yields the existence of a number $n_2 \in \mathbb{N}$ and clopen nonempty sets Y_{ij}, Z_{ij} ($i = 1, \ldots, n_1$, $j = 1, \ldots, n_2$) all of diameter $\leqslant \frac{1}{2}$ such that for each $i \in \{1, \ldots, n_1\}$, $Y_{i1}, \ldots, Y_{in_2}$ is a partition of Y_i, $Z_{i1}, \ldots, Z_{in_2}$ is a partition of Z_i. Inductively we arrive at the following. There is a sequence $n_1, n_2, \ldots$ of natural numbers such that for each $m \in \mathbb{N}$ and for each $i_1 \leqslant n_1, \ldots, i_m \leqslant n_m$ there are nonempty clopen sets $Y_{i_1 \ldots i_m}$, $Z_{i_1 \ldots i_m}$ of diameter $\leqslant 1/m$ such that

$$Y_{i_1 \ldots i_{m-1} 1}, Y_{i_1 \ldots i_{m-1} 2}, \ldots, Y_{i_1 \ldots i_{m-1} n_m} \text{ is a partition of } Y_{i_1 \ldots i_{m-1}}$$

$$Z_{i_1 \ldots i_{m-1} 1}, Z_{i_1 \ldots i_{m-1} 2}, \ldots, Z_{i_1 \ldots i_{m-1} n_m} \text{ is a partition of } Z_{i_1 \ldots i_{m-1}}$$

Now let $x \in Y$. There is a unique $i_1 \leqslant n_1$ such that $x \in Y_{i_1}$, there is a unique $i_2 \leqslant n_2$ such that $x \in Y_{i_1 i_2}$, etc. The intersection of $Z_{i_1}, Z_{i_1 i_2}, \ldots$ contains precisely one point which is, by definition, $f(x)$. It takes a standard reasoning to show that f is a homeomorphism.

COROLLARY 74.3. *Let K be locally compact.*
(i) *Each two nonempty compact open subsets of K are homeomorphic.*
(ii) *K is homeomorphic to $\mathbb{Q}_p$. In particular, for any prime q, $\mathbb{Q}_p$ and $\mathbb{Q}_q$ are homeomorphic* (Compare Exercise 33.B), *$\mathbb{Z}_p$ and $\mathbb{Z}_q$ are homeomorphic.*
(iii) (Peano curve) *For each $n \in \mathbb{N}$, $\mathbb{Q}_p$ and $\mathbb{Q}_p^n$ are homeomorphic and so are $\mathbb{Z}_p$ and $\mathbb{Z}_p^n$.*

How about diffeomorphisms? For example, do we have a diffeomorphism between $\mathbb{Z}_p$ and $T_p = \mathbb{Z}_p \backslash p\mathbb{Z}_p$? Yes, if $p = 2$ (of course!) but no if $p \neq 2$ as we will see below.

LEMMA 74.4. *Let U be a nonempty open compact subset of $\mathbb{Q}_p$. Let $U = B_1 \cup B_2 \cup \ldots \cup B_n = B_1' \cup B_2' \cup \ldots \cup B_m'$ be partitions of U into discs. Then $n \equiv m \pmod{p-1}$.*

Proof. There is a partition $S_1, \ldots, S_t$ of U consisting of discs of equal diameter that is a refinement of both the given partitions. Let k_j ($j \in \{1, \ldots, n\}$) be the number of those S_i that are contained in B_j. As these S_i cover B_j, the number k_j must be a power of p, whence $k_j \equiv 1 \pmod{p-1}$. Since $\sum_{j=1}^{n} k_j = t$ we get $t \equiv n \pmod{p-1}$. By the same token, $t \equiv m \pmod{p-1}$. The lemma follows.

Lemma 74.4 enables us to define the *type* of a nonempty open compact subset U of $\mathbb{Q}_p$ to be 'the number of discs U consists of, modulo $p-1$'. More precisely, if $U = B_1 \cup \ldots \cup B_n$ is a partition into discs then the type of U is the unique $s \in \{0, 1, \ldots, p-2\}$ for which $n \equiv s \pmod{p-1}$. Thus, if p is odd then $\mathbb{Z}_p$ has type 1. T_p has type 0.

THEOREM 74.5. *Two nonempty open compact subsets of $\mathbb{Q}_p$ are diffeomorphic if and only if they are of the same type.*

Proof. Let U, V be nonempty open compact subsets of $\mathbb{Q}_p$. If they are of the same type one can easily find a number $n \in \mathbb{N}$ such that both U and V are disjoint unions of n discs and we have an obvious (locally linear) diffeomorphism between U and V. Conversely, if $\sigma : U \to V$ is a diffeomorphism then $\sigma'(x) \neq 0$ for $x \in U$. By Theorem 72.1, for each $a \in U$ there is a disc B_a containing a such that $\sigma(B_a)$ is again a disc. By compactness there is a finite disjoint subcovering of $\{B_a : a \in U\}$, say $B_1, \ldots, B_n$. Then V is the disjoint union of the n discs $\sigma(B_1), \ldots, \sigma(B_n)$. Thus, U and V are of the same type.

COROLLARY 74.6. *$\mathbb{Z}_p$ and $\mathbb{Z}_p \backslash p\mathbb{Z}_p$ are not diffeomorphic if $p \neq 2$.*

Exercise 74.B. ('Type' for subsets of locally compact fields) Let K be locally compact. Prove Lemma 74.4 with K in place of $\mathbb{Q}_p$ and q in place of p, where q is the number of elements of the residue class field of K. Define, in the spirit of above, the type ($\in \{0, 1, \ldots, q-2\}$) of a nonempty compact open subset of K and prove that *two nonempty compact open subsets of K are diffeomorphic if and only if they are of the same type.*

Exercise 74.C. (A differentiable homeomorphism of T_p onto $\mathbb{Z}_p$) Let $p \neq 2$. We shall sketch a construction of a differentiable homeomorphism $T_p \to \mathbb{Z}_p$ whose derivative, according to Corollary 74.6, must have zeros.

(i) For $n \in \mathbb{N}$, let $B_n := \{x \in \mathbb{Z}_p : |x|_p < p^{-2n}\}$, $B_n' := \{x \in \mathbb{Z}_p : |x - p^{2n}|_p < p^{-2n}$. Prove that $B_{n+1} \cup B_{n+1}' \subset B_n$ for all n. Define $C_0 := \mathbb{Z}_p \backslash (B_1 \cup B_1')$ and $C_n := (B_n \backslash (B_{n+1} \cup B_{n+1}')) \cup B_n'$ for $n \in \mathbb{N}$. Show that

C_0 is of type $p-2$, that C_n is of type 0 for $n \in \mathbb{N}$, and that $C_0, C_1, \ldots$ form a partition of $\mathbb{Z}_p \setminus \{0\}$. (Draw a picture.)
(ii) Let $A_0 := T_p \setminus (1 + p\mathbb{Z}_p)$ and for $n \in \mathrm{N}$ let $A_n := (1 + p^n \mathbb{Z}_p) \setminus (1 + p^{n+1} \mathbb{Z}_p)$. Show that A_0 is of type $p-2$, that A_n is of type 0 for $n \in \mathbb{N}$ and that $A_0, A_1, \ldots$ form a partition of $T_p \setminus \{1\}$.
(iii) For each $n \in \mathbb{N} \cup \quad 0\}$ let $f_n : A_n \to C_n$ be a diffeomorphism. Show that together they define a differentiable bijection $f_\infty : T_p \setminus \{1\} \to \mathbb{Z}_p \setminus \{0\}$.
(iv) Extend f_∞ continuously to a homeomorphism $f : T_p \to \mathbb{Z}_p$. Show that f is differentiable and that $f'(0) = 0$.

Exercise 74.D. (On Peano curves, see also Exercises 75.F and 75.G)
(i) Show (for arbitrary K) that K and K^2 are homeomorphic and also that their 'closed' unit balls are homeomorphic.
(ii) Let $A_0 := T_p \setminus (1 + p\mathbb{Z}_p)$ and for $n \in \mathrm{N}$ let $A_n := (1 + p^n \mathbb{Z}_p) \setminus (1 + p^{n+1} \mathbb{Z}_p)$. Show that A_0 is of type $p-2$, that A_n is of type 0 for $n \in \mathbb{N}$ and that there do not exist C^1-maps of $\mathbb{Z}_p$ onto $\mathbb{Z}_p \times \mathbb{Z}_p$.
(iii) Let K be not locally compact. Does there exist a homeomorphism of $B_0(1)$ onto $B_0(1) \times B_0(1)$ that is C^1 in the style of (ii)?

75. Isometries

Are isometries $K \to K$ surjective? In this section we shall answer this question and also present several exercises that may illustrate the deviation from the archimedean theory.

PROPOSITION 75.1. *Let the residue class field of K be infinite. Then the 'closed' unit ball and the unit sphere are isometrically isomorphic.*
Proof. Let $j \leftrightarrow B_j$ ($j \in k$) be the 1–1 correspondence between the elements of the residue class field k of K and the additive cosets of $B_0(1^-)$ in $B_0(1)$. There is a bijection $\tau : k \to k^\times$. For each j, let σ_j be an isometry (translation) of B_j onto $B_{\tau(j)}$. Define $\sigma : B_0(1) \to B_0(1) \setminus B_0(1^-)$ by $\sigma(x) := \sigma_j(x)$ as soon as $x \in B_j$ for some $j \in k$. One easily checks that σ is a surjective isometry.

If k is finite then $B_0(1)$ and $B_0(1) \setminus B_0(1^-)$ are not isometrically isomorphic. The map σ constructed above is a C^∞-function, $\sigma' = 1$. If we extend it by defining $\sigma(x) := x$ if $|x| > 1$ we obtain an isometry $K \to \{x \in K : |x| \geqslant 1\}$.

COROLLARY 75.2. *If k is infinite then there is a non-surjective isometry $K \to K$.*

There is a second reason why $\mathbb{C}_p$ must have non-surjective isometries.

PROPOSITION 75.3. *Suppose that each increasing function $K \to K$ is surjective. Then K is spherically complete.*

Proof. Let K be not spherically complete; we construct a non-surjective increasing function $f : K \to K$. There is a nested sequence of balls $B_1 \supsetneqq B_2 \supsetneqq \ldots$ whose intersection is empty. The sets $A_0 := K \backslash B_1$, $A_1 := B_1 \backslash B_2$, $A_2 := B_2 \backslash B_3, \ldots$ form a partition of K into nonempty clopen sets. For each $n \in \mathbb{N}$ choose $a_n \in A_n$ and define $g(x) := a_{n+1}$ whenever $x \in A_n$. As g maps A_n into A_{n+1} it has no fixed point so that the map $f : K \to K$ given by

$$f(x) := x - g(x) \qquad (x \in K)$$

does not attain the value 0. We now show that f is increasing, which boils down to showing that $|g(x) - g(y)| < |x - y|$ for all $x, y \in K, x \neq y$. If $x \in A_n$, $y \in A_m$ for $n < m$ then $g(x) \in B_{n+1}$ and $y, g(y) \in B_m \subset B_{n+1}$, but $x \notin B_{n+1}$. Thus $|g(x) - g(y)| < d(x, B_{n+1}) = |x - y|$.

THEOREM 75.4. *The following conditions* (α) *and* (β) *on K are equivalent.*
(α) *K is spherically complete. Its residue class field is finite.*
(β) *Each isometry $K \to K$ is surjective.*

Proof. For (β) ⇒ (α) see above. To establish (α) ⇒ (β) it suffices to prove surjectivity of an isometry $f : B \to B$ where B is the 'closed' unit disc of K. Suppose we had an $a \in B \backslash f(B)$. By spherical completeness a has a best approximation in $f(B)$, i.e. there is a $b \in B$ such that $|a - f(b)| = \rho := d(a, f(B))$. Let $B' = B_b(\rho)$, $B'' := B_{f(b)}(\rho)$, $B''' := B_a(\rho^-)$. Then $f(B') \subset B''$, $B''' \subset B''$, $B''' \cap f(B) = \emptyset$. Let n be the number of elements of k. In B', hence in $f(B')$, we can find n equidistant points with distance ρ. Hence, each coset of B''' in B'' meets $f(B')$. In particular $B''' \cap f(B') \neq \emptyset$, a contradiction.

Exercise 75.A. (On surjectivity of isometries for locally compact K)
(i) Show that each isometry of a compact metric space into itself is surjective.
(ii) Use (i) to obtain a more down-to-earth proof of the surjectivity of an isometry $K \to K$ for locally compact K.

Exercise 75.B. Consider the sets $\mathbb{C}_p$, $\{x \in \mathbb{C}_p : |x|_p \leq 1\}$, $\{x \in \mathbb{C}_p : |x|_p < 1\}$. Which pairs are homeomorphic, isomorphic as additive groups, isometrically isomorphic?

**Exercise* 75.C. Let $X \subset K$ and let $f : X \to K$ satisfy a Lipschitz condition of order 1. Show that f is a K-linear combination of two isometries. (Hint. Let one of them be the identity.)

Exercise 75.D. Find a nowhere differentiable isometry $\mathbb{Z}_p \to \mathbb{Z}_p$. (Hint. The previous exercise might be helpful.)

Exercise 75.E. Let A, B be countable dense subsets of $\mathbb{Z}_p$. Then there exists an isometry of $\mathbb{Z}_p$ that maps A onto B. Prove this.

Exercise 75.F. (Isometrical Peano curves!) Let B be the 'closed' unit ball of K, let K^2 be normed as in Exercise 13.A. Prove the following.
(i) If the residue class field of K is finite then B and $B \times B$ are not isometrically isomorphic.
(ii) If the residue class field of K is infinite and the valuation of K is discrete then B and $B \times B$ are isometrically isomorphic. (Hint. Use the terminology of Theorem 12.1 and define $f : B \to B \times B$ by $f(\sum a_n \pi^n) = (\sum \sigma_1(a_n)\pi^n, \sigma_2(a_n)\pi^n)$ for a suitable choice of σ_1, σ_2.)

Exercise 75.G. (An isometry of $\mathbb{C}_p$ onto $\mathbb{C}_p^2$) The previous exercise does not cover the case $K = \mathbb{C}_p$. Thus we shall sketch a construction of an isometry between $\mathbb{C}_p$ and $\mathbb{C}_p^2$ whose restriction to the 'closed' unit ball B maps onto $B \times B$. We need the following fact. $\mathbb{C}_p$ contains a closed subfield K whose value group is $\{p^n : n \in \mathbb{Z}\}$ and which is infinite dimensional as a $\mathbb{Q}_p$-linear space. (One can prove that the closure of the smallest field containing Γ_u (see Section 33) will do. By way of exception we shall violate our principles and take this fact for granted.) Both the valuation on $\mathbb{C}_p$ and the norm on $\mathbb{C}_p^2$ are denoted $\| \quad \|$. Now proceed as follows.
(i) $\mathbb{C}_p$ and $\mathbb{C}_p \times \mathbb{C}_p$ as normed $\mathbb{Q}_p$-linear spaces have orthogonal bases $e_1, e_2, \ldots$ and $f_1, f_2, \ldots$ respectively (Corollary 50.10).
(ii) Let R be a complete set of representatives in $\mathbb{Q}$ of $\mathbb{Q}/\mathbb{Z}$ containing 0. We may assume that $\|e_n\|, \|f_n\| \in \{p^r : r \in R\}$ for all n.
(iii) For each $r \in R$ let V_r be the closure of $[\![e_j : \|e_j\| = p^r]\!]$ and let W_r be the closure of $[\![f_j : \|f_j\| \in p^r]\!]$. Then $\{e_j : \|e_j\| = p^r\}$ is an orthogonal base of V_r and $\{f_j : \|f_j\| = p^r\}$ is an orthogonal base of W_r.
(iv) If $r, s \in R$, $r \neq s$ then $V_r \perp V_s$ and $W_r \perp W_s$.
(v) $\{e_j : \|e_j\| = 1\}$ is a maximal orthogonal subset of $\{x \in \mathbb{C}_p : \|x\| \in \{p^n : n \in \mathbb{Z}\}\}$. Hence $V_0 \supset K$ and V_0 is infinite dimensional. In a similar way one proves that W_0 is infinite dimensional.
(vi) For each $r \in R$ the spaces V_r and W_r are infinite dimensional.
(vii) For each $r \in R$ define a surjective $\mathbb{Q}_p$-linear isometry $f_r : V_r \to W_r$.
(viii) Let $x \in \mathbb{C}_p$. Then $x = \sum_{r \in R} v_r$ where $v_r \in V_r$ for each $r \in R$. The formula $f(x) := \sum_{r \in R} f_r(v_r)$ defines a $\mathbb{Q}_p$-linear isometry of $\mathbb{C}_p$ onto $\mathbb{C}_p^2$.

Exercise 75.H. (Continuous endomorphisms of $\mathbb{C}_p$) At this stage one may start worrying about possible 'bad' behaviour of continuous endomorphisms of $\mathbb{C}_p$. However, the following is reassuring. Let $\sigma : \mathbb{C}_p \to \mathbb{C}_p$ be a nonzero continuous map satisfying $\sigma(x + y) = \sigma(x) + \sigma(y)$, $\sigma(xy) = \sigma(x)\,\sigma(y)$ for all $x, y \in \mathbb{C}_p$. Use the following steps to show that σ is in fact a surjective isometry.

(i) By Exercise 9.C the valuation $x \mapsto |\sigma(x)|_p$ is equivalent to $|\ \ |_p$.
(ii) σ maps $\mathbb{Q}_p$ into itself.
(iii) σ is a $\mathbb{Q}_p$-linear isometry of $\mathbb{C}_p$ into $\mathbb{C}_p$.
(iv) Let $f \in \mathbb{Q}_p[X]$ and let K be the smallest field containing $\mathbb{Q}_p$ and the roots of f. Then σ maps K onto K.
(v) σ is surjective.

76. Extension theorems

Let X be a closed subset of K. In order to extend a function $f : X \to K$ with a certain property to a function $g : K \to K$ with the same property it is sometimes helpful to define a suitable $\sigma : K \to K$ extending the identity on X and try $g = f \circ \sigma$.

LEMMA 76.1. *Let X be a closed nonempty subset of K. Let $\epsilon > 0$. There is a map $\sigma : K \to X$ such that*
(i) *$\sigma(x) = x$ for all $x \in X$,*
(ii) *σ is locally constant on $K \backslash X$,*
(iii) *$|\sigma(x) - \sigma(y)| \leqslant (1+\epsilon)|x-y|$ for all $x, y \in K$.*

Proof. The equivalence relation defined by

$$x \sim y \text{ if } [x, y] \subset K \backslash X$$

defines a partition of $K \backslash X$ into balls $\{B_j : j \in J\}$ for some indexing set J. We have $X \cap B_j = \emptyset$, hence $d(X, B_j) > 0$ for each j and we can find $a_j \in X$ such that $d(a_j, B_j) \leqslant (1+\epsilon)\, d(X, B_j)$. Define $\sigma : K \to X$ by the formula

$$\sigma(x) := \begin{cases} x \text{ if } x \in X \\ a_j \text{ if } x \in B_j \text{ for some } j \in J \end{cases}$$

Only (iii) requires a proof. It is trivial if $x, y \in X$. Let $x, y \notin X$. If $x, y \in B_j$ for some j then $\sigma(x) = \sigma(y)$. If $x \in B_i$, $y \in B_j$ for some $i \neq j$ then $|\sigma(x) - \sigma(y)| = |a_i - a_j| \leqslant \max(|a_i - x|, |x-y|, |y-a_j|)$. We have $|a_i - x| = d(a_i, B_i) \leqslant (1+\epsilon)\, d(X, B_i) \leqslant (1+\epsilon)\, d(X, x)$. Now $[x, y]$ intersects B_i so it properly contains B_i. By construction B_i is a maximal convex subset of $K \backslash X$ and therefore $[x, y] \cap X \neq \emptyset$ and we get $d(X, x) \leqslant |x-y|$ and $|a_i - x| \leqslant (1+\epsilon)|x-y|$ by the above. Similarly $|y - a_j| \leqslant (1+\epsilon)|x-y|$ and (iii) follows for this case. Let $x \in X$ and $y \notin X$. Then $y \in B_j$ for some j and $|\sigma(x) - \sigma(y)| = |x - a_j| \leqslant \max(|x-y|, |y-a_j|)$. As before $|y - a_j| \leqslant (1+\epsilon)\, d(X, B_j)$. Also $|x-y| \geqslant d(x, B_j) \geqslant d(X, B_j)$ so that $|\sigma(x) - \sigma(y)| \leqslant (1+\epsilon)|x-y|$.

Exercise 76.A. Assume that the set X of the above lemma is spherically complete as a metric space. Show that there exists a map $\sigma : K \to X$ satisfying (i), (ii) and (iii)' $|\sigma(x) - \sigma(y)| \leqslant |x - y|$ for all $x, y \in K$. (Use Theorem 21.2.) Observe that the above applies to the case $|K^\times|$ discrete, X closed.

THEOREM 76.2. *Let X be a closed subset of K, and let* $f : X \to K$. *Then f can be extended to a function* $g : K \to K$ *for which* $g(K) = f(X)$ *in such a way that*
(i) *if f is (uniformly) continuous then so is g,*
(ii) *if, for some positive constants M, a, f satisfies the Lipschitz condition*

$$|f(x) - f(y)| \leqslant M|x - y|^a \qquad (x, y \in X)$$

and if $M' > M$ *then*

$$|g(x) - g(y)| \leqslant M'|x - y|^a \qquad (x, y \in K)$$

(iii) *if X has no isolated points and f is differentiable,* $f' = 0$ *then* $g' = 0$,
(iv) *if X has no isolated points and* $f \in C^1(X \to K)$, $f' = 0$ *then* $g \in C^1(K \to K)$ *and* $g' = 0$.
Proof. For the cases (i), (iii), (iv) choose arbitrary $\epsilon > 0$; for case (ii) make sure that $M(1 + \epsilon)^a = M$. Define $g := f \circ \sigma$ where σ is as in Lemma 76.1. The rest is straightforward.

Exercise 76.B. Let char$(K) = 0$, let X be a closed subset of K. For some $n \in \mathbb{N}$ let $f \in C^n(X \to K)$, $f' = 0$. Use Theorem 29.12 to show that f can be extended to a C^n-function $g : K \to K$ for which $g' = 0$.

The map σ of Lemma 76.1 need not be differentiable so it is not clear whether $f \circ \sigma$ is differentiable for a differentiable $f : X \to K$. Yet, we have

THEOREM 76.3. *Let X be a closed subset of K without isolated points and let* $f : X \to K$ *be differentiable. Then f can be extended to a differentiable function* $K \to K$.
Proof. By Theorem 70.1, f' is of the first class of Baire so there are continuous $f_1, f_2, \ldots : X \to K$ for which $f' = \lim_{n\to\infty} f_n$. With $\sigma : K \to X$ as in Lemma 76.1 we then have $f' \circ \sigma = \lim_{n\to\infty} f_n \circ \sigma$ so that $f' \circ \sigma : K \to K$ is of the first class of Baire, and has an antiderivative $g : K \to K$. Obviously $(f - g)'(x) = 0$ for all $x \in X$, so, by Theorem 76.2(iii), $((f-g) \circ \sigma)' = 0$. The function $g + (f-g) \circ \sigma$ is a differentiable extension of f.
Remark. For a similar extension theorem for C^n-functions see Exercise 79.D and Corollary 81.4.

Exercise 76.C. According to Exercise 72.G a differentiable extension $\mathbb{C}_p \to \mathbb{Q}_p$ of the identity $\mathbb{Q}_p \to \mathbb{Q}_p$ must have zero derivative. Is this in conflict with Theorem 76.3?

Quite different techniques are needed for extensions of isometries (to isometries).

THEOREM 76.4. *The following conditions are equivalent.*
(α) *K is spherically complete. Its residue class field is finite.*
(β) *Each K-valued isometry defined on a subset of K can be extended to an isometry $K \to K$.*

Proof. (β) $\Rightarrow$ (α). By Theorem 75.4 it suffices to prove that an isometry $f: K \to K$ is surjective. By (β) we can extend $f^{-1} : f(K) \to K$ which is nonabsurd only if $f(K) = K$. To prove (α) $\Rightarrow$ (β) it suffices to extend a K-valued isometry f defined on a subset X of K to an isometry $\bar{f}: X \cup \{a\} \to K$ where $a \in K \setminus X$ (apply Zorn's lemma). Further, we may assume that X is closed so that $\rho := d(a, X) > 0$. We distinguish two cases.
(i) a has no best approximation in X. Then choose $x_1, x_2, \ldots \in X$ such that $|a - x_1| > |a - x_2| > \ldots$ and $\lim_{n \to \infty} |a - x_n| = \rho$. For each $n \in \mathbb{N}$ define

$$B_n := B_{f(x_n)}(|a - x_n|)$$

We claim that

$$B_1 \supset B_2 \supset \ldots$$

Indeed, if $z \in B_{n+1}$ then $|z - f(x_n)| \leqslant \max(|z - f(x_{n+1})|, |f(x_{n+1}) - f(x_n)|) \leqslant \max(|a - x_{n+1}|, |x_{n+1} - x_n|) \leqslant \max(|a - x_{n+1}|, |x_{n+1} - a|, |a - x_n|)$ so that $z \in B_n$. By spherical completeness $\bigcap_n B_n$ is not empty, extend f by letting $\bar{f}(a)$ be an arbitrary element of $\bigcap_n B_n$. We prove that $\bar{f}$ is an isometry, i.e. that $|\bar{f}(a) - f(x)| = |a - x|$ for all $x \in X$. Let $x \in X$. It is not a best approximation of a so $|a - x_n| < |a - x|$ for some $n \in \mathbb{N}$. Then $|f(x_n) - f(x)| = |x_n - x| = |a - x|$. As $\bar{f}(a) \in B_n$ we have $|\bar{f}(a) - f(x_n)| \leqslant |a - x_n| < |a - x|$. Thus, $|\bar{f}(a) - f(x)| = \max(|\bar{f}(a) - f(x_n)|, |f(x_n) - f(x)|) = |a - x|$.
(ii) a has a best approximation in X. Form the nonempty set

$$A_1 := \{x \in X : |a - x| = \rho\}$$

and let A_2 be a maximal subset of A_1 with the property '$x, y \in A_2, x \neq y$ then $|x - y| = \rho$'. Let the residue class field of K have q elements. In the 'annulus' $\{x \in K : |a - x| = \rho\}$ there are at most $q - 1$ points having distances ρ to one another, so A_2 is a finite set $\{r_1, \ldots, r_n\}$, where $n < q$. Then the n points of $f(A_2)$ also have distances ρ to one another. Since $n < q$ there is an element $\nu \in K$ such that $|\nu - f(r_1)| = |\nu - f(r_2)| = \ldots = |\nu - f(r_n)| = \rho$. Extend f by choosing $\bar{f}(a) := \nu$. To show that $\bar{f}$ is an isometry it suffices to check that

$$(*) \qquad |\nu - f(x)| = |a - x|$$

for all $x \in X$. If $x \in A_2$ then (*) holds. Let $x \in A_1$. There is a $j \in \{1, \ldots, n\}$ such that $|x - r_j| < \rho$. We have $|a - x| = \rho = |a - r_j| = |v - f(r_j)|$ and $|f(r_j) - f(x)| = |r_j - x| < \rho$. Hence, $|v - f(x)| = \max(|v - f(r_j)|, |f(r_j) - f(x)|) = \rho = |a - x|$. Finally, let $x \in X \backslash A_1$. Then $|a - x| > \rho$. Since $|a - r_1| = \rho$ we have $|x - r_1| = |a - x| > \rho$ so that $|f(x) - f(r_1)| > \rho$. We know that $|v - f(r_1)| = \rho$. Therefore, $|v - f(x)| = \max(|v - f(r_1)|, |f(r_1) - f(x)|) = |f(r_1) - f(x)| = |r_1 - x| = |a - x|$ and we are done.

For increasing functions (which are special isometries) we have the following characterization in the spirit of Theorems 75.4 and 76.4.

THEOREM 76.5. *The following conditions are equivalent.*
(α) *K is spherically complete.*
(β) *Each increasing function defined on a subset of K can be extended to an increasing function $K \to K$.*
(γ) *Each increasing function $K \to K$ is surjective.*

Proof. (α) $\Rightarrow$ (β). It suffices to extend an increasing function f defined on a subset X of K to an increasing function $\bar{f}: X \cup \{a\} \to K$ where $a \in K \backslash X$. This can be done if and only if there is an element $\bar{f}(a) \in K$ such that

$$\left| \frac{\bar{f}(a) - f(x)}{a - x} - 1 \right| < 1 \qquad (x \in X)$$

in other words

$$\bar{f}(a) \in B_{f(x)+a-x}(|a - x|^-) \qquad (x \in X)$$

So we are done if we can show that the collection of the balls $B_{f(x)+a-x}(|a - x|^-)$, where x runs through X, has a nonempty intersection. By Exercise 20.A(i) and spherical completeness it suffices to check nondisjointness of any two of these balls. Let $x, y \in X, x \neq y$. We have $|f(x) + a - x - (f(y) + a - y)| = |f(x) - f(y) - (x - y)| < |x - y| \leqslant \max(|a - x|, |a - y|)$. It follows that $B_{f(x)+a-x}(|a - x|^-) \cap B_{f(y)+a-y}(|a - y|^-) \neq \emptyset$. ($\beta$) $\Rightarrow$ (γ) is easy. (γ) $\Rightarrow$ (α) is just Proposition 75.3.

We can formulate Theorem 76.5 in a less mysterious way by using the concept of a pseudocontraction.

DEFINITION 76.6. Let (Y_1, d_1) and (Y_2, d_2) be metric spaces. A map $f: Y_1 \to Y_2$ is a *pseudocontraction* if $d_2(f(x), f(y)) < d_1(x, y)$ for all $x, y \in Y_1, x \neq y$.

Using the simple fact that a function f on K is increasing if and only if $f - x$ is a pseudocontraction we arrive easily at the following restatement of Theorem 76.5. See also Exercise 76.D.

COROLLARY 76.7. *The following conditions are equivalent.*
(α) K is spherically complete.
(β) For each subset X of K and each pseudocontraction $f : X \to K$ there is a pseudocontraction $K \to K$ extending f.
(γ) Each pseudocontraction $K \to K$ has a (unique) fixed point.

Exercise 76.D. Property (γ) of Corollary 76.7 suggests an analogy to Banach's contraction theorem (see Appendix A.1). In fact, prove the following. Let (Y, d) be a spherically complete ultrametric space and let $f : Y \to Y$ be a pseudocontraction. Then f has a unique fixed point. (Hint. Consider the balls $B_a := \{ x \in Y : d(x, f(a)) < d(a, f(a)) \}$ $(a \in Y)$.) However, in contrast to Banach's theorem iteration will not always lead to a fixed point, i.e. it may happen that $x, f(x), f(f(x)), \ldots$ is divergent for some $x \in Y$.

Exercise 76.E. Show that spherical completeness of K is also equivalent to the following. For every subset X of K and each $f : X \to K$ satisfying

$$|f(x) - f(y)| \leqslant |x - y| \qquad (x, y \in X)$$

there is an extension $\bar{f} : K \to K$ of f such that

$$|\bar{f}(x) - \bar{f}(y)| \leqslant |x - y| \qquad (x, y \in K)$$

(Compare Theorem 76.2(ii) and Theorem 76.4(β).)

PART 2: C^n-THEORY

In Sections 27–29 we already made a start with the theory of C^n-functions. In this part we shall continue our investigations and turn to fundamental questions that are still waiting for an answer, such as the following.

Do we have a local invertibility theorem for C^n-functions?
Is a composition of two C^n-functions again a C^n-function?
Is the derivative of a C^n-function a C^{n-1}-function?
Does a C^{n-1}-function have a C^n-antiderivative?

The corresponding statements for real valued functions defined on an interval are trivial or at least easy to prove by using the statement $f' \in C^{n-1} \Rightarrow f \in C^n$, which is false in the ultrametric theory. In Sections 77 and 78 we shall give an affirmative answer to the first three questions. Due to shortage of powerful tools the proofs are somewhat involved, but essentially they have nothing to do with ultrametric analysis and work equally well for real valued functions defined on a subset of $\mathbb{R}$ (for example $\mathbb{Q}$, the Cantor set, or even an interval for that matter). The fourth question is different. The answer is only partially 'yes' and we need ultrametric techniques for the proof (Sections

79–82). In Section 83 we shall discuss an alternative definition of a C^n -function involving only two variables, rather than $n+1$. Finally, in Section 84 we shall define C^1-and C^2-functions of two variables

THROUGHOUT PART 2 X IS A NONEMPTY SUBSET OF K WITHOUT ISOLATED POINTS

77. Local invertibility of C^n-functions

Our starting point is the following proposition. We leave the (easy) proof to the reader.

PROPOSITION. *Let $f \in C^1(X \to K)$ and let $f'(a) \neq 0$ for some $a \in X$. Then there is a neighbourhood U of a such that $f \mid U \cap X$ is injective. Its inverse $g : f(U \cap X) \to U \cap X$ is a C^1-function.*

Now let $n > 1$ and assume that the function f above is in $C^n(X \to K)$. We shall prove that $\Phi_n g$ can be expressed in terms of $\Phi_1 g, \ldots, \Phi_{n-1} g$, $\Phi_1 f, \ldots, \Phi_n f$. For example, one checks easily that for $(x_1, x_2, x_3) \in \nabla^3 f(U \cap X)$ we have

$$\Phi_2 g(x_1, x_2, x_3) = -\Phi_2 f(g(x_1), g(x_2), g(x_3))\, \Phi_1 g(x_1, x_2) \cdot \Phi_1 g(x_1, x_3)\, \Phi_1 g(x_2, x_3)$$

which is the case $n = 2$ of the following lemma.

LEMMA 77.1. *Let $f : X \to K$ be an injection and let $g : f(X) \to X$ be its inverse. Let $n \in \mathbb{N}$, $n \geqslant 2$. Let S_n be the set of the following functions defined on $\nabla^{n+1} f(X)$.*

$$(x_1, \ldots, x_{n+1}) \mapsto \Phi_1 g(x_{i_1}, x_{i_2}) \qquad (i_1 < i_2)$$
$$(x_1, \ldots, x_{n+1}) \mapsto \Phi_2 g(x_{i_1}, x_{i_2}, x_{i_3}) \qquad (i_1 < i_2 < i_3)$$
$$\vdots$$
$$(x_1, \ldots, x_{n+1}) \mapsto \Phi_{n-1} g(x_{i_1}, \ldots, x_{i_n}) \qquad (i_1 < i_2 < \ldots < i_n)$$

and

$$(x_1, \ldots, x_{n+1}) \mapsto \Phi_1 f(g(x_{i_1}), g(x_{i_2})) \qquad (i_1 < i_2)$$
$$(x_1, \ldots, x_{n+1}) \mapsto \Phi_2 f(g(x_{i_1}), g(x_{i_2}), g(x_{i_3})) \qquad (i_1 < i_2 < i_3)$$
$$\vdots$$
$$(x_1, \ldots, x_{n+1}) \mapsto \Phi_{n-1} f(g(x_{i_1}), \ldots, g(x_{i_n})) \qquad (i_1 < i_2 < \ldots < i_n)$$
$$(x_1, \ldots, x_{n+1}) \mapsto \Phi_n f(g(x_1), \ldots, g(x_{n+1}))$$

Let R_n be the ring generated by S_n. Then $\Phi_n g \in R_n$.

Proof. A simple inspection shows that if $h \in R_{n-1}$ then the function

$$(x_1, \ldots, x_{n+1}) \mapsto h(x_1, \ldots, x_n) \quad ((x_1, \ldots, x_{n+1}) \in \nabla^{n+1} f(X))$$

is an element of R_n. To prove the lemma it suffices to check the induction step from $n-1$ to n where $n \geq 3$. We have for $(x_1, \ldots, x_{n+1}) \in \nabla^{n+1} f(X)$

$$\Phi_n g(x_1, \ldots, x_{n+1}) = (x_1 - x_2)^{-1} (\Phi_{n-1} g(x_1, x_3, \ldots, x_{n+1}) - \Phi_{n-1} g(x_2, x_3, \ldots, x_{n+1}))$$

The induction hypothesis states that $\Phi_{n-1} g \in R_{n-1}$. So we are done if we can prove that $h \in R_{n-1}$ implies $\Delta h \in R_n$ where

$$\Delta h(x_1, \ldots, x_{n+1}) := (x_1 - x_2)^{-1} (h(x_1, x_3, \ldots, x_{n+1}) - h(x_2, x_3, \ldots, x_{n+1}))$$

In other words we must show that

$$B := \{h \in R_{n-1} : \Delta h \in R_n\}$$

equals R_{n-1}. We shall prove this by showing that B is a ring containing S_{n-1}. Δ is a linear map, so B is a group under addition. Let $h, t \in B$. We prove that $ht \in B$, i.e. that $\Delta(ht) \in R_n$ as follows. For $(x_1, \ldots, x_{n+1}) \in \nabla^{n+1} f(X)$ we have

$$\Delta(ht)(x_1, \ldots, x_{n+1}) = t(x_1, x_3, \ldots, x_{n+1}) \Delta h(x_1, \ldots, x_{n+1}) + h(x_2, x_3, \ldots, x_{n+1}) \Delta t(x_1, x_2, \ldots, x_{n+1})$$

By the first lines of this proof $(x_1, \ldots, x_{n+1}) \mapsto t(x_1, x_3, \ldots, x_{n+1})$ and $(x_1, x_2, \ldots, x_{n+1}) \mapsto h(x_2, x_3, \ldots, x_{n+1})$ are in R_n. By the definition of B we have $\Delta h \in R_n$ and $\Delta t \in R_n$. Since R_n is a ring, we get $\Delta(ht) \in R_n$. It follows that B is a ring. Finally we prove that $S_{n-1} \subset B$. Let $h \in S_{n-1}$ be a function of the 'first type', i.e.

$$h(x_1, \ldots, x_n) = \Phi_j g(x_{i_1}, \ldots, x_{i_{j+1}}) \quad (1 \leq j \leq n-2)$$

If $1 \notin \{i_1, \ldots, i_{j+1}\}$ then $\Delta h = 0$. Otherwise $i_1 = 1$ and for $(x, y, x_2, \ldots, x_n) \in \nabla^{n+1} f(X)$ we then have

$$h(x, x_2, \ldots, x_n) = \Phi_j g(x, x_{i_2}, \ldots, x_{i_{j+1}})$$
$$h(y, x_2, \ldots, x_n) = \Phi_j g(y, x_{i_2}, \ldots, x_{i_{j+1}})$$

so that

$$\Delta h(x, y, x_2, \ldots, x_n) = \Phi_{j+1} g(x, y, x_{i_2}, \ldots, x_{i_{j+1}})$$

Hence $\Delta h \in R_n$, which implies $h \in B$. Now let h be of the 'second type', i.e.

$$h(x_1, \ldots, x_n) = \Phi_j f(g(x_{i_1}), g(x_{i_2}), \ldots, g(x_{i_{j+1}})) \qquad (1 \leq j \leq n-1)$$

Again, $\Delta h = 0$ if $1 \notin \{i_1, \ldots, i_{j+1}\}$. If $i_1 = 1$ then for $(x, y, x_2, \ldots, x_n) \in \nabla^{n+1} f(X)$ we have

$$h(x, x_2, \ldots, x_n) = \Phi_j f(g(x), g(x_{i_2}), \ldots, g(x_{i_{j+1}}))$$
$$h(y, x_2, \ldots, x_n) = \Phi_j f(g(y), g(x_{i_2}), \ldots, g(x_{i_{j+1}}))$$

so that

$$\Delta h(x, y, x_2, \ldots, x_n) = \Phi_{j+1} f(g(x), g(y), g(x_{i_2}), \ldots, g(x_{i_{j+1}}))\, \Phi_1 g(x, y)$$

It follows that Δh is a product of two functions in S_n. Hence $\Delta h \in R_n$, which implies $h \in B$.

As a corollary we get

THEOREM 77.2. (Local invertibility of C^n-functions) *Let $f \in C^n(X \to K)$ and let $f'(a) \neq 0$ for some $a \in X$. Then there is a neighbourhood U of a such that $f \mid U \cap X$ is a bijection onto $f(U \cap X)$ whose inverse is a C^n-function.*

Proof. By induction on n. Suppose the theorem is true for $n-1$. Let $f \in C^n(X \to K)$, $f'(a) \neq 0$. By the induction hypothesis there is a neighbourhood U of a such that the local inverse

$$g : f(U \cap X) \to U \cap X$$

of f is a C^{n-1}-function. Apply the previous lemma to $f \mid U \cap X$ and $g \mid f(U \cap X)$. Then $\Phi_n g \in R_n$. But S_n, and therefore R_n, consists of functions that are continuously extendable to $f(U \cap X)^{n+1}$ Hence g is a C^n-function.

COROLLARY 77.3. *Let $f \in C^1(X \to K)$ and let $f'(a) \neq 0$ for some $a \in X$. Let U be a neighbourhood of a such that the local inverse $g : f(U \cap X) \to U \cap X$ is C^1. If $f \in C^n(X \to K)$ for some n then $g \in C^n(f(U \cap X) \to K)$.*

Remark.

If X is an open set (or a neighbourhood of a) we may improve the above corollary by stating that we can choose $U \subset X$ such that $f(U)$ contains a full neighbourhood of $f(a)$. This is simply Theorem 27.5. For arbitrary X it is not always true that for small neighbourhoods U of a the set $f(U \cap X)$ is a neighbourhood of $f(a)$ relative to $f(X)$ (i.e. $f(U \cap X) = V \cap f(X)$ for some neighbourhood V of $f(a)$), as is illustrated by the following exercise.

Exercise 77.A. Let $B_1, B_2, \ldots$ be disjoint balls in $\mathbb{Z}_p$ 'tending to 0', for example $B_n := \{x \in \mathbb{Z}_p : |x - p^n| < p^{-2n}\}$. 'Tear apart' $\mathbb{Z}_p$ by translating the balls B_n as follows. Let

$$X := (\mathbb{Z}_p \setminus \bigcup_{n=1}^{\infty} B_n) \cup (p^{-1} + B_1) \cup (p^{-2} + B_2) \cup \ldots$$

Then X is a closed subset of $\mathbb{Q}_p$ without isolated points. The function $f : X \to \mathbb{Z}_p$ that 'restores the damage' is defined by

$$f(x) := \begin{cases} x \text{ if } x \in \mathbb{Z}_p \setminus \bigcup_n B_n \\ -p^{-n} + x \text{ if } x \in p^{-n} + B_n \text{ for some } n \in \mathbb{N} \end{cases}$$

(i) Show that f is a C^∞-bijection of X onto $\mathbb{Z}_p$ and that $f' = 1$, but that f^{-1} is not continuous at 0. Is this in conflict with Theorem 77.2?

The techniques used to prove the local invertibility theorem also apply for compositions of C^n-functions. We content ourselves with stating a lemma in the spirit of Lemma 77.1 yielding Theorem 77.5 as a corollary.

LEMMA 77.4. *Let $g : X \to K$ be continuous, let $Y \supset g(X)$ have no isolated points and let, for some $n \in \mathbb{N}$, $f \in C^n(Y \to K)$. Let S_n' be the set of the following functions defined on $\nabla^{n+1} X$.*

$$(x_1, \ldots, x_{n+1}) \mapsto \Phi_1 g(x_{i_1}, x_{i_2}) \qquad (i_1 < i_2)$$
$$(x_1, \ldots, x_{n+1}) \mapsto \Phi_2 g(x_{i_1}, x_{i_2}, x_{i_3}) \qquad (i_1 < i_2 < i_3)$$
$$\vdots$$
$$(x_1, \ldots, x_{n+1}) \mapsto \Phi_n g(x_1, \ldots, x_{n+1})$$

and

$$(x_1, \ldots, x_{n+1}) \mapsto \bar{\Phi}_1 f(g(x_{i_1}), g(x_{i_2})) \qquad (i_1 < i_2)$$
$$(x_1, \ldots, x_{n+1}) \mapsto \bar{\Phi}_2 f(g(x_{i_1}), g(x_{i_2}), g(x_{i_3})) \qquad (i_1 < i_2 < i_3)$$
$$\vdots$$
$$(x_1, \ldots, x_{n+1}) \mapsto \bar{\Phi}_n f(g(x_1), \ldots, g(x_{n+1}))$$

Let R_n' be the ring generated by S_n. Then $\Phi_n(f \circ g) \in R_n'$.

THEOREM 77.5. (Composition of C^n-functions) *Let $g : X \to K$, let $Y \supset f(X)$ have no isolated points, and let $f : Y \to K$.*
(i) *If $f \in C^n(Y \to K)$, $g \in C^n(X \to K)$ for some $n \in \mathbb{N}$ then $f \circ g \in C^n(X \to K)$.*
(ii) *If $f \in C^\infty(Y \to K)$, $g \in C^\infty(X \to K)$ then $f \circ g \in C^\infty(X \to K)$.*

78. Differentiation $C^n \to C^{n-1}$

In this section we shall prove that D_j (see Definition 29.1) maps $C^n(X \to K)$ into $C^{n-j}(X \to K)$ $(0 \leqslant j \leqslant n)$, in particular that the derivative of a C^n-function is a C^{n-1}-function. With an eye on the nonzero characteristic case we prefer to work with $D_j f$ rather than $f^{(j)} = j! D_j f$ (Theorem 29.5). For notations and terminology we refer to Section 29.

LEMMA 78.1. *Let $n \in \mathbb{N}$ and $f \in C^{n-1}(X \to K)$. Then for all $(x_1, \ldots, x_{n+1}) \in \nabla^{n+1} X$ we have*

$$\Phi_j D_{n-j} f(x_1, \ldots, x_{j+1}) = \sum_{u \in S_{jn}} \tilde{\Phi}_n f(u) \qquad (1 \leqslant j \leqslant n)$$

where S_{jn} is the set of all $(x_{m_1}, x_{m_2}, \ldots, x_{m_{n+1}}) \in X^{n+1}$ for which $m_1 \leqslant m_2 \leqslant \ldots \leqslant m_{n+1}$ and $\{m_1, m_2, \ldots, m_{n+1}\} = \{1, 2, \ldots, j+1\}$. S_{jn} has $\binom{n}{j}$ elements.

Proof. An element $(x_1, x_1, \ldots, x_1, x_2, \ldots, x_2, x_3, \ldots, x_{j+1}, x_{j+1}, \ldots, x_{j+1})$ of S_{jn} is determined once we know at what spots x_i is followed by x_{i+1} $(1 \leqslant i \leqslant j)$. This is selecting j out of n available spots so S_{jn} has $\binom{n}{j}$ elements. We prove the formula by induction on n. The formula for $\Phi_1 D_{n-1} f$ is essentially established in the proof of Theorem 29.5. For the step from $\{1, 2, \ldots, n-1\}$ to n, let $f \in C^{n-1}(X \to K)$; we prove the formula for $\Phi_j D_{n-j} f$ for $2 \leqslant j \leqslant n$. By the definition of Φ_j and the induction hypothesis

$$\begin{aligned}
&\Phi_j D_{n-j} f(x_1, \ldots, x_{j+1}) \\
&\quad = (x_1 - x_2)^{-1} (\Phi_{j-1} D_{n-j} f(x_1, x_3, \ldots, x_{j+1}) - \\
&\quad\quad \Phi_{j-1} D_{n-j} f(x_2, x_3, \ldots, x_{j+1})) \\
&\quad = (x_1 - x_2)^{-1} \left(\sum_{u \in A} \overline{\Phi}_{n-1} f(u) - \sum_{u \in B} \overline{\Phi}_{n-1} f(u) \right)
\end{aligned}$$

where A and B are subsets of $\{(x_{m_1}, \ldots, x_{m_n}) \in X^n : m_1 \leqslant m_2 \leqslant \ldots \leqslant m_n\}$ determined by $\{m_1, m_2, \ldots, m_n\} = \{1, 3, 4, \ldots, j+1\}$ and $\{m_1, m_2, \ldots, m_n\} = \{2, 3, 4, \ldots, j+1\}$ respectively. By interchanging 1 and 2 we obtain a correspondence $u \leftrightarrow u'$ between A and B. Let $u = (u_1, u_2, \ldots, u_n) \in A$. There is $k \in \mathbb{N}$ such that $u_1 = u_2 = \ldots = u_k = x_1$, $x_{k+1} = x_2$. Then $u = (x_1, x_1, \ldots, x_1, u_{k+1}, \ldots, u_n)$ and its corresponding element u' of B equals $(x_2, x_2, \ldots, x_2, u_{k+1}, \ldots, u_n)$. By the extended rule (iii) of Lemma 29.2

$$\overline{\Phi}_{n-1} f(u) - \overline{\Phi}_{n-1} f(u') = (x_1 - x_2) \sum_{t \in A_u} \tilde{\Phi}_n f(t)$$

where A_u is the set of all $(x_{m_1}, x_{m_2}, \ldots, x_{m_{n+1}}) \in X^{n+1}$ such that

$$\begin{aligned}
&m_1 \leqslant m_2 \leqslant \ldots \leqslant m_{k+1} \\
&\{m_1, m_2, \ldots, m_{k+1}\} = \{1, 2\} \\
&(x_{m_{k+2}}, \ldots, x_{m_{n+1}}) = (u_{k+1}, \ldots, u_n)
\end{aligned}$$

The A_u where u runs through A form a partition of S_{jn}. Hence

$$\Phi_j D_{n-j} f(x_1, \dots, x_{j+1}) = \sum_{u \in A} \sum_{t \in A_u} \tilde{\Phi}_n f(t) = \sum_{t \in S_{jn}} \tilde{\Phi}_n f(t).$$

The following theorem obtains.

THEOREM 78.2. *Let $n \in \mathbb{N}$ and $f \in C^n(X \to K)$. Then $D_{n-j} f \in C^j(X \to K)$ and $D_j D_{n-j} f = \binom{n}{j} D_n f$ $(0 \leqslant j \leqslant n)$.*

Proof. The right-hand side of the formula of Lemma 78.1 defines a continuous function on X^{j+1} implying that $D_{n-j} f \in C^j(X \to K)$. After taking limits (for $(x_1, x_2, \dots, x_{j+1})$ tending to $(a, a, \dots, a)$) at both sides of the formula we obtain $D_j D_{n-j} f = \binom{n}{j} D_n f$.

We investigate (but now for a C^{n-1}-function f) what happens if in the formula of Lemma 78.1 we take limits for $(x_1, \dots, x_{j+1})$ tending to an element of $X^{j+1} \setminus \Delta$.

LEMMA 78.3. *Let $n \in \mathbb{N}$. For $f \in C^{n-1}(X \to K)$ and $(x, y) \in \nabla^2 X$, set*

$$\rho_1 f(x, y) := \tilde{\Phi}_n f(x, y, y, \dots, y)$$
$$\rho_2 f(x, y) := \tilde{\Phi}_n f(x, x, y, \dots, y)$$
$$\vdots$$
$$\rho_n f(x, y) := \tilde{\Phi}_n f(x, x, \dots, x, y)$$

Then $\rho_1, \dots, \rho_n$ are continuous on $\nabla^2 X$ and satisfy the relation

$$\begin{bmatrix} \tilde{\Phi}_1 D_{n-1} f(x, y) \\ \Phi_2 D_{n-2} f(x, y, y) \\ \vdots \\ \\ \tilde{\Phi}_{n-1} D_1 f(x, \dots, y) \end{bmatrix} = \begin{bmatrix} \binom{0}{0} & \binom{1}{0} & \cdot & \cdot & \cdot & \binom{n-1}{0} \\ 0 & \binom{1}{1} & \binom{2}{1} & \cdot & \cdot & \binom{n-1}{1} \\ 0 & 0 & \cdot & & & \vdots \\ & & & \cdot & & \vdots \\ & & & & \cdot & \vdots \\ 0 & 0 & \cdot & \cdot & 0\ \binom{n-2}{n-2} & \binom{n-1}{n-2} \end{bmatrix} \begin{bmatrix} \rho_n f(x, y) \\ \rho_{n-i} f(x, y) \\ \vdots \\ \\ \rho_1 f(x, y) \end{bmatrix}$$

for all $(x, y) \in \nabla^2 X$.

Proof. We only have to prove the matrix identity. Let $(x, y) \in \nabla^2 X$. If in the formula of Lemma 78.1 we let x_1 tend to x and $x_2, \dots, x_{j+1}$ to y then the left-hand side becomes $\tilde{\Phi}_j D_{n-j} f(x, y, y, \dots, y)$. For the right-hand side observe that, for $1 \leqslant s \leqslant n$, there are $\binom{n-s}{j-1}$ elements of S_{jn} that 'start' with precisely s times x_1 (i.e. elements of the form $(u_1, u_2, \ldots, u_{n+1})$ where $u_1 = x_1, u_2 = x_1, \ldots, u_s = x_1, u_{s+1} = x_2$). Their contribution to the sum is (after taking limits) $\binom{n-s}{j-1}\, \rho_s f(x, y)$ so the right-hand side becomes $\sum_{s=1}^{n} \binom{n-s}{j-1}\, \rho_s f(x, y) = \sum_{i=0}^{n-1} \binom{i}{j-1}\, \rho_{n-i} f(x, y)$ which was to be shown.

The result of the following exercise will be needed later.

**Exercise* 78.A. Let f be as in Lemma 78.3. Suppose in addition that $f' \in C^{n-1}(X \to K)$, that $\operatorname{char}(K) = 0$ and that $\lim_{x, y \to a} \rho_1 f(x, y) = f^{(n)}(a)/n!$ for some $a \in X$. Use the matrix identity to show successively that $\lim_{x, y \to a} \rho_j f(x, y) = f^{(n)}(a)/n!$ for $j = 2, 3, \ldots, n$.

79. Antiderivation $C \to C^1$

In Section 64 we constructed an antiderivation map $P : C(\mathbb{Z}_p \to K) \to C^1(\mathbb{Z}_p \to K)$. Our purpose is to generalize it and obtain an antiderivation map $C(X \to K) \to C^1(X \to K)$.

THROUGHOUT SECTIONS 79–84 WE FIX $\rho \in \mathbb{R}, 0 < \rho < 1$.

An *approximation of the identity on* X is a sequence $\sigma_0, \sigma_1, \ldots$ of maps $X \to X$ such that σ_0 is constant, $\sigma_m \circ \sigma_n = \sigma_n \circ \sigma_m = \sigma_n$ if $m \geqslant n$, and such that for all $n \in \mathbb{N}, x, y \in X$

$$|x-y| < \rho^n \text{ implies } \sigma_n(x) = \sigma_n(y)$$
$$|\sigma_n(x) - x| < \rho^n$$

The maps $x \mapsto x_n$, where $x_0, x_1, \ldots$ is the standard sequence of $x \in \mathbb{Z}_p$ defined in Section 62, form an obvious example. For every nonempty $X \subset K$ there exist approximations of the identity. (Choose $x_0 \in X$ and set $\sigma_0(x) := x_0$ for all $x \in X$; the equivalence relation given by $|x-y| < \rho$ yields a partition of X into 'open' balls (relative to X). Let R_1 be a full set of representatives, arrange that $x_0 \in R_1$ and define σ_1 by the conditions $\sigma_1(x) \in R_1$, $|\sigma_1(x) - x| < \rho$ $(x \in X)$ Next, choose a full set of representatives $R_2 \supset R_1$ of the relation $|x-y| < \rho^2$ and define σ_2 by $\sigma_2(x) \in R_2$, $|\sigma_2(x) - x| < \rho^2$. Et cetera.)

We now define an antiderivation map (compare Section 64).

DEFINITION 79.1. Let $\sigma_0, \sigma_1, \ldots$ be an approximation of the identity on X. For a continuous function $f : X \to K$ set

$$Pf(x) := \sum_{n=0}^{\infty} f(x_n)(x_{n+1} - x_n) \qquad (x \in X)$$

where $x_n := \sigma_n(x)$ $(x \in X, n \in \{0, 1, 2, \ldots\})$.

THEOREM 79.2. *Each continuous function $X \to K$ has a C^1-antiderivative. In fact, the map P of Definition* 79.1 *has the following properties.*

(i) *P is a linear map $C(X \to K) \to C^1(X \to K)$.*

(ii) *$(Pf)' = f$ for all $f \in C(X \to K)$.*

(iii) *If $f \in C(X \to K)$ is bounded then $\|\Phi_1 Pf\|_\infty \leqslant \|f\|_\infty$.*

Proof. First observe that $Pf(x)$ is well defined since f is continuous, hence bounded on the compact set $\{x_0, x_1, \ldots, x\}$ and $\lim_{n\to\infty}(x_{n+1} - x_n) = 0$. Now let $a \in X$ and $\epsilon > 0$. There is $m \in \mathbb{N}$ such that $|f(x) - f(a)| < \epsilon$ whenever $|x-a| < \rho^m$. Let $x, y \in X$ such that $x \neq y$, $|x-a| < \rho^m$, $|y-a| < \rho^m$. To prove (i) and (ii) it suffices to show

$$|\Phi_1 Pf(x, y) - f(a)| \leqslant \epsilon$$

There is a unique $s \in \mathbb{N}$ such that $\rho^{s+1} \leqslant |x-y| < \rho^s$. Then $x_0 = y_0, \ldots, x_s = y_s$, $x_{s+1} \neq y_{s+1}$. Further, we have $s \geqslant m$. Write $Pf(x) - Pf(y) - (x-y) f(a) = (f(x_s) - f(a))(x_{s+1} - y_{s+1}) + \sum_{n>s}(f(x_n) - f(a))(x_{n+1} - x_n) + \sum_{n>s}(f(y_n) - f(a))(y_{n+1} - y_n)$. Now $|x_n - a| < \rho^m$ for $n \geqslant s$ so that in the above formula $|f(x_s) - f(a)| < \epsilon$, $|f(x_n) - f(a)| < \epsilon$. Similarly, $|f(y_n) - f(a)| < \epsilon$. Further, $|x_{s+1} - y_{s+1}| \leqslant |x_{s+1} - x| \vee |x-y| \vee |y - y_{s+1}| \leqslant \rho^{s+1} \vee |x-y| = |x-y|$. For $n > s$, $|x_{n+1} - x_n| \leqslant |x_{n+1} - x| \vee |x - x_n| < \rho^n \leqslant \rho^{s+1} \leqslant |x-y|$. So we get

$$|Pf(x) - Pf(y) - (x-y) f(a)| \leqslant \epsilon|x-y|$$

which was to be shown. To prove (iii), let $x, y \in X$, $x \neq y$. Then there is a unique $s \in \{0, 1, 2, \ldots\}$ such that $\rho^{s+1} \leqslant |x-y| < \rho^s$. We write $Pf(x) - Pf(y) = f(x_s)(x_{s+1} - y_{s+1}) + \sum_{n>s} f(x_n)(x_{n+1} - x_n) + \sum_{n>s} f(y_n)(y_{n+1} - y_n)$. Again, the terms $|x_{s+1} - y_{s+1}|$, $|x_{n+1} - x_n|$, $|y_{n+1} - y_n|$ in this formula are all $\leqslant |x-y|$. We get $|Pf(x) - Pf(y)| \leqslant \|f\|_\infty |x-y|$, which is (iii).

Exercise 79.A. Show that $Pf(x) = x - x_0$ $(x \in X)$ if f is the constant function 1. More generally, show that Pf is locally linear if f is locally constant.

Exercise 79.B. (Compare Exercise 64.C) Show that for each $j \in \mathbb{N}$ the function

$$x \mapsto \sum_{n \geqslant j} f(x_n)(x_{n+1} - x_n) \qquad (x \in X)$$

is a C^1-antiderivative of $f \in C(X \to K)$. Deduce that for each $\epsilon > 0$ there exists a linear antiderivation map $Q : C(X \to K) \to C^1(X \to K)$ such that $\|Qf\|_\infty \leqslant \epsilon \|f\|_\infty$, $\|Qf\|_1 = \|f\|_\infty$ for each bounded continuous function $f : X \to K$. (Recall that $\|h\|_1 = \|h\|_\infty \vee \|\Phi_1 h\|_\infty$ $(h : X \to K)$.)

Exercise 79.C. (Mean value property of Pf, see Theorem 64.2) (i) Let X be compact or $|K^\times|$ be discrete. Show that there exists an antiderivation map

$P : C(X \to K) \to C^1(X \to K)$ satisfying (i), (ii), (iii) of Theorem 79.2 and (if $f \in C(X \to K)$ is bounded) that

$$\left|\frac{Pf(x) - Pf(y)}{x-y}\right| \leqslant \max\{|f(z)| : z \in [x, y]\}$$

for all $x, y \in X$ $(x \neq y)$.
(ii) For the general case, let $\epsilon > 0$. Show that there exists a map $P : C(X \to K) \to C^1(X \to K)$ satisfying (i), (ii), (iii) of Theorem 79.2 and

$$\left|\frac{Pf(x) - Pf(y)}{x-y}\right| \leqslant \sup\{|f(z)| : |z-x| \leqslant (1+\epsilon)|x-y|\}$$

for all $x, y \in X$, $x \neq y$.

Exercise 79.D. (Sequel to Theorem 76.3) Use Theorem 79.2 and the technique of Theorem 76.3 to prove the following. *Let X be closed and let $f : X \to K$ be a C^1-function. Then f can be extended to a C^1-function $K \to K$.* (See also Corollary 81.4.)

Exercise 79.E. Let $f \in C(X \to K)$ and let P be as in Theorem 79.2. Show that Pf is the unique continuous function $h : X \to K$ for which $h(x_0) = 0$ and $h(x_{n+1}) - h(x_n) = f(x_n)(x_{n+1} - x_n)$ for all $x \in X$ and $n \in \{0, 1, 2, \ldots\}$. (Compare Definition 64.1.)

80. Antiderivation $C^{n-1} \to C^n$. A candidate

The function $x \mapsto x$ does not have a C^2-antiderivative if char $(K) = 2$! In fact, by Corollary 29.6, for such an antiderivative g we would have $g'' = 0$, which is absurd. This fact makes it somewhat doubtful whether our map P of the previous section maps C^1-functions into C^2-functions even in the characteristic zero case. Indeed it goes wrong.

PROPOSITION 80.1. *Let $f \in C^1(X \to K)$. Let P be as in Definition* 79.1. *Then $Pf \in C^2(X \to K)$ implies $f' = 0$.*
Proof. We may assume char $(K) \neq 2$. By the definition of approximation of the identity we have $(x_m)_n = x_m$ if $n \geqslant m$, $(x_m)_n = x_n$ *if* $n \leqslant m$ so that $Pf(x_0) = 0$ and

$$Pf(x_m) = \sum_{n<m} f(x_n)(x_{n+1} - x_n) \qquad (m \in \mathbf{N}, x \in X)$$

It follows that

$$Pf(x_{m+1}) - Pf(x_m) = f(x_m)(x_{m+1} - x_m) \qquad (m \in \mathbf{N})$$

By Taylor's formula (Theorem 29.4) and Theorem 29.5 we have

$$Pf(x_{m+1}) - Pf(x_m) = f(x_m)(x_{m+1} - x_m) + \tfrac{1}{2}f'(x_m)(x_{m+1} - x_m)^2 + H_2(x_{m+1}, x_m)(x_{m+1} - x_m)^2$$

where H_2 is a continuous function on $X \times X$ vanishing on the diagonal. Hence

$$\tfrac{1}{2}f'(x_m) + H_2(x_{m+1}, x_m) = 0 \qquad (x \in X, x_{m+1} \neq x_m)$$

It is easy to see that each point of the diagonal can be approximated by elements of the form (x_{m+1}, x_m) where $x_{m+1} \neq x_m$. It follows that $f' = 0$.

To get a clue for finding an antiderivation map $C^1 \to C^2$ (if char $(K) \neq 2$) observe that P satisfies $Pf(x_{n+1}) - Pf(x_n) = f(x_n)(x_{n+1} - x_n) =$ (Taylor's formula) $= f(x_n)(x_{n+1} - x_n) + H_1(x_{n+1}, x_n)(x_{n+1} - x_n)$ where H_1 is a continuous function, zero on the diagonal. Thus, Pf is an antiderivative of f for which the rest term H_1 is zero at all points (x_{n+1}, x_n). Now let $f \in C^1(X \to K)$ and, by analogy, let us ask for an antiderivative F of f for which $F(x_0) = 0$ and $H_2(x_{n+1}, x_n) = 0$, i.e.

$$F(x_{n+1}) - F(x_n) = f(x_n)(x_{n+1} - x_n) + \tfrac{1}{2}f'(x_n)(x_{n+1} - x_n)^2$$

Such an F is uniquely determined because summation over n yields the formula

$$F(x) = \sum_{n=0}^{\infty} f(x_n)(x_{n+1} - x_n) + \tfrac{1}{2} \sum_{n=0}^{\infty} f'(x_n)(x_{n+1} - x_n)^2$$

We may hope that F is a C^2-antiderivative of f. More generally, we have the following candidate for antiderivation $C^{n-1} \to C^n$.

DEFINITION 80.2. Let char $(K) = 0$, let $n \in \mathbb{N}$, $f \in C^{n-1}(X \to K)$. With x_n as in Definition 79.1 set

$$P_n f(x) := \sum_{m=0}^{\infty} \sum_{j=0}^{n-1} \frac{f^{(j)}(x_m)}{(j+1)!}(x_{m+1} - x_m)^{j+1} \qquad (x \in X)$$

We shall prove (Theorem 81.3) that $P_n f$ is indeed a C^n-antiderivative of f. Notice that $P_n f$ is a well-defined function on X and that P_1 is what we have called P previously. In the next three exercises we assume char $(K) = 0$.

Exercise 80.A. Let n, $s \in \mathbb{N}$, $1 \leqslant s \leqslant n-1$. Let $f(x) = x^s$ $(x \in K)$. Show that $P_n f(x) = (x^{s+1} - x_0^{s+1}) / (s+1)$ $(x \in K)$.

**Exercise* 80.B. Let $f \in C^{n-1}(X \to K)$ for some $n \in \mathbb{N}$. Show that $P_n f$ is an antiderivative of f.

Exercise 80.C. Let $f \in C(X \to K)$. Prove that

$$2\,P_2\,P_1\,f(x) = 2\,x P_1 f(x) - \sum_{n=0}^{\infty} f(x_n)\,(x_{n+1}^2 - x_n^2) \qquad (x \in X)$$

Theorem 80.3 is a first step towards our goal.

THEOREM 80.3. (Taylor formula for $P_n f$) *Let K have characteristic* 0, *let $n \in \mathbb{N}, f \in C^{n-1}(X \to K)$. Then, with $P_n f$ as in Definition* 80.2,

$$P_n f(x) - P_n f(y) = \sum_{j=1}^{n} \frac{(x-y)^j}{j!} f^{(j-1)}(y) + (x-y)^n R_n(x,y) \quad (x, y \in X)$$

where R_n is a continuous function vanishing on the diagonal.

Proof. For $x, y \in X, x \neq y$ we define $R_n(x,y)$ by means of the above formula. We have to show that $\lim_{x,\,y \to a} R_n(x,y) = 0$ for any $a \in X$. Let $x, y \in X$. We have

$$P_n f(x) = \sum_{m=0}^{\infty} \sum_{j=0}^{n-1} \frac{1}{(j+1)!} f^{(j)}(x_m)(x_{m+1} - x_m)^{j+1}$$

For each $j \in \{0, 1, \ldots, n-1\}$, $f^{(j)} \in C^{n-j-1}(X \to K)$ (Corollary 78.2). So there are continuous functions Λ_j $(0 \leqslant j \leqslant n-1)$, zero on the diagonal, such that

$$f^{(j)}(x_m) = \sum_{s=0}^{n-j-1} \frac{(x_m - y)^s}{s!} f^{(j+s)}(y) + (x_m - y)^{n-j-1} \Lambda_j(x_m, y)$$

Substitution in the formula for $P_n f(x)$ and some elementary computation yields

$$P_n f(x) = \sum_{\nu=0}^{n-1} ((x-y)^\nu - (x_0 - y)^\nu) \frac{1}{(\nu+1)!} f^{(\nu)}(y) + \sum_{m=0}^{\infty} \sum_{j-0}^{n-1} \frac{1}{(j+1)!} (x_m - y)^{n-j-1} (x_{m+1} - x_m)^{j+1} \Lambda_j(x_m, y)$$

Similarly,

$$P_n f(y) = \sum_{\nu=0}^{n-1} ((y-y)^\nu - (y_0 - y)^\nu) \frac{1}{(\nu+1)!} f^{(\nu)}(y) + \sum_{m=0}^{\infty} \sum_{j=0}^{n-1} \frac{1}{(j+1)!} (y_m - y)^{n-j-1} (y_{m+1} - y_m)^{j+1} \Lambda_j(y_m, y)$$

Subtraction yields (remember that $x_0 = y_0$)

$$(*)\ (x-y)^n R_n(x,y) = \sum_{m=0}^{\infty} \sum_{j=0}^{n-1} \frac{1}{(j+1)!} \{ (x_m - y)^{n-j-1} \cdot (x_{m+1}-x_m)^{j+1} \Lambda_j(x_m, y) - (y_m - y)^{n-j-1} (y_{m+1}-y_m)^{j+1} \cdot \Lambda_j(y_m, y) \}$$

Let $\epsilon > 0$. There is s such that $|u-a| < \rho^s$, $|v-a| < \rho^s$ implies

$$|\Lambda_j(u,v)| < \epsilon\rho^n\ |n!| \qquad (j = 0, 1, \ldots, n-1)$$

Let $|x-a| < \rho^s$, $|y-a| < \rho^s$, $x \neq y$. We shall prove that $|R_n(x,y)| \leqslant \epsilon$. First observe that $x_0 = y_0, \ldots, x_t = y_t$, $x_{t+1} \neq y_{t+1}$ for some $t \geqslant s$. Then $\rho^{t+1} \leqslant |x-y| < \rho^t$. For $m \geqslant t$ (the terms with $m < t$ in (*) vanish) we have $|x_m - y| \leqslant |x_m - x| \vee |x-y| \leqslant \rho^m \vee \rho^t = \rho^t \leqslant \rho^{-1}\ |x-y|$. Similarly, the terms $|x_{m+1} - x_m|$, $|y_m - y|$, $|y_{m+1} - y_m|$ are all $\leqslant \rho^{-1}\ |x-y|$. We find

$$|(x-y)^n R_n(x,y)| \leqslant |n!|^{-1} \rho^{-n} |x-y|^n \epsilon \rho^n |n!| = \epsilon |x-y|^n.$$

Exercise 80.D. Define $P_n f$ for the case $\mathrm{char}(K) = p$, $f \in C^{n-1}(X \to K)$, $n < p$ formally as in Definition 80.2 and extend Theorem 80.3 to this case.

From Theorem 80.3 it does not follow at once that $P_n f \in C^n(X \to K)$. Indeed, the following lemma remains to be shown.

LEMMA 80.4. *Let K have characteristic* 0, *let $n \in \mathbb{N}$, $f \in C^{n-1}(X \to K)$. Suppose F is an antiderivative of f such that there exists a continuous function $S : X \times X \to K$, zero on the diagonal, for which*

$$F(x) = F(y) + (x-y)f(y) + (x-y)^2 \frac{f'(y)}{2!} + \ldots + (x-y)^n \frac{f^{(n-1)}(y)}{n!} + (x-y)^n S(x,y) \quad (x, y \in X)$$

for all $x, y \in X$. Then $F \in C^n(X \to K)$.

Stated in a different form Lemma 80.4 reads as follows.

LEMMA 80.5. *Let K have characteristic* 0, *let $n \in \mathbb{N}$, let $f : X \to K$ be differentiable such that $f' \in C^{n-1}(X \to K)$. Suppose*

$$f(x) = f(y) + \sum_{j=1}^{n} (x-y)^j \frac{f^{(j)}(y)}{j!} + (x-y)^n R_n(x,y) \ (x, y \in X)$$

where R_n is a continuous function, zero on the diagonal. Then $f \in C^n(X \to K)$.

Section 81 is fully devoted to proving Lemma 80.5.

81. Surjectivity of differentiation $C^n \to C^{n-1}$

LEMMA 81.1. *Let the characteristic of K be 0, let $n \in \mathbb{N}$, let $f \in C^{n-1}(X \to K)$ such that f' is also in $C^{n-1}(X \to K)$. Suppose in addition that*

$$f(x) = f(y) + \sum_{j=1}^{n} (x-y)^j \frac{f^{(j)}(y)}{j!} + (x-y)^n R_n(x, y) \quad (x, y \in X)$$

where R_n is a continuous function, zero on the diagonal. Then for each $a \in X$

$$\lim_{x,\, y \to a} \rho_j f(x, y) = \frac{f^{(n)}(a)}{n!} \qquad (1 \leqslant j \leqslant n)$$

where $\rho_1, \ldots, \rho_n$ are as in Lemma 78.3.

Proof. For $x, y \in X$ we have, by Taylor's formula,

$$f(x) = f(y) + \sum_{j=1}^{n-1} (x-y)^j \frac{f^{(j)}(y)}{j!} + (x-y)^{n-1} \cdot$$
$$(\bar{\Phi}_{n-1} f(x, y, y, \ldots, y) - D_{n-1} f(y))$$

It follows that

$$\frac{f^{(n)}(y)}{n!} + R_n(x, y) = (x-y)^{-1} (\bar{\Phi}_{n-1} f(x, y, y, \ldots, y) - D_{n-1} f(y))$$
$$= \tilde{\Phi}_n f(x, y, y, \ldots, y) = \rho_1 f(x, y)$$

By continuity

$$\lim_{x,\, y \to a} \rho_1 f(x, y) = \frac{f^{(n)}(a)}{n!}$$

That also

$$\lim_{x,\, y \to a} \rho_j f(x, y) = \frac{f^{(n)}(a)}{n!} \qquad (2 \leqslant j \leqslant n)$$

follows from Exercise 78.A.

LEMMA 81.2. *Let $n \in \mathbb{N}$, $f \in C^{n-1}(X \to K)$. Let B, S be balls in K. Suppose for $j \in \{1, \ldots, n\}$*

$$\rho_j f(x, y) \in S \qquad (x, y \in B \cap X, x \neq y)$$

Then

$$\tilde{\Phi}_n f(x_1, \ldots, x_{n+1}) \in S$$

for all $(x_1, \ldots, x_{n+1}) \in X^{n+1} \setminus \Delta$ for which $x_i \in B$ for each i.

Proof. Let $\Lambda := \{(x_1, \ldots, x_{n+1}) \in X^{n+1} \setminus \Delta : x_i \in B$ for each $i\}$. For $2 \leqslant j \leqslant n+1$, let $\Lambda_j := \{(x_1, \ldots, x_{n+1}) \in \Lambda : \{x_1, \ldots, x_{n+1}\}$ has j elements$\}$. Then $\Lambda = \cup_{j=2}^{n+1} \Lambda_j$. From the assumption on ρ_j it follows at once that $\tilde{\Phi}_n f(\Lambda_2) \subset S$. It suffices to prove that each element of $\tilde{\Phi}_n f(\Lambda_j)$ is a 'convex' combination of elements of $\tilde{\Phi}_n f(\Lambda_{j-1})$ $(3 \leqslant j \leqslant n+1)$. So let $u \in \Lambda_j$ for some $j \geqslant 3$. By symmetry of $\tilde{\Phi}_n f$ we may assume that u has the form

$$u = (x_1, \ldots, x_1, x_2, \ldots, x_2, x_3, \ldots)$$

where x_1, x_2, x_3 are distinct and $|x_1 - x_2| \geqslant \max(|x_1 - x_3|, |x_2 - x_3|)$. Suppose x_1 occurs k times, x_2 occurs l times. For simplicity of notation we write

$$u = (x_1^k, x_2^l, x_3, \ldots)$$

From

$$\tilde{\Phi}_n f(x_1^k, x_2^l, x_3, \ldots) = \frac{x_1 - x_3}{x_1 - x_2} \tilde{\Phi}_n f(x_1^k, x_2^{l-1}, x_3, \ldots)$$

$$+ \frac{x_3 - x_2}{x_1 - x_2} \tilde{\Phi}_n f(x_1^{k-1}, x_2^l, x_3, \ldots)$$

we see that we can write $\tilde{\Phi}_n f(u)$ as a convex combination (value of coefficients $\leqslant 1$, sum of coefficients $= 1$) of $\tilde{\Phi}_n f(x_1^k, x_2^{l-1}, x_3, \ldots)$ and $\tilde{\Phi}_n f(x_1^{k-1}, x_2^l, x_3, \ldots)$. If $l - 1 \geqslant 1$ we can continue by writing the first expression as a convex combination of $\tilde{\Phi}_n f(x_1^k, x_2^{l-2}, x_3, \ldots)$ and $\tilde{\Phi}_n f(x_1^{k-1}, x_2^{l-1}, x_3, \ldots)$. We can treat the second expression in a similar way if $k-1 \geqslant 1$. Going on this way we obtain $\tilde{\Phi}_n f(u)$ expressed as a convex combination of elements of the form $\tilde{\Phi}_n f(v)$ where the coordinates of v are in $\{x_1, x_2, x_3, \ldots\}$ but where x_1 or x_2 is missing, i.e. $v \in \Lambda_{j-1}$.

After this preparatory work we are ready to prove

THEOREM 81.3. *Let K have characteristic* 0, *let $n \in \mathbb{N}$. Then each $f \in C^{n-1}(X \to K)$ has a C^n-antiderivative. In fact the function $P_n f$ of Definition* 80.2 *is a C^n-antiderivative of f.*

Proof. If suffices to prove Lemma 80.5, which is trivial for $n = 1$. Suppose Lemma 80.5 is true for $n-1$ in place of n. Let f satisfy the conditions of Lemma 80.5. By the induction hypothesis we have $f \in C^{n-1}(X \to K)$. Then Lemma 81.1 tells that $\lim_{x,\,y \to a} \rho_j f(x, y) = f^{(n)}(a)/n!$ for each $a \in X$ and $1 \leqslant j \leqslant n$. From Lemma 81.2 it then follows that also $\tilde{\Phi}_n f(x_1, \ldots, x_n)$ tends to $f^{(n)}(a)/n!$ if $(x_1, \ldots, x_{n+1}) \in X^{n+1} \setminus \Delta$ tends to $(a, a, \ldots, a)$. By Theorem 29.9, $f \in C^n(X \to K)$.

COROLLARY 81.4. (Extension theorem for C^n- functions) *Let* char $(K) = 0$, *let X be a closed subset of K, let $n \in \mathbb{N}$ and $f \in C^n (X \to K)$. Then f can be extended to a C^n-function $K \to K$.*

Proof. First assume that $f' = 0$. Let $\sigma : K \to X$ be as in Lemma 76.1 and set $\bar{f} := f \circ \sigma$. For $x, y \in K$ $(x \neq y)$ we have

$$\frac{\bar{f}(x) - \bar{f}(y)}{(x-y)^n} = \begin{cases} 0 \text{ if } \sigma(x) = \sigma(y) \\ \dfrac{f(\sigma(x)) - f(\sigma(y))}{(\sigma(x) - \sigma(y))^n} \cdot \left(\dfrac{\sigma(x) - \sigma(y)}{x-y} \right)^n \text{ if } \sigma(x) \neq \sigma(y) \end{cases}$$

By Theorem 29.12 applied to f, $(\bar{f}(x) - \bar{f}(y)) / (x-y)^n$ tends to zero if x, y approach some $a \in K$ (observe that σ satisfies a Lipschitz condition). Hence, by the same theorem, $\bar{f} \in C^n (K \to K)$ is a required extension of f. For the general case we proceed by induction on n. Let $f \in C^n (X \to K)$. Then $f' \in C^{n-1} (X \to K)$ and can be extended to a C^{n-1}-function $h : K \to K$. By Theorem 81.3, h has a C^n-antiderivative g. Then $f - g \mid X$ is in $C^n (X \to K)$, has vanishing derivative. By the first part of this proof $(f-g) \circ \sigma$ is a C^n-extension of $f - g \mid X$. Then $\bar{f} := g + (f-g) \circ \sigma$ is a C^n-extension of f.

Exercise 81.A. Let $\text{char}(K) = p$, let $n \in \mathbb{N}$, $n < p$. Show that each $f \in C^{n-1} (X \to K)$ has a C^n-antiderivative.

82. Surjectivity of differentiation $C^\infty \to C^\infty$

If $\text{char}(K) = 0$ each C^∞-function f has a C^n-antiderivative $P_n f$ for each n. To prove that f has a C^∞-antiderivative one might try $\lim_{n \to \infty} P_n f$. However, we shall have to make some modifications if only for the reason that $\lim_{n \to \infty} P_n f$ does not always exist.

THEOREM 82.1. *Let K have characteristic 0. Then each $f \in C^\infty (X \to K)$ has a C^∞-antiderivative.*

Proof. Let $j \in \{0, 1, 2, \ldots\}$. $f^{(j)}$ is continuous, hence locally bounded and there exists a partition of X into 'closed' balls B_{ji} (relative to X) of radius < 1, where i runs through some indexing set I_j such that $f^{(j)}$ is bounded on each B_{ji}. For each $i \in I_j$ we can choose $m_{ji} \in \mathbb{N}$ such that (recall that $0 < \rho < 1$)

$$(*) \qquad \rho^{m_{ji}} \leqslant d(B_{ji}) < 1, \ |f^{(j)}(x)| \rho^{m_{ji}} < |(j+1)!| \rho^j \quad (x \in B_{ji})$$

Define $F_j : X \to K$ as follows. If $x \in X$ then $x \in B_{ji}$ for precisely one $j \in I_j$. Set

$$F_j(x) := \sum_{m \geqslant m_{ji}} \frac{f^{(j)}(x_m)}{(j+1)!}(x_{m+1} - x_m)^{j+1}$$

We shall prove that $F := \sum_{j=0}^{\infty} F_j$ is a C^∞-antiderivative of f by means of the following steps.

(i) Each F_j is well defined.

(ii) For each $j \in \{0, 1, 2, \ldots\}$ and for all $i \in I_j$

$$|F_j(x)| \leqslant \rho^{jm_{ji}+j} \qquad (x \in B_{ji})$$

so that F is well defined.

(iii) $\sum_{j=0}^{n} F_j$ is a C^n-antiderivative of f for each $n \in \mathbb{N}$.

(iv) For each n, $\sum_{j=n+1}^{\infty} F_j$ is a C^n-function with zero derivative.

Proof of (i). $f^{(j)}$ is bounded on B_{ji} and $\lim_{m\to\infty}(x_{m+1} - x_m) = 0$.

Proof of (ii). Let $x \in B_{ji}$ and $m \geqslant m_{ji}$. Then by (*)

$$|x_{m+1} - x_m| \leqslant |x - x_m| \leqslant \rho^m \leqslant \rho^{m_{ji}} \leqslant d(B_{ji})$$

from which it follows that $x_m \in B_{ji}$ and $|x_{m+1} - x_m| \leqslant \rho^{m_{ji}}$. Applying the second formula of (*) with x replaced by x_m we get

$$\left| \frac{f^{(j)}(x_m)}{(j+1)!}(x_{m+1} - x_m)^{j+1} \right| \leqslant \rho^j \rho^{-m_{ji}} \rho^{m_{ji}(j+1)} = \rho^{jm_{ji}+j}$$

and (ii) is proved.

Proof of (iii). The functions F_j and $x \mapsto \sum_{m=0}^{\infty} f^{(j)}(x_m)(x_{m+1} - x_m)^{j+1}/(j+1)!$ differ (on each B_{ji}, hence globally) by a locally constant function. Summation from $j = 0$ to $j = n$ shows that $\sum_{j=0}^{n} F_j - P_{n+1}f$ is locally constant. By Theorem 81.3, $\sum_{j=0}^{n} F_j \in C^{n+1}(X \to K) \subset C^n(X \to K)$ and $(\sum_{j=0}^{n} F_j)' = f$.

Proof of (iv). Set $H := \sum_{j=n+1}^{\infty} F_j$. We shall prove that $|H(x) - H(y)| \leqslant |x - y|^{n+1}$ for all $x, y \in X$ which, by Theorem 29.12, implies (iv). To obtain the inequality it suffices to prove

$$(**) \qquad |F_j(x) - F_j(y)| \leqslant |x - y|^{n+1} \qquad (x, y \in X)$$

for each $j \geqslant n + 1$. We consider several cases.

(*a*) $x \in B_{ji}$, $y \in B_{ji'}$ where $i \neq i'$. Then by (*) $|x - y| \geqslant d(B_{ji}) \geqslant \rho^{m_{ji}}$ so that $|x - y|^{n+1} \geqslant \rho^{m_{ji}(n+1)}$. By (ii), $|F_j(x)| \leqslant \rho^{jm_{ji}+j}$. As $jm_{ji} + j \geqslant (n+1)m_{ji}$ we have $|F_j(x)| \leqslant |x - y|^{n+1}$. By symmetry, $|F_j(y)| \leqslant |x - y|^{n+1}$ and (**) follows.

(*b*) There is i such that $x, y \in B_{ji}$. We may assume $x \neq y$, there exists an $s \in \mathbb{N} \cup \{0\}$ such that (recall that $d(B_{ji}) < 1$)

$$\rho^{s+1} \leqslant |x-y| < \rho^s$$

Then $|x-y|^{n+1} \geqslant \rho^{(s+1)(n+1)}$. Consider two subcases.

(*b*. 1) $s < m_{ji}$. Then by (ii) $|F_j(x)| \leqslant \rho^{jm_{ji}+j}$ and since $jm_{ji}+j \geqslant (n+1)\cdot(s+1)+j \geqslant (s+1)(n+1)$ we have $|F_j(x)| \leqslant |x-y|^{n+1}$. By symmetry $|F_j(y)| \leqslant |x-y|^{n+1}$ and (**) follows.

(*b*. 2) $s \geqslant m_{ji}$. Then since $x_0 = y_0, \ldots, x_s = y_s$ we have for $m = m_{ji}, \ldots, s-1$

$$\frac{f^{(j)}(x_m)}{(j+1)!}(x_{m+1}-x_m)^{j+1} = \frac{f^{(j)}(y_m)}{(j+1)!}(y_{m+1}-y_m)^{j+1}$$

so that

$$F_j(x) - F_j(y) = \sum_{m \geqslant s} \frac{f^{(j)}(x_m)}{(j+1)!}(x_{m+1}-x_m)^{j+1} - \sum_{m \geqslant s} \frac{f^{(j)}(y_m)}{(j+1)!}(y_{m+1}-y_m)^{j+1}$$

If $m \geqslant s$ we have by (*) (observe that $x_m \in B_{ji}$)

$$\left| \frac{f^{(j)}(x_m)}{(j+1)!}(x_{m+1}-x_m)^{j+1} \right| \leqslant \rho^{j-m_{ji}+m(j+1)}$$

$$\left| \frac{f^{(j)}(y_m)}{(j+1)!}(y_{m+1}-y_m)^{j+1} \right| \leqslant \rho^{j-m_{ji}+m(j+1)}$$

and we find $|F_j(x) - F_j(y)| \leqslant \rho^{j-m_{ji}+s(j+1)}$. Using the fact that $j \geqslant n+1$ and our assumption $s \geqslant m_{ji}$ we obtain $j - m_{ji} + s(j+1) = (s+1)j + s - m_{ji} \geqslant (s+1)(n+1)$. In consequence

$$|F_j(x) - F_j(y)| \leqslant \rho^{(s+1)(n+1)} \leqslant |x-y|^{n+1}$$

which finishes the proof.

Remark. The above construction does not give us a *linear* antiderivation map $C^\infty(X \to K) \to C^\infty(X \to K)$.

83. C^3-functions

In Theorem 29.4 we proved Taylor's formula. If $f \in C^n(X \to K)$ then $f(x) = f(y) + \sum_{j=1}^{n-1}(x-y)^j D_j f(y) + (x-y)^n \bar{\Phi}_n f(x, y, y, \ldots, y)$ $(x, y \in X)$. In this section we shall consider the following 'converse'.

PROBLEM 83.1. Let $f : X \to K$, $n \in \mathbb{N}$. Suppose there are continuous functions $\lambda_1, \ldots, \lambda_{n-1} : X \to K$ and $\Lambda_n : X \times X \to K$ such that

$$(*) \quad f(x) = f(y) + \sum_{j=1}^{n-1} (x-y)^j \lambda_j(y) + (x-y)^n \Lambda_n(x, y) \qquad (x, y \in X)$$

Does it follow that $f \in C^n(X \to K)$?

This problem is of interest because a positive answer would enable us to give an alternative definition of a C^n-function involving only two variables rather than the $n+1$ variables occurring in Definition 29.1. For $n = 1, 2$ the answer to the problem is 'yes' (Proposition 27.2 (γ) and Proposition 28.4). Surprisingly for $n = 3$ we have the following (see also Exercise 83.A).

EXAMPLE 83.2. *Problem* 83.1 *has a negative answer for* $n = 3$.
Proof. Let $X := \{\sum_{n=0}^{\infty} a_n p^{n!} \in \mathbb{Z}_p : a_n \in \{0, 1\}$ for all $n\}$ and define $f : X \to \mathbb{Z}_p$ by

$$f\left(\sum_{n=0}^{\infty} a_n p^{n!}\right) = \sum_{n=0}^{\infty} a_n p^{3n!} \qquad \left(\sum_{n=0}^{\infty} a_n p^{n!} \in X\right)$$

We shall prove that f satisfies the conditions of Problem 83.1, but $f \notin C^3(X \to \mathbb{Q}_p)$. First observe that X is a closed subset of $\mathbb{Z}_p$ without isolated points. Further, $|f(x) - f(y)|_p = |x-y|_p^3$ for all $x, y \in X$ so that by Theorem 29.12 $f \in C^2(X \to \mathbb{Q}_p)$ and $f' = 0$ but certainly $f \notin C^3(X \to \mathbb{Q}_p)$. We now prove that

$$\lim_{(x, y) \to (a, a)} \frac{f(x) - f(y)}{(x-y)^3} = 1$$

for each $a \in X$. (Then, with $\lambda_1 = \lambda_2 = 0$ and Λ_3 equal to the continuous extension of $(x, y) \mapsto (f(x) - f(y)) / (x-y)^3$, f satisfies (*) of Problem 83.1 and we are done.) Let $k \in \mathbb{N}$. We shall prove that $x, y \in X$, $|x-y|_p = p^{-k!}$ implies $|(x-y)^{-3}(f(x) - f(y)) - 1|_p \leqslant p^{-k \cdot k!}$. Write $x = \sum_{n=0}^{\infty} a_n p^{n!}$, $y = \sum_{n=0}^{\infty} b_n p^{n!}$. Then $a_j = b_j$ for $j < k$, $a_k \neq b_k$. One verifies immediately that

$$\begin{aligned} f(x) - f(y) &= (a_k - b_k) p^{3k!} + u_k \\ (x-y)^3 &= (a_k - b_k)^3 p^{3k!} + v_k \end{aligned}$$

where $|u_k|_p \leqslant p^{-3(k+1)!}$, $|v_k|_p \leqslant p^{-3(k+1)!}$ so that $\max(|u_k|_p, |v_k|_p) \leqslant p^{-(k+3)\,k!}$. Since $a_k, b_k \in \{0, 1\}$ we have

$$(a_k - b_k)^3 = a_k - b_k$$

and we get

$$|f(x) - f(y) - (x-y)^3|_p = |u_k - v_k|_p \leqslant p^{-(k+3)\,k!} = |x-y|_p^3\, p^{-k \cdot k!}$$

which finishes the proof.

Exercise 83.A. Show that the above example provides a negative answer to Problem 83.1 for each $n \in \mathbb{N}$, $n \geqslant 3$. (If $n > 3$ choose $\lambda_j := 0$ if $j \neq 3$, $\lambda_3 := 1$, $\Lambda_n(a, a) := 0$ for each $a \in X$.)

We shall now formulate an extra (but quite reasonable) assumption on the domain X in order that the answer to Problem 83.1 is 'yes'. To get a clear picture of what is going on we shall concentrate on the crucial case $n = 3$.

DEFINITION 83.3. Let C be a real number, $C \geqslant 1$. A triple $\{x_1, x_2, x_3\} \subset X$ is a *C-regular triangle* if $\max(|x_1 - x_2|, |x_2 - x_3|, |x_3 - x_1|) \leqslant C \min(|x_1 - x_2|, |x_2 - x_3|, |x_3 - x_1|)$. X has *property B_3* if there is a $C \geqslant 1$ such that each pair $\{x_1, x_2\}$ $(x_1 \neq x_2)$ in X can be extended to a C-regular triangle $\{x_1, x_2, x_3\}$.

**Exercise* 83.B
(i) Show that discs in K have property B_3. Let $C \geqslant 1$ be such that each set $\{x_1, x_2\}$ of two elements in $B_0(1) \subset K$ can be extended to a C-regular triangle in $B_0(1)$. Discuss how 'small' we can choose C in each of the following cases. (1) The residue class field k of K has more than two elements. (2) k has two elements and the valuation on K is dense. (3) $K = \mathbb{Q}_2$.
(ii) Show that the set X of Example 83.2 does not have property B_3.

THEOREM 83.4. (A positive answer to Problem 83.1) *Let X have property B_3, let $f: X \to K$. Suppose f has the following property. There are continuous functions $\lambda_1, \lambda_2: X \to K$ and $\Lambda_3: X \times X \to K$ such that*

$$f(x) = f(y) + (x-y)\lambda_1(y) + (x-y)^2\lambda_2(y) + (x-y)^3\Lambda_3(x, y) \qquad (x, y \in X)$$

Then $f \in C^3(X \to K)$ (and $\lambda_1 = D_1 f$, $\lambda_2 = D_2 f$, $\Lambda_3(x, y) = \overline{\Phi}_3 f(x, y, y, y)$ for all $x, y \in X$).

Proof. By Proposition 28.4 we have $f \in C^2(X \to K)$ and $\lambda_1 = D_1 f$, $\lambda_2 = D_2 f$. According to Lemma 29.7 the function $\tilde{\Phi}_3 f$ is defined as a continuous function on $X^4 \setminus \Delta$ and it is easily seen that $\Lambda_3(x, y) = \tilde{\Phi}_3 f(x, y, y, y)$ $(x, y \in$

$X, x \neq y$). To obtain the theorem it suffices to prove that the (symmetric) function $h : X^4 \to K$ defined by the formula

$$h(x_1, x_2, x_3, x_4) := \begin{cases} \Lambda_3(x_1, x_1) \text{ if } x_1 = x_2 = x_3 = x_4 \\ \\ \tilde{\Phi}_3 f(x_1, x_2, x_3, x_4) \text{ otherwise} \end{cases}$$

is continuous at each point of Δ. By property B_3 there is a $C \geqslant 1$ such that every pair of two elements of X can be extended to a C-regular triangle. Let $a \in X$, $\epsilon > 0$. There is $\delta > 0$ such that

(*) $$x, y \in B_a(\delta) \cap X \Rightarrow |\Lambda_3(x, y) - \Lambda_3(a, a)| \leqslant C^{-2}\epsilon$$

We shall prove that

$$U := \{(x_1, x_2, x_3, x_4) \in X^4 : |h(x_1, x_2, x_3, x_4) - h(a, a, a, a)| < \epsilon\}$$

contains $B_a(\delta C^{-1})^4 \cap X^4$ (which implies continuity of h at (a, a, a, a)). Thus, let $(x_1, x_2, x_3, x_4) \in B_a(\delta C^{-1})^4 \cap X^4$. To see that this element is in U we use the following steps.

(i) If (x_1, x_2, x_3, x_4) is of the form (x, y, y, y) then by (*) and the formula $\Lambda_3(x, y) = \tilde{\Phi}_3 f(x, y, y, y)$ we have $(x_1, x_2, x_3, x_4) \in U$. By symmetry of h we may conclude that $(x_1, x_2, x_3, x_4) \in U$ as soon as among x_1, x_2, x_3, x_4 at least three elements coincide.

(ii) (The crucial step) Let (x_1, x_2, x_3, x_4) have the form (x, y, x, y) where $x \neq y$. Then there is an element $z \in X$ such that $\{x, y, z\}$ is a C-regular triangle so that certainly $z \in B_a(\delta) \cap X$. A continuous extension of rule (i) of Lemma 29.2 shows that

$$(x - y)\, h(x, y, x, y) = (x - z)\, h(x, z, x, y) + (z - y)\, h(z, y, x, y)$$

If in this formula we substitute the further decompositions

$$(x - y)\, h(z, y, x, x) = (z - x)\, h(z, x, x, x) + (x - y)\, h(x, y, x, x)$$
$$(z - x)\, h(z, x, y, y) = (z - y)\, h(z, y, y, y) + (y - x)\, h(y, x, y, y)$$

we find (after applying (*) and using the symmetry of h) that $h(x, y, x, y)$ can be written as $\sum_{j=1}^{4} \mu_j b_j$ ($|\mu_j| \leqslant C^2$, $|b_j - \Lambda_3(a, a)| < C^{-2}\epsilon$ for each j, $\sum_{j=1}^{4} \mu_j = 1$). Hence $|h(x, y, x, y) - h(a, a, a, a)| < \epsilon$, i.e. $(x_1, x_2, x_3, x_4) \in U$.

(iii) From (i) and (ii) it follows that if $\{x_1, x_2, x_3, x_4\}$ has one or two elements then $(x_1, x_2, x_3, x_4) \in U$.

(iv) Let $\{x_1, x_2, x_3, x_4\}$ have three elements. By symmetry we may suppose that (x_1, x_2, x_3, x_4) has the form (x, y, z, z) where $|x - y| \geqslant \max(|x - z|, |z - y|)$. From the formula

$$(x - y)\, h(x, y, z, z) = (x - z)\, h(x, z, z, z) + (z - y)\, h(z, y, z, z)$$

and (iii) we infer that $h(x, y, z, z) = \mu_1 b_1 + \mu_2 b_2$ where $|\mu_j| \leqslant 1$, $|b_j - \Lambda_3(a, a)| < \epsilon$ for $j = 1, 2$ and $\mu_1 + \mu_2 = 1$. We see that $(x_1, x_2, x_3, x_4) \in U$.
(v) Finally, let $\{x_1, x_2, x_3, x_4\}$ have four elements. As in (iv) we may suppose that $(x_1, x_2, x_3, x_4) = (x, y, z, t)$ and $|x - y| \geqslant \max(|x - z|, |z - y|)$. The problem is reduced to case (iv) by means of the formula

$$(x-y)\, h(x, y, z, t) = (x - z)\, h(x, z, z, t) + (z - y)\, h(z, y, z, t)$$

which also finishes the proof.

A solution of Problem 83.1 for $n > 3$ in the style of Theorem 83.4 requires no new ideas but only takes some patience. We simply state the result and refer to W. H. Schikhof (*Non-archimedean calculus*, Report 7812, Mathematisch Instituut, Nijmegen, the Netherlands (1978)) for a detailed proof.

Let C be a real number, $C \geqslant 1$. A finite subset $\{x_1, x_2, \ldots, x_n\}$ of X is a *C-regular polygon* if $|x_i - x_j| \leqslant C\, |x_k - x_m|$ $(i, j, k, m \in \{1, 2, \ldots, n\}$, $k \neq m)$. Let $n \in \mathbb{N}$, $n \geqslant 3$. X has *property* B_n if there is a $C \geqslant 1$ such that each pair $\{x_1, x_2\}$ $(x_1 \neq x_2)$ in X can be extended to a C-regular polygon $\{x_1, x_2, \ldots, x_n\}$ in X. Open sets in K have property B_n for each $n \in \mathbb{N}$.

THEOREM 83.5. *Let $n \in \mathbb{N}$, $n \geqslant 3$, let X have property B_n, let $f : X \to K$. Suppose there exist continuous functions $\lambda_1, \ldots, \lambda_{n-1} : X \to K$ and Λ_n : $X \times X \to K$ such that*

$$f(x) = f(y) + \sum_{j=1}^{n-1} (x-y)^j \lambda_j(y) + (x-y)^n \Lambda_n(x, y) \quad (x, y \in X)$$

Then $f \in C^n(X \to K)$.

Remark. Theorem 83.5 can be used to prove an ultrametric version of Borel's theorem: *Let $\lambda_0, \lambda_1, \ldots$ be any sequence in K. Then there exists a C^∞-function $f : K \to K$ such that $D_n f(0) = \lambda_n$ for all $n \in \mathbb{N} \cup \{0\}$*. (See the reference given above.)

84. Functions of two variables

Although our interest lies in functions of one variable we shall make a brief excursion to calculus of several variables. For simplicity we work with functions $K^2 \to K$ but the results of this section can easily be extended to functions defined on an open subset of K^2. The partial derivatives $\partial f/\partial x$ and $\partial f/\partial y$ are defined just as in the real case.

In the spirit of Definition 27.1 we say that $f : K^2 \to K$ is a *C^1-function* if the difference quotients $\Phi_1^{(1)} f$, $\Phi_1^{(2)} f$ given by the formulas

$$\Phi_1^{(1)} f(x, x', y) := \frac{f(x, y) - f(x', y)}{x - x'} \qquad (x, x', y \in K, x \neq x')$$

$$\Phi_1^{(2)} f(x, y, y') := \frac{f(x, y) - f(x, y')}{y - y'} \qquad (x, y, y' \in K, y \neq y')$$

can be extended to continuous functions $\overline{\Phi}_1^{(1)} f$, $\overline{\Phi}_1^{(2)} f$ respectively, defined on K^3. We have obviously

$$\frac{\partial f}{\partial x}(x, y) = \overline{\Phi}_1^{(1)} f(x, x, y), \qquad \frac{\partial f}{\partial y}(x, y) = \overline{\Phi}_1^{(2)} f(x, y, y)$$

for all $x, y \in K$.

The following theorem states that 'every continuous vector field has a potential'.

THEOREM 84.1. *Let $f, g : K^2 \to K$ be continuous. Then there exists a C^1-function $F : K^2 \to K$ such that*

$$\frac{\partial F}{\partial x} = f \text{ and } \frac{\partial F}{\partial y} = g$$

Proof. Let $x \mapsto x_n$ ($n \in \{0, 1, 2, \ldots\}$) be an approximation of the identity in the sense of Section 79. Set

$$F(x, y) := \sum_{n=0}^{\infty} f(x_n, y_n)(x_{n+1} - x_n) + \sum_{n=0}^{\infty} g(x_n, y_n)(y_{n+1} - y_n) \qquad (x, y \in K)$$

Using the techniques of the proof of Theorem 79.2 one can prove easily that F satisfies the requirements. We leave the details to the reader.

COROLLARY 84.2. *There is a (C^1-)function $f : K^2 \to K$ whose four second partial derivatives exist and are continuous but for which*

$$\frac{\partial^2 f}{\partial x \, \partial y} = 1 \text{ and } \frac{\partial^2 f}{\partial y \, \partial x} = 0$$

Proof. By Theorem 84.1 there is a C^1-function f such that $\partial f/\partial y = x$ and $\partial f/\partial x = 0$. Differentiation yields the result.

To translate the theorem stating that the mixed partial derivatives $\partial^2 f/\partial x \partial y$ and $\partial^2 f/\partial y \partial x$ are equal if f is a C^2-function $\mathbb{R}^2 \to \mathbb{R}$ we shall define a C^2-function by requiring that second order difference quotients are continuously extendable. First observe that $\Phi_1^{(1)} f(x, x', y)$ is symmetric in x, x' (and that $\Phi_1^{(2)} f(x, y, y')$ is symmetric in y, y'). So we have essentially two difference quotients of $\Phi_1^{(1)} f$ defined by the formulas

$$\Phi_2^{(11)} f(x,x', x'', y) := \frac{\Phi_1^{(1)} f(x, x', y) - \Phi_1^{(1)} f(x, x'', y)}{x' - x''}$$

$$\Phi_2^{(21)} f(x, x', y, y') := \frac{\Phi_1^{(1)} f(x, x', y) - \Phi_1^{(1)} f(x, x', y')}{y - y'}$$

Similarly we have the two difference quotients $\Phi_2^{(12)} f$ and $\Phi_2^{(22)} f$ of $\Phi_1^{(2)} f$.

A function $f : K^2 \to K$ is a C^2-*function* if the four second order difference quotients $\Phi_2^{(ij)} f$ $(i, j \in \{1, 2\})$ can be extended to continuous functions on K^4. With this definition the following theorem is almost trivial.

THEOREM 84.3. *If* $f : K^2 \to K$ *is a* C^2*-function then*

$$\frac{\partial^2 f}{\partial x\, \partial y} = \frac{\partial^2 f}{\partial y\, \partial x}$$

Proof. This follows easily from the following three formulas.

$$\frac{\partial^2 f}{\partial y\, \partial x}(x, y) = \lim_{y' \to y} \lim_{x' \to x} \Phi_2^{(21)} f(x, x', y, y')$$

$$\frac{\partial^2 f}{\partial x\, \partial y}(x, y) = \lim_{x' \to x} \lim_{y' \to y} \Phi_2^{(12)} f(x, x', y, y')$$

$$\Phi_2^{(21)} f = \Phi_2^{(12)} f$$

Exercise 84.A. Let $p \equiv 3 \pmod 4$ and let $K := \mathbf{Q}_p(\sqrt{-1})$. The map $(x, y) \mapsto x + iy$ $(x, y \in \mathbf{Q}_p)$ where i is a root of $X^2 + 1$ is a bijection of $\mathbf{Q}_p^2$ onto K. A function $f : K \to K$ induces maps $u, v : \mathbf{Q}_p^2 \to \mathbf{Q}_p$ via $f(x + iy) = u(x, y) + iv(x, y)$ $(x, y \in \mathbf{Q}_p)$. Show that for differentiable f the functions u, v satisfy the familiar Cauchy-Riemann relations $\partial u/\partial x = \partial v/\partial y$ and $\partial u/\partial y = -\partial v/\partial x$.

PART 3: MONOTONE FUNCTIONS

In the last part of this chapter we discuss several definitions of 'monotonicity' using the ideas of Section 24. The theory of monotone functions and sequences is speculative in the sense that it has not yet been proved to be of substantial interest to other parts of p-adic analysis. We shall only touch upon the basic notions and theorems and restrict ourselves to $K = \mathbf{Q}_p$ whenever this may simplify things. For more details we refer to W. H. Schikhof (*Non-archimedean monotone functions,* Report 7916, Mathematisch Instituut, Nijmegen, the Netherlands (1979)).

85. Sides of 0 in K

In this section we shall have a closer look at the group of signs $\Sigma = K^\times / K^+$. We shall use the notations and terminology of Section 24. Let $x \in K^\times, a \in \Sigma$. Instead of sgn $x = a$ we shall sometimes write $x \in a$ (where a is considered as a subset of K rather than an element of the abstract group Σ).

We define a *multiplication* $K^\times \times \Sigma \to \Sigma$ by

$$x\, a := (\operatorname{sgn} x)\, a \qquad (x \in K^\times, a \in \Sigma)$$

In particular *the opposite sign* $-a$ of $a \in \Sigma$ is defined by $-a := (-1)\, a$. We have obviously $-a = \{-x : x \in a\}$. Observe that $a = -a$ if char$(k) = 2$.

The *absolute value* of a sign a is the real number $|a|$ defined by

$$|a| = |x| \qquad (x \in a)$$

(If $x, y \in a$ then $|x| = |y|$ so that the definition is meaningful.)

We also introduce a (restricted) *addition* $\oplus$ between elements of Σ as follows. Let $a, \beta \in \Sigma$. The set $a + \beta := \{x + y : x \in a, y \in \beta\}$ contains 0 if $a = -\beta$ so $a + (-a)$ is not a sign. If $a \neq -\beta$ choose $a \in a, b \in \beta$. By assumption $|a + b| = |a| \vee |b|$ and we get $a + \beta = B_a(|a|^-) + B_b(|b|^-) = B_{a+b}((|a| \vee |b|)^-) = B_{a+b}(|a+b|^-)$ which is indeed a sign (in fact, sgn $(a+b)$). Define

$$a \oplus \beta := a + \beta \qquad (a, \beta \in \Sigma, a \neq -\beta)$$

PROPOSITION 85.1. *Let $a, \beta, \gamma \in \Sigma$.*

(i) $|a\beta| = |a|\,|\beta|$, $|a^{-1}| = |a|^{-1}$.

(ii) *If $a \oplus \beta$ is defined then so is $\beta \oplus a$ and $a \oplus \beta = \beta \oplus a$.*

(iii) *If $(a \oplus \beta) \oplus \gamma$ and $a \oplus (\beta \oplus \gamma)$ are defined then $(a \oplus \beta) \oplus \gamma = a \oplus (\beta \oplus \gamma)$.*

(iv) *If $a \oplus \beta$ is defined then $\gamma(a \oplus \beta) = \gamma a \oplus \gamma\beta$.*

(v) *If $a \oplus \beta$ is defined then $|a \oplus \beta| = |a| \vee |\beta|$.*

(vi) *If $|s| = |a| \vee |\beta|$ for some $s \in a + \beta$ then $a \oplus \beta$ is defined.*

(vii) $|a| < |\beta|$ *if and only if $a \oplus \beta = \beta$.*

Proof. Direct verification.

As an illustration we shall describe the group of signs of $\mathbb{Q}_p$. Let θ be a primitive $(p-1)$ th root of unity, fixed from now on. Set

$$\Sigma' := \left\{p^n \theta^j : n \in \mathbb{Z}, j \in \{0, 1, \ldots, p-2\}\right\}$$

Σ is a multiplicative subgroup of $\mathbb{Q}_p^\times$, its elements form a complete set of representatives in $\mathbb{Q}_p^\times$ modulo $\mathbb{Q}_p^+$. Thus, Σ' is isomorphic to Σ in a natural way. We prefer to work with Σ' rather than with the abstract group Σ The following statements are easy to prove.

PROPOSITION 85.2. *Let $x \in \mathbb{Q}_p^\times$ have the Teichmüller expansion $\Sigma_{n \geq k} a_n p^n$ for some $k \in \mathbb{Z}, a_k \neq 0$* (see Exercise 27.I). *Then*

$$\operatorname{sgn} x = a_k p^k$$

If $a = p^m \theta^i, \beta = p^n \theta^j \in \Sigma'$ then

$$\begin{aligned} x\, a &= a_k\, \theta^i p^{m+k} \\ |a| &= p^{-m} \\ a \oplus \beta &= \beta \text{ if } m > n \\ a \oplus \beta &= p^n\, \omega_p\, (\theta^i + \theta^j) \text{ if } m = n \text{ and } |\theta^i + \theta^j|_p = 1 \end{aligned}$$

where ω_p is the Teichmüller character (see Definition 33.3).

Exercise 85.A. Try to interpret the definitions of Σ, multiplication of field elements and signs, opposite sign, absolute value of a sign, $\oplus$, for the real case.

Exercise 85.B. (Other formulas for sgn in $\mathbb{Q}_p$) Prove the following formulas.

$$\operatorname{sgn} x = p^{\operatorname{ord}_p x}\, \omega_p\, (p^{-\operatorname{ord}_p x}\, x) \qquad (x \in \mathbb{Q}_p^\times)$$

$$\operatorname{sgn} x = x\, (\exp \log_p x)^{-1} \qquad (x \in \mathbb{Q}_p^\times, p \neq 2)$$

Let $a \in \Sigma$. The relation $>_a$ in K is defined by the formula

$$x >_a y \text{ if } x - y \in a \qquad (x, y \in K)$$

PROPOSITION 85.3. *Let $x, y, z \in K$.*

(i) *If $x \neq y$ then $x >_a y$ for precisely one $a \in \Sigma$. If $x = y$ then $x >_a y$ for no $a \in \Sigma$.*

(ii) *If $x >_a y$ for some $a \in \Sigma$ then $x + z >_a y + z$.*

(iii) *If $x >_a y$, $z >_\beta 0$ for some $a, \beta \in \Sigma$ then $xz >_{a\beta} yz$.*

(iv) *If $x >_a y$, $y >_\beta z$ for some $a, \beta \in \Sigma$, $a \neq -\beta$ then $x >_{a \oplus \beta} z$.*

Proof. Obvious.

86. Monotone functions of type σ

DEFINITION 86.1. Let $\sigma : \Sigma \to \Sigma$ be a bijection. A function $f : K \to K$ is *monotone of type σ* if $x >_a y$ implies $f(x) >_{\sigma(a)} f(y)$ for all $x, y \in K$ and $a \in \Sigma$.

Observe that increasing functions and monotone functions of type $a \in \Sigma$ are special cases (choose for σ the identity, the multiplication by a, respectively, see Definition 24.6).

Not every bijection $\Sigma \to \Sigma$ can 'occur as a type'. In fact we have the following.

THEOREM 86.2. *Let $\sigma : \Sigma \to \Sigma$ be a bijection for which there exists a function $K \to K$, monotone of type σ. Then $\sigma(-a) = -\sigma(a)$ and $\sigma(a \oplus \beta) = \sigma(a) \oplus \sigma(\beta)$ for all $a, \beta \in \Sigma$, $a \neq -\beta$.*

Proof. Let $f : K \to K$ be monotone of type σ. Let $a \in \Sigma$ and $x >_a y$. Then $y >_{-a} x$ so that $f(y) >_{\sigma(-a)} f(x)$. We also have $f(x) >_{\sigma(a)} f(y)$, hence $f(y) >_{-\sigma(a)} f(x)$. Thus, $\sigma(-a) = -\sigma(a)$. Now let $a, \beta \in \Sigma$, $a \neq -\beta$. Then $\sigma(a) \neq \sigma(-\beta) = -\sigma(\beta)$ so that $\sigma(a) \oplus \sigma(\beta)$ is defined. Choose $x, y, z \in K$ for which $x >_a y$, $y >_\beta z$. Then $x >_{a \oplus \beta} z$ so that $f(x) >_{\sigma(a \oplus \beta)} f(z)$. We also have $f(x) >_{\sigma(a)} f(y), f(y) >_{\sigma(\beta)} f(z)$ so that $f(x) >_{\sigma(a) \oplus \sigma(\beta)} f(z)$. It follows that $\sigma(a \oplus \beta) = \sigma(a) \oplus \sigma(\beta)$.

**Exercise* 86.A. Let $\sigma : \Sigma \to \Sigma$ be a bijection satisfying $\sigma(-a) = -\sigma(a)$ and $\sigma(a \oplus \beta) = \sigma(a) \oplus \sigma(\beta)$ for all $a, \beta \in \Sigma$, $a \neq -\beta$. Prove that $|\sigma(a)| < |\sigma(\beta)|$ if and only if $|a| < |\beta|$.

For spherically complete K we have a converse of Theorem 86.2.

THEOREM 86.3. *Let K be spherically complete. Let $\sigma : \Sigma \to \Sigma$ be a bijection satisfying $\sigma(-a) = -\sigma(a)$ and $\sigma(a \oplus \beta) = \sigma(a) \oplus \sigma(\beta)$ for all $a, \beta \in \Sigma$, $a \neq -\beta$. Then there exists a function $f : K \to K$ that is monotone of type σ.*

Proof. We shall prove the following. Let $Y \subset K$, $a \in K \backslash Y$. Suppose $f : Y \to K$ is monotone of type σ (i.e. $x >_a y$ implies $f(x) >_{\sigma(a)} f(y)$ for all $x, y \in Y$ and $a \in \Sigma$). Then f can be extended to a function $\bar{f}$ on $Y \cup \{a\}$ that is monotone of type σ. (Then we can finish the proof by choosing $Y := \{0\}$, $f : Y \to K$ arbitrary and by extending f with the help of Zorn's lemma to a function $K \to K$, monotone of type σ.) We have to choose an element $\bar{f}(a) \in K$ such that $f(x) - \bar{f}(a) \in \sigma(\operatorname{sgn}(x - a)), \bar{f}(a) - f(x) \in \sigma(\operatorname{sgn}(a - x))$ for all $x \in Y$. It suffices to consider only the second condition. Thus we must show that the discs $B_x := f(x) + \sigma(\operatorname{sgn}(a - x))$, where x runs through Y, have a nonempty intersection. By spherical completeness we are done if we can show that $B_x \cap B_y \neq \emptyset$ if $x, y \in Y$, $x \neq y$. So let $x, y \in Y$, $x \neq y$, set $a := \operatorname{sgn}(a - x)$, $\beta := \operatorname{sgn}(a - y)$ and choose $b \in \sigma(a)$, $c \in \sigma(\beta)$. We shall prove that the distance between $f(x) + b$ and $f(y) + c$ is strictly less than max $(d(B_x), d(B_y)) = |\sigma(a)| \vee |\sigma(\beta)|$. We consider two cases.

(i) $a = \beta$. Then $a - x \in a, a - y \in a$ so that $|x - y| < |a - x| = |a|$ and $|\operatorname{sgn}(x - y)| < |a|$. By Exercise 86.A we have $|\operatorname{sgn}(f(x) - f(y))| = |\sigma(\operatorname{sgn}(x - y))| < |\sigma(a)|$. Hence, $|f(x) - f(y)| < |\sigma(a)|$. We have also $b, c \in \sigma(a)$ so $|b - c| < |\sigma(a)|$. Hence, $|f(x) + b - (f(y) + c)| < |\sigma(a)|$.

(ii) $a \neq \beta$. Then $x - y = a - y - (a - x) \in \beta \oplus -a$ so $f(x) - f(y) + b - c \in$

$\sigma(\beta \oplus - a) + \sigma(a) + \sigma(-\beta) = \sigma(\beta \oplus - a) + \sigma(a \oplus - \beta) = \sigma(\beta \oplus - a) - \sigma(\beta \oplus - a)$. Thus, $|f(x) - f(y) + b - c| < |\sigma(\beta \oplus - a)| = |\sigma(\beta) \oplus \sigma(-a)| = |\sigma(\beta)| \vee |\sigma(a)|$.

As an example we consider the case $K = \mathbb{Q}_p$. Let Σ' be as in Section 85.

THEOREM 86.4. *Let $\sigma : \Sigma' \to \Sigma'$ be a bijection for which there exists a function $\mathbb{Q}_p \to \mathbb{Q}_p$, monotone of type σ. Then σ has the form*

$$(*) \qquad p^n \theta^j \mapsto p^{n+c} \theta^{j+s(n)} \qquad (n \in \mathbb{Z}, j \in \{0, 1, \ldots, p-2\})$$

for some $c \in \mathbb{Z}$ and some $s : \mathbb{Z} \to \{0, 1, \ldots, p-2\}$. Conversely, let $\sigma : \Sigma' \to \Sigma'$ have the form (*). Then

$$\sum_n a_n p^n \mapsto \sum_n a_n \theta^{s(n)} p^{n+c}$$

(where $\Sigma_n a_n p^n$ is the Teichmüller representation) defines a function $f : \mathbb{Q}_p \to \mathbb{Q}_p$ that is monotone of type σ.

Proof. To prove the first part observe that, by Exercise 86.A, $|\sigma(a)|_p$ is a strictly increasing function of $|a|_p$. By the surjectivity of σ we must have that $n \mapsto |\sigma(p^n)|_p$ is a decreasing bijection of $\mathbb{Z}$ onto $|\mathbb{Q}_p^\times| \simeq \mathbb{Z}$. It follows easily that there is a $c \in \mathbb{Z}$ such that $|\sigma(p^n)|_p = p^{-n-c} = |p^{n+c}|_p$ for all $n \in \mathbb{Z}$. For each $n \in \mathbb{Z}$ there is therefore an element $s(n) \in \{0, 1, \ldots, p-2\}$ such that $\sigma(p^n) = p^{n+c}\theta^{s(n)}$ Let $j \in \{0, 1, \ldots, p-2\}$. Then there is a $t \in \{1, \ldots, p-1\}$ such that $|\theta^j - t|_p < 1$. It follows easily from Theorem 86.2 that $\sigma(ta) = t\sigma(a)$ for all $a \in \Sigma'$ and we have $\sigma(\theta^j p^n) = \sigma(\text{sgn } tp^n) = \sigma(tp^n) = t\sigma(p^n) = (\text{sgn } \theta^j)\, \sigma(p^n) = \theta^j \sigma(p^n) = p^{n+c}\theta^{j+s(n)}$. We see that σ has the required form (*). To prove the second part, let $x, y \in \mathbb{Q}_p, x >_a y$ for some $a \in \Sigma'$. We show that $f(x) >_{\sigma(a)} f(y)$. Let $x = \Sigma_n a_n p^n$, $y = \Sigma_n b_n p^n$ (Teichmüller representations) and let $a = p^m \theta^j$ for some $m \in \mathbb{Z}$, $j \in \{0, 1, \ldots, p-2\}$. From $x >_a y$ we obtain $|x - y - p^m \theta^j|_p < |x-y|_p$ so that $a_n = b_n$ if $n < m$, $|a_m - b_m - \theta^j|_p < 1$. Hence $\text{sgn }(a_m - b_m) = \theta^j$. Now we have

$$f(x) - f(y) = \sum_{n \geqslant m} (a_n - b_n)\theta^{s(n)} p^{n+c} = (a_m - b_m)\theta^{s(m)} p^{m+c} + r$$

where $|r|_p < |f(x) - f(y)|_p$. Thus, $\text{sgn }(f(x) - f(y)) = \text{sgn }((a_m - b_m)\theta^{s(m)} p^{m+c}) = \theta^{s(m)+j} p^{m+c} = \sigma(p^m \theta^j) = \sigma(a)$, i.e. $f(x) >_{\sigma(a)} f(y)$.

For the general theory the following theorem is of interest.

THEOREM 86.5. *Let $f : K \to K$ be monotone of some type $\sigma : \Sigma \to \Sigma$. Then there exists a strictly increasing bijection $\varphi : |K| \to |K|$ such that*

$|f(x)-f(y)| = \varphi(|x-y|)$ for all $x, y \in K$. In particular, f is uniformly continuous.

Proof. From Definition 86.1 it follows that f is injective. Let $x, y, z, t \in K$ be such that $|x-y| < |z-t|$. We prove that $|f(x)-f(y)| < |f(z)-f(t)|$. By injectivity of f we may assume $x \neq y$. Then $x >_a y$, $z >_\beta t$ for some $a, \beta \in \Sigma$, $|a| < |\beta|$. Then $|f(x)-f(y)| = |\sigma(a)| < |\sigma(\beta)| = |f(z)-f(t)|$. From what we have proved so far it follows that $|f(x)-f(y)|$ is a strictly increasing function of $|x-y|$, i.e. $|f(x)-f(y)| = \varphi(|x-y|)$ for some strictly increasing $\varphi : |K| \to |K|$. Since σ is surjective the set $\{f(x)-f(y) : x, y \in K, x \neq y\}$ meets every sign, hence $\{|f(x)-f(y)| : x, y \in K\} = |K|$ so that φ is surjective. Since 0 is an accumulation point of $|K|$, φ is continuous at 0 implying uniform continuity of f.

COROLLARY 86.6. *Let K have discrete valuation and let $f : K \to K$ be monotone of type $\sigma : \Sigma \to \Sigma$. Then f is a scalar multiple of an isometry.*

Proof. Let φ be as in Theorem 86.5. The value group of K is (order-) isomorphic to $\mathbb{Z}$, hence φ induces an increasing bijection $\mathbb{Z} \to \mathbb{Z}$. The latter are all of the form $n \mapsto n + a$ $(n \in \mathbb{Z})$ for some $a \in \mathbb{Z}$. It follows that $|f(x)-f(y)| = \varphi(|x-y|) = c\,|x-y|$ for some $c \in |K^\times|$.

Exercise 86.B. Let K have discrete valuation. Show that Σ is isomorphic (as a group) to $|K^\times| \times k^\times$ (k is the residue class field of K).

Exercise 86.C. Describe the class of all bijections $\sigma : \Sigma' \to \Sigma'$ for which there exists a differentiable function $\mathbb{Q}_p \to \mathbb{Q}_p$, monotone of type σ.

Exercise 86.D. (Nonmonotone isometries) Let $p \neq 2$. Choose a permutation τ of $\{0, 1, 2, \ldots, p-1\}$ with $\tau(0) = 0$ such that the isometry f defined by $f(\sum_n a_n p^n) = \sum_n \tau(a_n) p^n$ $(\sum_n a_n p^n \in \mathbb{Q}_p)$ is monotone of type σ for no bijection $\sigma : \Sigma' \to \Sigma'$.

87. Monotonicity without type

Instead of 'sides of 0' we may take 'betweenness' as a starting point leading to a little more general notion of monotonicity. For a function $f : \mathbb{R} \to \mathbb{R}$ we have equivalence of the following conditions (a) and (β).

(a) f is monotone (in the non-strict sense).

(β) If x is between y and z then $f(x)$ is between $f(y)$ and $f(z)$ $(x, y, z \in \mathbb{R})$.

With Definition 24.1 in mind this leads us to the following definition (for simplicity we consider functions $\mathbb{Z}_p \to \mathbb{Q}_p$).

DEFINITION 87.1. A function $f : \mathbb{Z}_p \to \mathbb{Q}_p$ is *monotone* if $z \in [x, y]$ implies $f(z) \in [f(x), f(y)]$ for all $x, y, z \in \mathbb{Z}_p$.
In other words, f is monotone if for all $x, y, z \in \mathbb{Z}_p$

$$|z-x|_p \leqslant |x-y|_p \Rightarrow |f(z)-f(x)|_p \leqslant |f(x)-f(y)|_p$$

It follows that isometries (compare Exercise 86.D) and (restrictions to $\mathbb{Z}_p$ of) monotone functions $\mathbb{Q}_p \to \mathbb{Q}_p$ of a certain type are monotone in the sense of Definition 87.1 (Theorem 86.5). A C^1-function $f : \mathbb{Z}_p \to \mathbb{Q}_p$ is 'locally monotone' at a if $f'(a) \neq 0$. There are also non-injective examples. In fact, for each $n \in \mathbb{N}$ the locally constant function

$$x \mapsto x_n = p^n\,[p^{-n}x]_p = \sum_{j=0}^{n-1} a_j p^j \qquad (x = \sum_{j=0}^{\infty} a_j p^j \in \mathbb{Z}_p)$$

is easily seen to be monotone (see the beginning of Section 62). The following exercise contains some immediate consequences of the definition.

**Exercise* 87.A. Prove the following properties (compare the real case).
(i) $f : \mathbb{Z}_p \to \mathbb{Q}_p$ is monotone if and only if for each convex (Definition 24.1) subset C of $\mathbb{Q}_p$ the set $f^{-1}(C)$ is convex.
(ii) The set of all monotone functions $\mathbb{Z}_p \to \mathbb{Q}_p$ is closed under pointwise limits and scalar multiplication.
(iii) If $f : \mathbb{Z}_p \to \mathbb{Q}_p$ is monotone and $f(a) = f(b)$ for some $a, b \in \mathbb{Z}_p$ then f is constant on $[a, b]$.

LEMMA 87.2. *Let $f : \mathbb{Z}_p \to \mathbb{Q}_p$ be monotone. If $a, b, c \in \mathbb{Z}_p$, $|a-b|_p < |a-c|_p$ and $f(a) \neq f(c)$ then $|f(a)-f(b)|_p < |f(a)-f(c)|_p$.*
Proof. Let $B := [a, c]$. By monotony $f(B) \subset [f(a), f(c)]$. Define an equivalence relation $\sim$ on B by

$$x \sim y \text{ if } |f(x)-f(y)|_p < |f(a)-f(c)|_p \qquad (x, y \in B)$$

The ball $[f(a), f(c)]$ contains at most p elements that have distances $|f(a) - f(c)|$ to one another hence B decomposes into n equivalence classes $B_1, \ldots, B_n$ where $n \leqslant p$. Since a and c are not equivalent we have also $n \geqslant 2$. By monotony of f each B_j is convex. But the only way to cover B by means of n nonempty convex mutually disjoint subsets, where also $2 \leqslant n \leqslant p$, is the one by means of p balls of radius $p^{-1}\ |a-c|_p$. Hence, each B_j is a ball of radius $p^{-1}\ |a-c|_p$ and since $|a-b|_p \leqslant p^{-1}\ |a-c|_p$ there is a j for which both a and b are elements of B_j. That is, $a \sim b$ so that $|f(a)-f(b)|_p < |f(a) - f(c)|_p$.

We use this lemma to prove the following.

THEOREM 87.3. *Let $f : \mathbb{Z}_p \to \mathbb{Q}_p$ be monotone. Then $f \in \mathrm{Lip}_1 (\mathbb{Z}_p \to \mathbb{Q}_p)$. In particular, monotone functions are continuous.*

Proof. Let $M := \max \{|f(i) - f(j)|_p : i, j \in \{0, 1, \ldots, p-1\}\}$. Let $a \in \mathbb{Z}_p$. We prove by induction on n that $x \in \mathbb{Z}_p$, $|x-a|_p = p^{-n}$ implies $|f(x) - f(a)|_p \leqslant p^{-n} M$. First let $n = 0$, i.e. $|x-a|_p = 1$. We can find $i, j \in \{0, 1, \ldots p-1\}$, $i \neq j$ such that $|x-i|_p < 1$, $|a-j|_p < 1$. Then $|x-a|_p = |i-a|_p = |i-j|_p$ so that by monotony $|f(x) - f(a)|_p \leqslant |f(i) - f(a)|_p \leqslant |f(i) - f(j)|_p \leqslant M$. To prove the step from $n-1$ to n let $|x-a|_p = p^{-n}$. Then $|x-a|_p < |a+p^{n-1} - a|_p$. If $f(a+p^{n-1}) \neq f(a)$ then by the previous lemma $|f(x) - f(a)|_p < |f(a+p^{n-1}) - f(a)|_p$ which is $\leqslant p^{-(n-1)} M$ by the induction hypothesis. Hence $|f(x) - f(a)|_p \leqslant p^{-1} p^{-(n-1)} M = p^{-n} M$. If $f(a+p^{n-1}) = f(a)$ then by monotony $|f(x) - f(a)|_p \leqslant |f(a+p^{n-1}) - f(a)|_p = 0$ and we have trivially $|f(x) - f(a)|_p \leqslant p^{-n} M$.

Exercise 87.B. Let us call a function $f : \mathbb{C}_p \to \mathbb{C}_p$ monotone if $x, y, z \in \mathbb{C}_p$, $|z-x|_p \leqslant |x-y|_p$ implies $|f(z) - f(x)|_p \leqslant |f(x) - f(y)|_p$. One can prove that there exist noncontinuous monotone functions $\mathbb{C}_p \to \mathbb{C}_p$. Prove however that a monotone function $f : \mathbb{C}_p \to \mathbb{C}_p$ has at most countably many points of discontinuity. (Hint. If a is a point of discontinuity then $f(a)$ is an isolated point of $f(\mathbb{C}_p)$.)

Exercise 87.C. Let $f : \mathbb{Z}_p \to \mathbb{Q}_p$ be monotone. Show that for each $n \in \mathbb{N}$ the function $f_n : \mathbb{Z}_p \to \mathbb{Q}_p$ defined by $f_n(x) := p^n [p^{-n} f(x)]_p$ $(x \in \mathbb{Z}_p)$ (see Section 62) is locally constant and monotone. Further, prove that $\lim_{n\to\infty} f_n = f$ uniformly. *Thus, every monotone function $\mathbb{Z}_p \to \mathbb{Q}_p$ can uniformly be approximated by locally constant monotone functions.*

Exercise 87.D. Let $f : \mathbb{Z}_p \to \mathbb{Q}_p$. Prove the equivalence of the following conditions.

(α) $f \in \mathrm{Lip}_1 (\mathbb{Z}_p \to \mathbb{Q}_p)$.

(β) f is the difference of two monotone functions.

(γ) f is a linear combination of two isometries.

Remark. For a real valued function f defined on a closed bounded subinterval of $\mathbb{R}$ condition (β) is equivalent to 'f is of bounded variation'.

The following theorem is surprising at first sight.

THEOREM 87.4. *Let $f : \mathbb{Z}_p \to \mathbb{Z}_p$ be monotone and surjective. Then f is an isometry.*

Proof. From Theorem 87.3 and its proof it follows that f satisfies the Lipschitz condition $|f(x) - f(y)|_p \leqslant |x-y|_p$ $(x, y \in \mathbb{Z}_p)$. If it can be shown that f is injective then we may conclude from Lemma 87.2 that for all triples $x, y, z \in \mathbb{Z}_p$

$$|x-y|_p \leqslant |y-z|_p \text{ if and only if } |f(x)-f(y)|_p \leqslant |f(y)-f(z)|_p$$

implying monotony of $f^{-1} : \mathbb{Z}_p \to \mathbb{Z}_p$. Again, by Theorem 87.4 we then have $|f^{-1}(x)-f^{-1}(y)|_p \leqslant |x-y|_p$ $(x, y \in \mathbb{Z}_p)$. Hence f is an isometry. Thus, we shall finish the proof by showing that f is injective. Let $f(a)=f(b)$ for some $a, b \in \mathbb{Z}_p$. There is a sequence $x_1, x_2, \ldots$ in $\mathbb{Z}_p$ such that $\lim_{n\to\infty} f(x_n) = f(a)$ and $f(x_n) \neq f(a)$ for all n. By compactness we may assume that $x_1, x_2, \ldots$ converges. Set $c := \lim_{n\to\infty} x_n$. Suppose $c \neq a$. Then $|x_n - c|_p < |c-a|_p$ for large n so that by monotony $|f(x_n)-f(c)|_p \leqslant |f(c)-f(a)|_p$ for large n. Now f is continuous so $f(c) = \lim_{n\to\infty} f(x_n) = f(a)$. We see that $f(x_n) = f(c) = f(a)$ for large n, a contradiction. It follows that $c = a$. A similar reasoning with a replaced by b yields $c = b$. Thus, $a = b$ and f is injective.

Exercise 87.E. (p-adic Darboux continuity) A function $[0, 1] \to \mathbb{R}$ is Darboux continuous if and only if it maps convex subsets of the closed unit interval onto convex subsets of $\mathbb{R}$. Thus (see also Exercise 87.A (i)) we try the following definition. A function $f : \mathbb{Z}_p \to \mathbb{Z}_p$ is *Darboux continuous* if for each convex set $S \subset \mathbb{Z}_p$ the image $f(S)$ is convex in $\mathbb{Z}_p$. The term 'local Darboux continuity' is defined in an obvious way. Prove the following and compare the real case.

(i) A Darboux continuous function $\mathbb{Z}_p \to \mathbb{Z}_p$ need not be continuous.
(ii) A continuous function $\mathbb{Z}_p \to \mathbb{Z}_p$ need not be Darboux continuous (!)
(iii) A bijective Darboux continuous function $\mathbb{Z}_p \to \mathbb{Z}_p$ is an isometry.
(iv) A C^1-function $f : \mathbb{Z}_p \to \mathbb{Z}_p$ is locally Darboux continuous at points a where $f'(a) \neq 0$.
(v) There exist differentiable functions $\mathbb{Z}_p \to \mathbb{Z}_p$ that are nowhere locally Darboux continuous.

Finally we introduce the notion of a p-adic monotone sequence.

DEFINITION 87.5. A sequence $a_1, a_2, \ldots$ in $\mathbb{Q}_p$ is *monotone* if for each triple $m, s, n \in \mathbb{N}$ for which $m \leqslant s \leqslant n$ we have $a_s \in [a_m, a_n]$.

In the next exercise we consider some immediate consequences and a few examples.

**Exercise* 87.F. Prove the following.
(i) A sequence $a_1, a_2, \ldots$ in $\mathbb{Q}_p$ is monotone if and only if $|a_n - a_m|_p = \max\{|a_{j+1} - a_j|_p : m \leqslant j < n\}$ $(m, n \in \mathbb{N}, m < n)$.
(ii) If $a_1, a_2, \ldots$ is monotone and if $a_m = a_n$ for some $m, n \in \mathbb{N}$ then $a_m = a_s = a_n$ for all s between m and n.

(iii) Subsequences of monotone sequences are monotone.
(iv) A monotone function $\mathbb{Z}_p \to \mathbb{Q}_p$ maps monotone sequences into monotone sequences.
(v) A sequence for which $|a_1 - a_2|_p > |a_2 - a_3|_p > \ldots$ is monotone. The converse is not true.
(vi) For each $x = \sum_{j=0}^{\infty} a_j p^j \in \mathbb{Z}_p$ the sequence $n \mapsto \sum_{j=0}^{n-1} a_j p^j$ is monotone.
(vii) The sequence $1, 1, p, p, p^2, p^2, \ldots$ is monotone.
(viii) The sequence $n \mapsto n - n_-$ (see Definition 47.3) is monotone.

The following theorem shows that p-adic monotone sequences behave very much like real monotone sequences.

THEOREM 87.6. (Properties of p-adic monotone sequences)
(i) *If a sequence $a_1, a_2, \ldots$ is monotone and convergent with limit a then* $|a - a_1|_p \geqslant |a - a_2|_p \geqslant \ldots$
(ii) *A monotone sequence having a convergent subsequence is itself convergent.*
(iii) *A bounded monotone sequence is convergent.*
(iv) *If $a_1, a_2, \ldots$ is unbounded and monotone then* $\lim_{n\to\infty} |a_n|_p = \infty$.
(v) *Each sequence has a monotone subsequence.*
Proof. Let $j \in \mathbb{N}$. For $n > j$ we have $|a_n - a_j|_p \geqslant |a_n - a_{j+1}|_p$. After taking limits for $n \to \infty$ we obtain $|a - a_j|_p \geqslant |a - a_{j+1}|_p$ which is (i). To prove (ii) let $b_1, b_2, \ldots$ be a convergent subsequence of a monotone sequence $a_1, a_2, \ldots$ Set $b := \lim_{n\to\infty} b_n$ and let $\epsilon > 0$. There is a $j \in \mathbb{N}$ such that $|b - b_j|_p < \epsilon$. Now $b_j = a_m$ for some m. For $n \geqslant m$ we have $|a_n - a_m|_p \leqslant |b_s - b_j|_p$ for some large enough $s > j$ and $|b_s - b_j|_p \leqslant |b - b_s|_p \vee |b - b_j|_p \doteq |b - b_j|_p < \epsilon$, using (i). It follows that $|b - a_n|_p \leqslant |b - b_j|_p \vee |a_m - a_n|_p < \epsilon$ and $\lim_{n\to\infty} a_n = b$. Properties (iii) and (iv) are direct consequences of (ii) and the local compactness of $\mathbb{Q}_p$. Finally we prove (v). Let $a_1, a_2, \ldots$ be a sequence in $\mathbb{Q}_p$. If it is unbounded we simply take a subsequence $b_1, b_2, \ldots$ for which $|b_1|_p < |b_2|_p < \ldots$ Then $b_1, b_2, \ldots$ is monotone. Let $a_1, a_2, \ldots$ be bounded, assume also that $n \neq m$ implies $a_n \neq a_m$. By compactness it has a subsequence that converges to a, say. By taking a further suitable subsequence $b_1, b_2, \ldots$ we can arrange that $|a - b_1|_p > |a - b_2|_p > \ldots$ Then also $|b_1 - b_2|_p > |b_2 - b_3|_p > \ldots$ and $b_1, b_2, \ldots$ is monotone by Exercise 87.F (v).

Exercise 87.G. (p-adic sequences whose partial sums form a monotone sequence) For a real sequence $a_1, a_2, \ldots$ the sequence $n \mapsto \sum_{j=1}^{n} a_j$ is monotone if and only if either $a_2, a_3, \ldots$ are all $\geqslant 0$ or $a_2, a_3, \ldots$ are all $\leqslant$

0. The p-adic translation reads as follows. Let $a_1, a_2, \ldots$ be a p-adic sequence. Then $n \mapsto \sum_{j=1}^{n} a_j$ is monotone if and only if

$$(*) \quad |a_n + a_{n+1} + \ldots + a_m|_p = \max(|a_n|_p, |a_{n+1}|_p, \ldots, |a_m|_p) \qquad (2 \leqslant n \leqslant m)$$

Prove this. Furthermore, show that if $a_1, a_2, \ldots$ is a bounded sequence for which (*) holds then $\lim_{n\to\infty} a_n = 0$.

Exercise 87.H. (Monotone sequences of functions) A sequence of functions $f_1, f_2, \ldots : \mathbb{Z}_p \to \mathbb{Q}_p$ is *monotone* if for each $x \in \mathbb{Z}_p$ the sequence $f_1(x)$, $f_2(x), \ldots$ is monotone.

(i) Prove that every continuous function $\mathbb{Z}_p \to \mathbb{Q}_p$ is the uniform limit of a monotone sequence of locally constant functions.

(ii) Limits of monotone sequences of continuous functions need not be continuous. In fact, prove that for every closed (or open) subset Y of $\mathbb{Z}_p$ the function ξ_Y is the limit of a monotone sequence of characteristic functions of clopen sets.

(iii) On the other hand, if $f_1, f_2, \ldots$ are continuous functions $\mathbb{Z}_p \to \mathbb{Q}_p$ such that for all $x \in \mathbb{Z}_p$

$$(*) \qquad |f_1(x) - f_2(x)|_p > |f_2(x) - f_3(x)|_p > \ldots$$

(this condition implies monotony of $f_1, f_2, \ldots$) then $f := \lim_{n\to\infty} f_n$ exists uniformly and f is continuous.

(iv) Can you reach the same conclusion if in (iii) condition (*) is replaced by the weaker condition

$$|f_1(x) - f_2(x)|_p \geqslant |f_2(x) - f_3(x)|_p \geqslant \ldots,$$
$$\lim_{n\to\infty} |f_n(x) - f_{n+1}(x)|_p = 0?$$

(v) Prove the following p-adic Dini theorem. *If $f_1, f_2, \ldots$ is a monotone sequence of continuous functions $\mathbb{Z}_p \to \mathbb{Q}_p$ and if $\lim_{n\to\infty} f_n(x) = 0$ for each $x \in \mathbb{Z}_p$ then $\lim_{n\to\infty} f_n = 0$ uniformly.*

APPENDIX A

Aspects of functional analysis

A.1. Two theorems on metric spaces

We shall prove Banach's contraction theorem and the category theorem of Baire. This section contains no *ultra*metric theory.

A map f from a metric space $X = (X, d)$ into itself is a *contraction* if there exists a real number c $(0 < c < 1)$ such that $d(f(x), f(y)) \leqslant cd(x, y)$ for all $x, y \in X$. A *fixed point* of f is an element $a \in X$ for which $f(a) = a$.

THEOREM. (Banach's contraction theorem) *Let X be a complete metric space and let $f : X \to X$ be a contraction. Then f has a unique fixed point a. For every $x \in X$ the sequence $x, f(x), f(f(x)), \ldots$ converges to a.*

Proof. Choose $x \in X$. Define $x_0 := x$ and $x_{n+1} = f(x_n)$ $(n \in \mathbb{N} \cup \{0\})$. Since $d(x_{n+1}, x_n) \leqslant c\, d(x_n, x_{n-1}) \leqslant \ldots \leqslant c^n\, d(x_1, x_0)$ we have for $m, n \in \mathbb{N}$, $m > n$ that $d(x_m, x_n) \leqslant d(x_m, x_{m-1}) + d(x_{m-1}, x_{m-2}) + \ldots + d(x_{n+1}, x_n) \leqslant (c^{m-1} + c^{m-2} + \ldots + c^n)\, d(x_1, x_0)$. It follows that $\lim_{n, m \to \infty} d(x_m, x_n) = 0$. By completeness, $a := \lim_{n \to \infty} x_n$ exists. By continuity, $f(a) = \lim_{n \to \infty} f(x_n) = \lim_{n \to \infty} x_{n+1} = a$. Thus, a is a fixed point. If $a, b \in X$ are fixed points then $d(a, b) = d(f(a), f(b)) \leqslant c\, d(a, b)$ so that $a = b$.

A subset A of a metric space is *a G_δ-set* if A is the intersection of countably many open sets, *nowhere dense* if the interior of the closure $\bar{A}$ of A is empty, *meagre* if A is a countable union of nowhere dense sets.

PROPOSITION. *The following conditions on a metric space are equivalent.*

(α) *No ball is meagre.*

(β) *The complement of a meagre set contains a dense G_δ-set.*

(γ) *The intersection of countably many dense G_δ-sets is itself a dense G_δ-set.*

Proof. (α) $\Rightarrow$ (γ). It suffices to prove that if $U_1, U_2, \ldots$ are dense open sets then $\bigcap_{n=1}^{\infty} U_n$ is dense. Let S be a closed ball; we prove that $\bigcap_{n=1}^{\infty} U_n$ meets S. The sets $S \backslash U_1, S \backslash U_2, \ldots$ are closed nowhere dense subsets of S. Since S is not meagre their union is not all of S so there is $x \in S$ such that $x \notin \bigcup_{n=1}^{\infty} (S \backslash U_n)$, i.e. $x \in \bigcap_{n=1}^{\infty} U_n$.

$(\gamma) \Rightarrow (\beta)$. Let M be a meagre subset of the metric space X. Then $M = \bigcup_{n=1}^{\infty} A_n \subset \bigcup_{n=1}^{\infty} \bar{A}_n$ where the interior of $\bar{A}_n$ is empty for each n. Thus, $X \setminus M \supset \bigcap_{n=1}^{\infty} (X \setminus \bar{A}_n)$. Each $X \setminus \bar{A}_n$ is open (hence a G_δ -set) and dense. By (γ), $\bigcap_{n=1}^{\infty} (X \setminus \bar{A}_n)$ is a dense G_δ -set.
$(\beta) \Rightarrow (\alpha)$ is evident.

A *Baire space* is a metric space satisfying one (or all) of the conditions $(\alpha), (\beta), (\gamma)$ of above.

THEOREM. (Category theorem of Baire) *A complete metric space is a Baire space.*
Proof. We prove that if $U_1, U_2, \ldots$ are open dense subsets then $\bigcap_{n=1}^{\infty} U_n$ is dense. Let B_0 be an open ball; we prove that $\bigcap_{n=1}^{\infty} U_n$ meets B_0. The set U_1 meets B_0, $U_1 \cap B_0$ is open so there is an open ball B_1 whose closure $\bar{B}_1$ is contained in $U_1 \cap B_0$. We may assume that the diameter of B_1 is $\leqslant 1$. The set U_2 meets B_1 and by the same token there is an open ball B_2, of diameter $\leqslant 1/2$, whose closure is contained in $U_2 \cap B_1$. Going on this way we find open balls $B_1 \supset B_2 \supset \ldots$ for which $\lim_{n \to \infty} d(B_n) = 0$ and $\bar{B}_n \subset U_n \cap B_0$ for each n. By completeness there is an element $a \in \bigcap_{n=1}^{\infty} \bar{B}_n$. Then clearly $a \in B_0$, $a \in \bigcap_{n=1}^{\infty} U_n$.

COROLLARY. (Other form of the category theorem) *If a complete metric space is the union of countably many closed sets then at least one of these sets contains a ball.*
Proof. A complete metric space is not meagre.

We shall prove a statement that looks much stronger than the theorem above.

THEOREM. (Generalization of the category theorem) *Let Y be a G_δ -subset of a complete metric space X. Then Y is a Baire space.*
Proof. Let $U_1, U_2, \ldots$ be subsets of Y that are open in Y and dense in Y. We prove that $\bigcap_{n=1}^{\infty} U_n$ is dense in Y. Let $Z := \bar{Y}$. Then Z is complete, hence a Baire space. The sets $U_1, U_2, \ldots$ are also dense in Z and G_δ -subsets of Z by the lemma below. $\bigcap_{n=1}^{\infty} U_n$ is dense in Z, hence in Y.

LEMMA. *Let $P \subset Q \subset R$ be subsets of a metric space. If P is a G_δ -subset of Q and if Q is a G_δ -subset of R then P is a G_δ -subset of R.*
Proof. There are subsets $U_1, U_2, \ldots$, open in Q, such that $P = \bigcap_{n=1}^{\infty} U_n$. There are subsets $V_1, V_2, \ldots$, open in R, such that $Q = \bigcap_{n=1}^{\infty} V_n$. For each $n \in \mathbb{N}$ there is a subset U_n', open in R, such that $U_n' \cap Q = U_n$. Then $P = \bigcap_{n=1}^{\infty} U_n' \cap Q = \bigcap_{n=1}^{\infty} \bigcap_{m=1}^{\infty} U_n' \cap V_m$ is a G_δ-subset of R.

A.2. Functions of the first class of Baire

IN THIS SECTION X IS A SUBSET OF K

DEFINITION. A function $f : X \to K$ is of the *first class of Baire* if there are continuous functions $f_1, f_2, \ldots : X \to K$ such that $\lim_{n\to\infty} f_n = f$ (pointwise). The set of functions $X \to K$ of the first class of Baire is denoted $B^1(X \to K)$.

This section contains some (painless) translations of 'classical' results.

**Exercise.* Under pointwise operations the space $B^1(X \to K)$ is a K-vector space (even a K-algebra) containing $C(X \to K)$.

**Exercise.* $\xi_{\{0\}} \in B^1(K \to K)$. More generally, a function $X \to K$ with only finitely many discontinuities is in $B^1(X \to K)$.

THEOREM. *$B^1(X \to K)$ is uniformly closed, i.e. if $f_1, f_2, \ldots \in B^1(X \to K)$ and $f = \lim_{n\to\infty} f_n$ uniformly then $f \in B^1(X \to K)$.*

Proof. We may assume that f and all f_n are bounded. Set $h_n := f_{n+1} - f_n$ ($n \in \mathbb{N}$). Then $f = f_1 + \sum_{n=1}^{\infty} h_n$ where $\lim_{n\to\infty} \|h_n\|_\infty = 0$. It suffices to show that $\sum_{n=1}^{\infty} h_n \in B^1(X \to K)$. Since $h_n \in B^1(X \to K)$ there is a sequence $s_{n1}, s_{n2}, \ldots$ of continuous functions converging to h_n pointwise. We may suppose that $\|s_{nj}\|_\infty \leqslant \|h_n\|_\infty$ for all j (define $\tilde{s}_{nj}(x) := s_{nj}(x)$ if $|s_{nj}(x)| \leqslant \|h_n\|_\infty$, $\tilde{s}_{nj}(x) := 0$ if $|s_{nj}(x)| > \|h_n\|_\infty$ and take $\tilde{s}_{nj}$ instead of s_{nj}). Then for each j the series $\sum_n s_{nj}$ is uniformly convergent so that $t_j := \sum_{n=1}^{\infty} s_{nj}$ is continuous for each j. We shall prove that $\lim_{j\to\infty} t_j = \sum_{n=1}^{\infty} h_n$ (pointwise). Let $x \in X$, $\epsilon > 0$. There is an N such that $\|h_n\|_\infty < \epsilon$ for $n \geqslant N$. Then also $|s_{nj}(x) - h_n(x)| < \epsilon$ for all j and $n \geqslant N$. For $n = 1, 2, \ldots, N-1$ we have $\lim_{j\to\infty} s_{nj}(x) = h_n(x)$ so that

$$\left| t_j(x) - \sum_{n=1}^{\infty} h_n(x) \right| = \left| \sum_{n=1}^{\infty} (s_{nj}(x) - h_n(x)) \right| \leqslant \max_{1 \leqslant n < N} |s_{nj}(x) - h_n(x)| \vee \epsilon < \epsilon$$

for sufficiently large j.

We now prove that a function $X \to K$ of the first class of Baire must have points of continuity (at least if X is a G_δ-subset of K). First we have a lemma that has nothing to do with $B^1(X \to K)$.

LEMMA. *Let $f : X \to K$ be any function. Then the points of continuity of f form a G_δ-subset of X.*

Proof. Define the oscillation function $\omega : X \to [0, \infty]$ by the formula

$$\omega(a) = \lim_{n\to\infty} \sup \{|f(x)-f(y)| : x, y \in B_a(1/n) \cap X\}$$

For each $r > 0$ the set $\{x \in X : \omega(x) < r\}$ is open in X. Hence $\{x \in X : \omega(x) = 0\} = \bigcap_{n=1}^{\infty} \{x \in X : \omega(x) < 1/n\}$ is a G_δ-set. Further, f is continuous at a if and only if $\omega(a) = 0$, which finishes the proof.

THEOREM. (Points of continuity for a function of the first class of Baire) *Let X be a G_δ-subset of K, let $f \in B^1(X \to K)$. Then the points of continuity of f form a dense G_δ-subset of X.*

Proof. Let ω be as in the previous lemma. X is a Baire space so by Appendix A.1 it suffices to show that for each $r > 0$ the set $\{x \in X : \omega(x) < r\}$ is dense. By restricting the problem to relatively open subsets of X we see that a proof of $\{x \in X : \omega(x) < r\} \neq \emptyset$ will do. Now let $f_1, f_2, \ldots$ be continuous functions on X for which $\lim_{n\to\infty} f_n = f$. For each n the set $E_n := \{x \in X : |f_j(x) - f_m(x)| > r/2$ for some $j, m \geqslant n\}$ is open in X. Also we have $\bigcap_{n=1}^{\infty} E_n = \emptyset$. By the category theorem E_n is not dense in X for some n so there is a nonempty open subset U of X with $U \cap E_n = \emptyset$. Hence $x \in U$ implies $|f_j(x) - f_m(x)| \leqslant r/2$ for all $j, m \geqslant n$. By choosing $m = n$ and letting $j \to \infty$ we arrive at $|f(x) - f_n(x)| \leqslant r/2$ for all $x \in U$. Let $a \in U$. By continuity there is a relatively open V, $a \in V \subset U$ such that $|f_n(x) - f_n(y)| \leqslant r/2$ for all $x, y \in V$. The inequality

$$|f(x)-f(y)| \leqslant |f(x)-f_n(x)| \vee |f_n(x)-f_n(y)| \vee |f_n(y)-f(y)| \qquad (x, y \in V)$$

yields $|f(x) - f(y)| \leqslant r/2$ for all $x, y \in V$. Hence, $\omega(x) \leqslant r/2 < r$ for all $x \in V$. In particular, $\omega(a) < r$ and we are done.

A. 3. Orthonormal bases of $C(X \to K)$

In Section 62 we met an orthonormal base of $C(\mathbb{Z}_p \to K)$ consisting of characteristic functions of clopen sets. We now generalize this result.

LEMMA. *Let X be a compact ultrametric space, let $B_1, B_2, \ldots$ be balls in X.*
(i) If $\xi_{B_1}, \xi_{B_2}, \ldots$ are linearly independent over K then $\{\xi_{B_1}, \xi_{B_2}, \ldots\}$ is an orthonormal set in $(C(X \to K), \|\ \|_\infty)$.
(ii) If $\{\xi_{B_1}, \xi_{B_2}, \ldots\}$ is a maximal linearly independent set of characteristic functions of balls then $\xi_{B_1}, \xi_{B_2}, \ldots$ is an orthonormal base of $(C(X \to K), \|\ \|_\infty)$.

Proof (i). We may assume that $d(B_1) \geqslant d(B_2) \geqslant \ldots$. Let $\lambda_1, \ldots, \lambda_n \in K (n \in \mathbb{N})$. By Proposition 50.4(iii) it suffices to prove

$$(*) \qquad \left\| \sum_{j=m}^{n} \lambda_j \xi_{B_j} \right\|_\infty \geq |\lambda_m|$$

for $m \in \{1, 2, \ldots, n-1\}$. Let $m = 1$. We shall find an $x \in B_1$ that is not in any of the other balls $B_2, B_3, \ldots$ (then $\| \Sigma_{j=1}^{n} \lambda_j \xi_{B_j} \|_\infty \geq | \Sigma_{j=1}^{n} \lambda_j \xi_{B_j}(x)| = |\lambda_1|$ and (*) is proved for $m = 1$). Suppose that B_1 could be covered by $B_2, B_3, \ldots$ For each j we have either $B_j \subset B_1$ or $B_j \cap B_1 = \emptyset$ (since $d(B_j) \leq d(B_1)$). Using this and compactness we find that B_1 is a disjoint union of finitely many balls $\in \{B_2, B_3, \ldots\}$ Thus, ξ_{B_1} can be written as a finite sum of certain ξ_{B_j} $(j \neq 1)$, which contradicts the linear independence. To prove (*) for $m = 2$ we simply repeat the above reasoning for B_2, $B_3, \ldots$ in place of $B_1, B_2, \ldots$ Et cetera.
(ii) Let $E := [\![\xi_{B_1}, \xi_{B_2}, \ldots]\!]$. By maximality $\xi_B \in E$ for each ball $B \subset X$. A clopen set $U \subset X$ is a disjoint union of balls, hence $\xi_U \in E$. Every locally constant function is in E. Thus, E is dense in $C(X \to K)$. Now apply (i) and Theorem 50.7.

THEOREM. *Let X be a compact ultrametric space. Then $C(X \to K)$ has an orthonormal base consisting of characteristic functions of balls. In fact, let $B_1, B_2, \ldots$ be an enumeration of the collection of balls in X* (Theorem 19.3). *Then one obtains an orthonormal base of $C(X \to K)$ by cancelling in $\xi_{B_1}, \xi_{B_2}, \ldots$ those ξ_{B_n} that are in $[\![\xi_{B_1}, \ldots, \xi_{B_{n-1}}]\!]$.*

COROLLARY. *Let X, Y, be infinite compact ultrametric spaces. Then $C(X \to K)$ and $C(Y \to K)$ are isomorphic as K-Banach spaces (to c_0).*

A. 4. The ultrametric Stone-Weierstrass theorem

Kaplansky's theorem (Theorem 43.3) admits the following generalization.

THEOREM. (Ultrametric Stone-Weierstrass theorem) *Let X be a compact ultrametric space, let $A \subset C(X \to K)$ have the following properties.*
(i) *A is a K-linear subspace of $C(X \to K)$ and closed for products.*
(ii) *A contains the constant functions.*
(iii) *For each $a, b \in X$, $a \neq b$ there is an $f \in A$ such that $f(a) \neq f(b)$.*
(iv) *A is uniformly closed.*
Then $A = C(X \to K)$.
(Observe that if we let X be a compact subset of K and choose 'the uniform closure of the polynomial functions' for A we obtain Kaplansky's theorem as a special case. It should be noticed however that we use Kaplansky's result in the proof below.)

Proof. Let $\Omega(X)$ be the collection of open compact subsets of X, and let $\Omega' := \{U \in \Omega(X) : \xi_U \in A\}$. It suffices to show that $\Omega' = \Omega(X)$. It follows from (i) and (ii) that finite intersections and unions of elements of Ω' are in Ω' and that Ω' is closed for complementation. Now let $a, b \in X$, $a \neq b$. We use (iii) to show that there is $U \in \Omega'$ such that $a \in U$, $b \notin U$ as follows. Let $f \in A$ be such that $f(a) \neq f(b)$. Then $V := \{t \in f(X) : |t - f(a)| < |f(b) - f(a)|\}$ is a clopen subset of the compact set $f(X) \subset K$. By Kaplansky's theorem there is a sequence $P_1, P_2, \ldots$ of polynomial functions such that $P_n \to \xi_V$ uniformly on $f(X)$. By (i) and (ii) $P_n \circ f \in A$ for each n and $P_n \circ f \to \xi_U$ uniformly, where $U = f^{-1}(V)$. Hence $\xi_U \in A$, i.e. $U \in \Omega'$. Clearly $a \in U$, $b \notin U$. Now let $T \in \Omega(X)$; we prove that $T \in \Omega'$. Let $y \in X \backslash T$. For each $x \in T$ there is $U_x \in \Omega'$ such that $x \in U_x$, $y \notin U_x$ A compactness argument yields the existence of $U^y \supset T$ such that $U^y \in \Omega'$, $y \notin U^y$. Now the complements of the sets U^y are in $X \backslash T$ and cover $X \backslash T$. Hence there are $y_1, \ldots, y_n \in X \backslash T$ such that $\bigcup_{i=1}^n (X \backslash U^{y_i}) = X \backslash T$. In other words $\bigcap_{i=1}^n U^{y_i} = T$. It follows that $T \in \Omega'$.

COROLLARY. *Let X and Y be compact ultrametric spaces. Then each $f \in C(X \times Y \to K)$ can uniformly be approximated by functions of the form $(x, y) \mapsto \sum_{j=1}^n f_j(x) g_j(y)$ $((x, y) \in X \times Y)$ where $n \in \mathbb{N}$, $f_j \in C(X \to K)$, $g_j \in C(Y \to K)$ $(1 \leqslant j \leqslant n)$.*

Exercise. A *character* of $\mathbb{Z}_p$ is a continuous function $a : \mathbb{Z}_p \to \mathbb{C}_p^\times$ for which $a(x + y) = a(x)\, a(y)$ for all $x, y \in \mathbb{Z}_p$. Prove that each continuous function $f : \mathbb{Z}_p \to \mathbb{C}_p$ can uniformly be approximated by $\mathbb{C}_p$-linear combinations of characters.

A. 5. Integration on compact spaces

In this section we shall sketch an ultrametric integration theory on compact spaces. For a wide generalization, see van Rooij (1978).

In the classical set up of integration on (locally) compact spaces one encounters two approaches both leading eventually to the same theory. One way is to start with a σ-additive measure defined on the collection of the Borel sets. The other way is to take an integral (= a positive linear function) on the space of continuous functions as a fundamental notion. We shall follow the second approach, but first we explain why in the p-adic theory the first approach leads to trivialities.

Let $\mathscr{B}$ be the collection of the Borel sets of $\mathbb{Z}_p$ (i.e. $\mathscr{B}$ is the smallest collection of subsets of $\mathbb{Z}_p$ that is closed for countable unions and intersections and that contains all open subsets of $\mathbb{Z}_p$). A *σ-additive measure on* $\mathbb{Z}_p$ is a map $\mu : \mathscr{B} \to \mathbb{Q}_p$ such that

$$\mu\Big(\bigcup_{j=1}^{\infty} B_j\Big) = \sum_{j=1}^{\infty} \mu(B_j)$$

whenever $B_1, B_2, \ldots$ are mutually disjoint Borel sets.

A trivial example is the *Dirac measure* δ_x at some point $x \in \mathbb{Z}_p$ defined by $\delta_x(B) := 1$ if $x \in B$, $\delta_x(B) := 0$ if $x \in \mathbb{Z}_p \backslash B$ for all $B \in \mathscr{B}$. If $x_1, x_2, \ldots \in \mathbb{Z}_p$ and $\lambda_1, \lambda_2, \ldots \in \mathbb{Q}_p$ such that $\lim_{n\to\infty} \lambda_n = 0$ we formally define $\sum_{n=1}^{\infty} \lambda_n \delta_{x_n}$ by the formula

$$\Big(\sum_{n=1}^{\infty} \lambda_n \delta_{x_n}\Big)(B) := \sum_{n=1}^{\infty} \lambda_n \delta_{x_n}(B) = \sum_{x_n \in B} \lambda_n$$

for all $B \in \mathscr{B}$. One easily checks that such 'convergent linear combinations of Dirac measures' are again σ-additive measures on $\mathbb{Z}_p$. We have the following surprising result.

THEOREM. *Each σ-additive measure on $\mathbb{Z}_p$ is a convergent linear combination of Dirac measures. More precisely, if μ is a σ-additive measure on $\mathbb{Z}_p$ then there are $x_1, x_2, \ldots \in \mathbb{Z}_p$ and $\lambda_1, \lambda_2, \ldots \in \mathbb{Q}_p$ such that* $\lim_{n\to\infty} \lambda_n =$ 0 *and* $\mu = \sum_{n=1}^{\infty} \lambda_n \delta_{x_n}$.

Proof. For each $r > 0$ the set $\{x \in \mathbb{Z}_p : |\mu(\{x\})|_p \geqslant r\}$ is finite. (If not, take mutually distinct p-adic integers $x_1, x_2, \ldots$ for which $|\mu(\{x_n\})|_p \geqslant r$. Then $\mu(\{x_1, x_2, \ldots\})$ must be equal to $\sum_{n=1}^{\infty} \mu(\{x_n\})$ which is meaningless.) It follows that $\{x \in \mathbb{Z}_p : \mu(\{x\}) \neq 0\}$ is at most countable; extend it to an infinite set with enumeration $x_1, x_2, \ldots$, say. Set $\lambda_n := \mu(\{x_n\})$ for all n. By considering $\mu(\{x_1, x_2, \ldots\}) = \sum_{n=1}^{\infty} \lambda_n$ we see that $\lim_{n\to\infty} \lambda_n = 0$. The measure $\mu_1 := \sum_{n=1}^{\infty} \lambda_n \delta_{x_n}$ is well defined and so is $\nu := \mu - \mu_1$. The latter has the property $\nu(\{x\}) = 0$ for all $x \in \mathbb{Z}_p$. So we shall be done once we have proved the following lemma.

LEMMA. *Let μ be a σ-additive measure on $\mathbb{Z}_p$ such that $\mu(\{a\}) = 0$ for all $a \in \mathbb{Z}_p$. Then $\mu = 0$.*

Proof. Let A be a Borel set, let $\epsilon > 0$. We shall prove that $|\mu(A)|_p \leqslant \epsilon$. If $B_1, \ldots, B_n$ is a partition of $\mathbb{Z}_p$ into balls and if $|\mu(A \cap B_j)|_p \leqslant \epsilon$ for each j then $|\mu(A)|_p = |\sum_{j=1}^{n} \mu(A \cap B_j)|_p \leqslant \epsilon$. Therefore it suffices to find, for each $a \in \mathbb{Z}_p$, a ball $B_a \ni a$ for which $|\mu(A \cap B_a)|_p \leqslant \epsilon$. Thus, let $a \in \mathbb{Z}_p$. Set $R_n := \{x \in \mathbb{Z}_p : |x - a|_p = p^{-n}\}$ $(n \in \mathbb{N} \cup \{0\})$. Then $\{a\}, R_0, R_1, \ldots$ is a disjoint covering of $\mathbb{Z}_p$ so that $A \cap \{a\}, A \cap R_0, A \cap R_1, \ldots$ is a partition of A into Borel sets. We have $\mu(A) = \sum_{n=0}^{\infty} \mu(A \cap R_n)$ hence

$\lim_{n\to\infty} \mu(A \cap R_n) = 0$. There is an m such that $|\mu(A \cap R_n)|_p \leqslant \epsilon$ for $n \geqslant m$. Now choose $B_a := B_a(p^{-m})$. We have

$$B_a = \{a\} \cup R_m \cup R_{m+1} \cup \ldots$$

so that

$$|\mu(A \cap B_a)|_p = \left| 0 + \sum_{n \geqslant m} \mu(A \cap R_n) \right|_p \leqslant \epsilon$$

which is what we intended to prove.

In the rest of this section X is a compact ultrametric space. For notations and terminology, see Section 13.

DEFINITION. An *integral* on $C(X \to K)$ is an element of the dual space $C(X \to K)'$ of $C(X \to K)$. A K-valued *measure* on X is a function $\mu : \Omega(X) \to K$, where $\Omega(X)$ denotes the collection of open compact subsets of X, such that
(i) if $U, V \in \Omega(X)$, $U \cap V = \emptyset$ then $\mu(U \cup V) = \mu(U) + \mu(V)$ (*additivity*),
(ii) $\|\mu\| := \sup\{|\mu(V)| : V \in \Omega(X)\} < \infty$ (*boundedness*).

The K-valued measures on X form a normed vector space $M(X \to K)$ under the obvious operations and with the norm $\| \quad \|$ defined in (ii).

The following statement can be viewed as the ultrametric analogue of Riesz' representation theorem (although the latter is a much deeper theorem).

THEOREM. *For each* $\varphi \in C(X \to K)'$

$$\mu_\varphi : U \mapsto \varphi(\xi_U) \qquad (U \in \Omega(X))$$

is a measure. The map $\varphi \mapsto \mu_\varphi$ *is a K-linear isometry of* $C(X \to K)'$ *onto* $M(X \to K)$.

Proof. It is clear that μ_φ is a measure and that $\varphi \mapsto \mu_\varphi$ is K-linear. Also we have $\|\mu_\varphi\| = \sup\{|\varphi(\xi_U)| : U \in \Omega(X)\} \leqslant \|\varphi\|$. To prove the opposite inequality, let $f \in C(X \to K)$ be locally constant, $\|f\|_\infty \leqslant 1$. There exist mutually disjoint $U_1, \ldots, U_n \in \Omega(X)$ and $\lambda_1, \ldots, \lambda_n \in K$ such that

$$f = \sum_{j=1}^{n} \lambda_j \xi_{U_j}, \qquad \max_{1 \leqslant j \leqslant n} |\lambda_j| \leqslant 1$$

Hence, $|\varphi(f)| \leqslant \max\{|\lambda_j|\,|\varphi(\xi_{U_j})| : 1 \leqslant j \leqslant n\} \leqslant \max\{|\mu_\varphi(U_j)| : 1 \leqslant j \leqslant n\} \leqslant \|\mu_\varphi\|$. It follows readily that $|\varphi(f)| \leqslant \|\mu_\varphi\|$ for all $f \in C(X \to K)$, $\|f\|_\infty \leqslant 1$ so that $\|\varphi\| \leqslant \|\mu_\varphi\|$. Thus, $\varphi \mapsto \mu_\varphi$ is an isometry; we prove surjectivity. Let $\mu \in M(X \to K)$. To find $\varphi \in C(X \to K)'$ such that $\mu_\varphi = \mu$, let $f : X \to K$ be locally constant. Then there is a partition $U_1, \ldots, U_n$ of X into clopen sets such that for every choice of $a_j \in U_j$

$$f = \sum_{j=1}^{n} f(a_j)\, \xi_{U_j}$$

Now set

$$(*) \qquad \varphi(f) := \sum_{j=1}^{n} f(a_j)\, \mu(U_j)$$

To see that this definition makes sense, let also $f = \sum_{j=1}^{m} f(b_j)\, \xi_{V_j}$ ($b_j \in V_j$; $V_1, \ldots, V_m$ is a clopen partition of X) and consider a refinement $W_1, \ldots, W_k$ of both $U_1, \ldots, U_n$ and $V_1, \ldots, V_m$. Choose $w_j \in W_j$ for each j. Then

$$\sum_{j=1}^{n} f(a_j)\, \mu(U_j) = \sum_{j=1}^{k} f(w_j)\, \mu(W_j) = \sum_{j=1}^{m} f(b_j)\, \mu(V_j)$$

(A similar device gives us the K-linearity of φ on the space of the locally constant functions.) For our locally constant function f we have by (*)

$$|\varphi(f)| = \left| \sum_{j=1}^{n} f(a_j)\, \mu(U_j) \right| \leqslant \|f\|_\infty \max_{1 \leqslant j \leqslant n} |\mu(U_j)| \leqslant \|f\|_\infty \, \|\mu\|$$

Hence φ can be extended to an element (again called φ) of $C(X \to K)'$. By (*) we have trivially $\varphi(\xi_U) = \mu(U)$ ($U \in \Omega(X)$) and the theorem is proved.

Exercise. Find a measure on $\mathbb{Z}_p$ that is not a convergent linear combination of Dirac measures. (Hint. Consider the integral $\sum_{n=0}^{\infty} a_n e_n \mapsto \sum_{n=0}^{\infty} a_n$ where $e_0, e_1, \ldots$ is van der Put's base.)

Exercise. (Description of the space of all integrals) Let X be infinite and let $e_1, e_2, \ldots$ be an orthonormal base (see Appendix A. 3) of $C(X \to K)$. Show that each integral on $C(X \to K)$ has the form

$$(*) \qquad \sum_{n=1}^{\infty} \lambda_n e_n \mapsto \sum_{n=1}^{\infty} \lambda_n \xi_n \qquad ((\lambda_1, \lambda_2, \ldots) \in c_0)$$

for some bounded sequence $\xi_1, \xi_2, \ldots$ in K and that, conversely, for each $(\xi_1, \xi_2, \ldots) \in l^\infty$ formula (*) defines an integral on $C(X \to K)$.

Exercise. Let μ be a measure on X and denote its corresponding integral by $f \mapsto \int_X f(x)\, d\,\mu(x)$. This notation suggests that 'f is integrated with respect to the measure μ'. Show that this is indeed true in the following sense. Let $f \in C(X \to K)$ and $\epsilon > 0$. Then there is a $\delta > 0$ such that for each partition $U_1, \ldots, U_n$ of X into clopen sets for which $\max_j d(U_j) < \delta$ and every choice of $a_j \in U_j$ ($1 \leqslant j \leqslant n$) we have $|\int_X f(x)\, d\,\mu(x) - \sum_{j=1}^{n} f(a_j)\, \mu(U_j)| < \epsilon$. (Compare Exercise 55.D.)

Exercise. (Fubini theorem) Let μ, ν be measures on the compact ultrametric spaces X and Y respectively. Let $f: X \times Y \to K$ be continuous. Use Appendix A.4 to prove

$$\int_X \int_Y f(x, y)\, \mathrm{d}\mu(x)\, \mathrm{d}\nu(y) = \int_X \int_Y f(x, y)\, \mathrm{d}\nu(y)\, \mathrm{d}\mu(x)$$

(notation as in the previous exercise).

Exercise. Let $z \in \mathbb{C}_p \setminus \mathbb{C}_p^+$. For $n \in \{0, 1, 2, \ldots\}$ define

$$\mu_z(a + p^n \mathbb{Z}_p) = \frac{z^a}{1 - z^{p^n}} \qquad (a \in \{0, 1, \ldots, p^n - 1\})$$

Show that μ_z extends uniquely to a $\mathbb{C}_p$-valued measure on $\mathbb{Z}_p$ and that $\|\mu_z\| \leqslant 1$ (such measures are used in Koblitz (1980) to define the p-adic zeta functions).

The second half of this section is devoted to extending an integral on $C(X \to K)$ to a larger class of ('μ-integrable') functions. To this end we shall introduce a seminorm $\|\ \ \|_\mu$ replacing $f \mapsto \mu(|f|)$ of the real theory and declare a function to be μ-integrable if there are $f_1, f_2, \ldots \in C(X \to K)$ for which $\lim_{n\to\infty} \|f - f_n\|_\mu = 0$.

Let μ be a K-valued measure on X and let its corresponding integral also be denoted by μ. We define the seminorm $\|\ \ \|_\mu$ on $C(X \to K)$ by

$$\|f\|_\mu := \sup \left\{ \frac{|\mu(fg)|}{\|g\|_\infty} : g \in C(X \to K), g \neq 0 \right\} \qquad (f \in C(X \to K))$$

(It is not hard to see that a similar definition for the case of a real valued integral μ on $C(X \to \mathbb{R})$ leads to $\|f\|_\mu = \mu(|f|)$.) Observe that $\|f\|_\mu$ is the norm of the linear function $g \mapsto \mu(fg)$ and that $|\mu(f)| \leqslant \|f\|_\mu \leqslant \|\mu\|\, \|f\|_\infty$ $(f \in C(X \to K))$.

THEOREM. *There is an upper semicontinuous function $N_\mu : X \to [0, \infty)$ such that*

$$(*) \qquad \|f\|_\mu = \sup_{x \in X} |f(x)| N_\mu(x) \qquad (f \in C(X \to K))$$

In fact, we may choose

$$N_\mu(x) := \inf \{\|\xi_U\|_\mu : x \in U \in \Omega(X)\} \qquad (x \in X)$$

Proof. We first prove the formula

$$\|\xi_U\|_\mu = \sup \{|\mu(V)| : V \subset U, V \in \Omega(X)\}$$

for each $U \in \Omega(X)$. Let $V \in \Omega(X)$, $V \subset U$. Then $|\mu(V)| = |\mu(\xi_U \xi_V)| \leqslant \|\xi_U\|_\mu \|\xi_V\|_\infty = \|\xi_U\|_\mu$. It follows that

$$\|\xi_U\|_\mu \geqslant \sup \{|\mu(V)| : V \subset U, V \in \Omega(X)\}$$

To prove the opposite inequality, let g be a locally constant function, $\|g\|_\infty \leqslant 1$. Write g as a finite sum

$$g = \sum_{j=1}^{n} \lambda_j \xi_{T_j}$$

where $|\lambda_j| \leqslant 1$ for all j and where $T_1, \ldots, T_n$ is a clopen partition of X. We have $|\mu(\xi_U g)| = |\mu(\sum_{j=1}^{n} \lambda_j \xi_{T_j \cap U})| = |\sum_{j=1}^{n} \lambda_j \mu(T_j \cap U)| \leqslant \max_{1 \leqslant j \leqslant n} |\mu(T_j \cap U)| \leqslant \sup \{|\mu(V)| : V \subset U, V \in \Omega(X)\}$. By taking the supremum over all locally constant g with $\|g\|_\infty \leqslant 1$ we arrive at

$$\|\xi_U\|_\mu \leqslant \sup \{|\mu(V)| : V \subset U, V \in \Omega(X)\}$$

and the announced formula is proved. Now define N_μ as above. Clearly $N_\mu : X \to [0, \infty)$, N_μ is bounded so that the formula

$$\|f\|_\mu' := \sup_{x \in X} |f(x)| N_\mu(x)$$

defines a seminorm on $C(X \to K)$. Both $\| \ \|_\mu'$ and $\| \ \|_\mu$ are continuous with respect to the norm $\| \ \|_\infty$ so to prove (*) it suffices to check that $\|f\|_\mu' = \|f\|_\mu$ for locally constant functions f. First, let f be the characteristic function of a clopen set U. By the definition of N_μ we have $\|\xi_U\|' = \sup_{x \in U} N_\mu(x) \leqslant \|\xi_U\|_\mu$. We proceed to prove that $\|\xi_U\|_\mu \leqslant \|\xi_U\|'$, i.e. that

$$\|\xi_U\|_\mu \leqslant \sup_{x \in U} N_\mu(x)$$

Let $\epsilon > 0$. For each $x \in U$ we have for sufficiently small clopen neighbourhoods $V \subset U$ of x that

$$\|\xi_V\|_\mu \leqslant N_\mu(x) + \epsilon$$

(Observe that $S, T \in \Omega(X)$, $S \subset T$ implies $\|\xi_S\|_\mu \leqslant \|\xi_T\|_\mu$, which follows from the formula we have proved at the beginning of this proof.) A standard compactness argument yields the existence of $x_1, \ldots, x_n \in U$ and a partition $V_1, \ldots, V_n$ of U into clopen sets such that $x_j \in V_j$ for each j and

$$\|\xi_{V_j}\|_\mu \leqslant N_\mu(x_j) + \epsilon \qquad (j \in \{1, \ldots, n\})$$

Then $\|\xi_U\|_\mu \leqslant \max_{1 \leqslant j \leqslant n} \|\xi_{V_j}\|_\mu \leqslant \max_{1 \leqslant j \leqslant n} N_\mu(x_j) + \epsilon \leqslant \sup_{x \in U} N_\mu(x) + \epsilon$ which finishes the proof of $\|f\|_\mu' = \|f\|_\mu$ for f a characteristic function. Now let f be a locally constant function, i.e.

$$f = \sum_{j=1}^{n} \lambda_j \xi_{U_j}$$

where $\lambda_1, \ldots, \lambda_n \in K$ and $U_1, \ldots, U_n$ is a clopen partition of X. We have $\|f\|_\mu \leqslant \max_{1 \leqslant j \leqslant n} |\lambda_j| \, \|\xi_{U_j}\|_\mu = \max_{1 \leqslant j \leqslant n} |\lambda_j| \, \|\xi_U\|'_\mu = \sup_{x \in X} |f(x)| \, N_\mu(x) = \|f\|'_\mu$. On the other hand, we have for each j, $V \subset U_j$, $V \in \Omega(X)$

$$\|f\|_\mu \geqslant |\mu(f\xi_V)| = |\lambda_j| \, |\mu(V)|$$

Thus, by the formula at the beginning of this proof,

$$\|f\|_\mu \geqslant |\lambda_j| \sup \{|\mu(V)| : V \subset U_j, V \in \Omega(X)\} = |\lambda_j| \, \|\xi_{U_j}\|_\mu$$

so that

$$\|f\|_\mu \geqslant \max_{1 \leqslant j \leqslant n} |\lambda_j| \, \|\xi_{U_j}\|'_\mu = \|f\|'_\mu$$

This completes the proof of (*). Finally we prove that N_μ is upper semicontinuous. Let $a \in X$, $\epsilon > 0$. There is a clopen neighbourhood U of a such that $\|\xi_U\|_\mu \leqslant N_\mu(a) + \epsilon$. For all $x \in U$ we have $N_\mu(x) \leqslant \|\xi_U\|_\mu \leqslant N_\mu(a) + \epsilon$.

Formula (*) of the above theorem enables us to define $\|f\|_\mu$ for *any* function $X \to K$.

DEFINITION. Let μ be an integral on X. For any $f : X \to K$ set

$$\|f\|_\mu := \sup_{x \in X} |f(x)| \, N_\mu(x)$$

$$E_\mu := \{f : X \to K : \|f\|_\mu < \infty\}$$

f is *μ-negligible* if $\|f\|_\mu = 0$. A subset A of X is *μ-negligible* if $\|\xi_A\|_\mu = 0$. Define

$$X_+ := \{x \in X : N_\mu(x) > 0\}$$

$$X_0 := \{x \in X : N_\mu(x) = 0\}$$

The following statements are easy to prove.

THEOREM. *The seminorm* $\| \;\; \|_\mu$ *induces a Banach space norm on* $E_\mu / \{f : \|f\|_\mu = 0\}$. X_0 *is a* G_δ *-set. A subset of* X *is* μ*-negligible if and only if it is contained in* X_0. X_0 *is the largest* μ*-negligible set. A function* $X \to K$ *is* μ*-negligible if and only if it vanishes on* X_+.

We now define μ-integrability.

DEFINITION. A function $f : X \to K$ is *μ-integrable* if there exists a sequence $f_1, f_2, \ldots$ in $C(X \to K)$ such that $\lim_{n \to \infty} \|f - f_n\|_\mu = 0$. Set

$$\mathscr{L}^1(X, \mu) := \{f : X \to K : f \text{ is } \mu\text{-integrable}\}$$

$$L^1(X, \mu) := \mathscr{L}^1(X, \mu) / \sim$$

where $\sim$ is the equivalence relation defined by

$$f \sim g \text{ if } f - g \text{ is } \mu\text{-negligible}$$

We have the following immediate consequences.

THEOREM. *$\mathscr{L}^1(X, \mu)$ is a linear space containing $C(X \to K)$ and the set of all μ-negligible functions $X \to K$. The integral μ on $C(X \to K)$ extends uniquely to a K-linear function $\bar{\mu}$ on $\mathscr{L}^1(X, \mu)$ for which $|\bar{\mu}(f)| \leqslant \|f\|_\mu$ for all $f \in \mathscr{L}^1(X, \mu)$. $L^1(X, \mu) = \mathscr{L}^1(X, \mu)/\{f : X \to K : \|f\|_\mu = 0\}$ is a K-Banach space with respect to the norm induced by $\|\ \ \|_\mu$.*

Exercise. Show that for the $\mathbb{Q}_q$-valued Haar integral μ on $C(\mathbb{Z}_p \to \mathbb{Q}_q)$, introduced in Exercise 62.G for $q \neq p$, we have $N_\mu(x) = 1$ for all $x \in \mathbb{Z}_p$, so that $\|\ \ \|_\mu = \|\ \ \|_\infty$. Thus $L^1(\mathbb{Z}_p, \mu) = C(\mathbb{Z}_p \to \mathbb{Q}_q)$, there are no μ-negligible sets other than the empty set.

Remark. For further theory (μ-measurability, convergence theorems, Fubini theorem, more general measure spaces, . . .) we refer to van Rooij (1978).

A.6. Measures and distributions on $\mathbb{Z}_p$

THROUGHOUT A.6 WE ASSUME $K \supset \mathbb{Q}_p$

In Section 55 we have met the Volkenborn integral, which is not an integral in the sense of Appendix A.5 since it is an element of $C^1(\mathbb{Z}_p \to K)'$ rather than $C(\mathbb{Z}_p \to K)'$. In this section we shall describe the dual space of $C^n(\mathbb{Z}_p \to K)$ for $n \in \{0, 1, 2, \ldots\}$. For $\mu \in C^n(\mathbb{Z}_p \to K)', f \in C^n(\mathbb{Z}_p \to K)$ we sometimes write $\int f(x) \mathrm{d}\mu(x)$ instead of $\mu(f)$.

First we consider the case $n = 0$. The fact that $\mathbb{Z}_p$ is a group enables us to define a convolution multiplication in $C(\mathbb{Z}_p \to K)'$.

THEOREM. *The formula*

$$(\mu * \nu)(f) = \int \left(\int f(x+y)\, \mathrm{d}\mu(x)\right) \mathrm{d}\nu(y) \quad (\mu, \nu \in C(\mathbb{Z}_p \to K)')$$

defines a multiplication (convolution) in $C(\mathbb{Z}_p \to K)'$ making it into a commutative K-Banach algebra with identity $\delta_0 : f \mapsto f(0)$.

Proof. A straightforward argument yields continuity of $h : y \mapsto \int_{\mathbb{Z}_p} f(x+y)\, \mathrm{d}\mu(x)$ so that $(\mu * \nu)(f)$ is well defined. The latter depends linearly on f and $|(\mu * \nu)(f)| \leqslant \|\nu\|\, \|h\|_\infty \leqslant \|\nu\|\, \|\mu\|\, \|f\|_\infty$. Thus $\mu * \nu \in C(\mathbb{Z}_p \to K)'$ and $\|\mu * \nu\| \leqslant \|\mu\|\, \|\nu\|$. Associativity and commutativity of convolution can be proved, for example by checking the formulas when applied to elements of the Mahler base and then by using linear and continuous extensions.

Henceforth we shall omit unnecessary brackets and write $\int\int f(x+y)\mathrm{d}\mu(x)\,\mathrm{d}\nu(y)$ $(= \int\int f(x+y)\mathrm{d}\nu(y)\mathrm{d}\mu(x))$ instead of $\int(\int\int(x+y)\mathrm{d}\mu(x))\mathrm{d}\nu(y)$. Also we write $\mu\binom{*}{n}$ instead of $\mu(\binom{*}{n})$.
The map

$$\mu \mapsto (\mu\tbinom{*}{0}, \mu\tbinom{*}{1}, \ldots) \qquad (\mu \in C(\mathbb{Z}_p \to K)')$$

is a bijective linear isometry between $C(\mathbb{Z}_p \to K)'$ and l^∞ (the space of all bounded sequences in K, with the supremum norm). To see how the convolution multiplication looks in terms of l^∞ observe that for $n \in \mathbb{N} \cup \{0\}$

$$(\mu * \nu)\tbinom{*}{n} = \int\int\tbinom{x+y}{n}\,\mathrm{d}\,\mu(x)\mathrm{d}\,\nu(y) = \sum_{j=0}^{n} \mu\tbinom{*}{j}\,\nu\tbinom{*}{n-j}$$

so that convolution induces the following multiplication in l^∞

$$(\xi_0, \xi_1, \ldots)\,(\eta_0, \eta_1, \ldots) = (\xi_0\,\eta_0, \xi_0\,\eta_1 + \xi_1\,\eta_0, \xi_0\,\eta_2 + \xi_1\,\eta_1 + \xi_2\,\eta_0, \ldots)$$

$((\xi_0, \xi_1, \ldots), (\eta_0, \eta_1, \ldots) \in l^\infty)$ which is essentially the multiplication law for the power series $\Sigma\, \xi_n x^n$ and $\Sigma\, \eta_n x^n$. Thus, we define $K\langle X\rangle$ to be the ring of those formal power series (see Appendix B) in the variable X whose coefficients are bounded and norm it by

$$\left|\sum_{j=0}^{\infty} a_j X^j\right| := \sup_j |a_j|$$

The following is evident.

PROPOSITION. *The map*

$$\mu \mapsto \sum_{j=0}^{\infty} \mu\tbinom{*}{j} X^j \qquad (\mu \in C(\mathbb{Z}_p \to K)')$$

is an isometrical isomorphism between the K-Banach algebras $C(\mathbb{Z}_p \to K)'$ *and* $K\langle X\rangle$.

This isomorphism reveals a curious property of $C(\mathbb{Z}_p \to K)'$.

THEOREM. *The norm on* $C(\mathbb{Z}_p \to K)'$ *is multiplicative, i.e.*

$$\|\mu * \nu\| = \|\mu\|\,\|\nu\| \qquad (\mu, \nu \in C(\mathbb{Z}_p \to K)')$$

Proof. We show that the norm $|\ \ |$ on $K\langle X\rangle$ is multiplicative. Being aware that K may have a dense valuation we introduce the norms $|\ \ |_\rho$ $(0 < \rho < 1)$ on $K\langle X\rangle$ by

$$\left| \sum_{j=0}^{\infty} a_j X^j \right|_\rho := \sup_j |a_j| \rho^j = \max_j |a_j| \rho^j$$

We prove that $| \ |_\rho$ is multiplicative. Let $f = \Sigma\, a_j X^j$, $g = \Sigma\, b_j X^j$ be non-zero elements of $K\langle X\rangle$. There are m, n such that

$$|f|_\rho = |a_m| \rho^m \ ; \ |a_i| \rho^i < |f|_\rho \text{ for } i < m$$
$$|g|_\rho = |b_n| \rho^n ; \ |b_j| \rho^j < |g|_\rho \text{ for } j < n.$$

The coefficient c_{m+n} of X^{m+n} in the product fg equals

$$\sum_{i+j=m+n} a_i b_j$$

If $i < m$ then $|a_i b_j| \rho^{i+j} < |a_m| \rho^m \ |g|_\rho = |f|_\rho \ |g|_\rho$. Similarly, if $j < n$ then $|a_i b_j| \rho^{i+j} < |f|_\rho \ |g|_\rho$. Since

$$|a_m b_n| \rho^{n+m} = |f|_\rho \ |g|_\rho$$

we obtain

$$|c_{m+n}| \rho^{n+m} = \max_{i+j=m+n} |a_i b_j| \rho^{n+m} = |f|_\rho \ |g|_\rho$$

It follows that $|fg|_\rho \geq |c_{m+n}| \rho^{n+m} = |f|_\rho \ |g|_\rho$. The opposite inequality is trivial. Now

$$|f| = \lim_{\rho \uparrow 1} |f|_\rho \qquad (f \in K\langle X\rangle)$$

yields the multiplicativity of $| \ |$.

The above theorem implies that there are no idempotent measures on $\mathbb{Z}_p$ other than 0 and the Dirac measure δ_0, a phenomenon that does not have an analogue in classical harmonic analysis.

Next we move to the dual space of $C^1(\mathbb{Z}_p \to K)$. As above we assign to every $\mu \in C^1(\mathbb{Z}_p \to K)'$ the formal power series

$$\sum_{j=0}^{\infty} \mu\binom{*}{j} X^j$$

THEOREM. *The map* $\mu \mapsto \Sigma_{j=0}^{\infty} \mu\binom{*}{j} X^j$ *is a* 1-1 *correspondence between the elements of* $C^1(\mathbb{Z}_p \to K)'$ *and the formal power series* $\Sigma\, b_j X^j$ *for which* $\{|b_j|/j : j \in \mathbb{N}\}$ *is bounded.*

Proof. For $j \geq 1$ we have by Theorem 53.5 (ii) and Lemma 53.3 (ii)

$$|\mu\binom{*}{j}| \leq \|\mu\| \ \|\binom{*}{j}\|_1 = \|\mu\| \ |\gamma_j|_p^{-1} \leq \|\mu\| j$$

where $\|\mu\|$ is the norm of μ in $C^1(\mathbb{Z}_p \to K)'$. Hence $\{|\mu\binom{x}{j}|/j : j \in \mathbb{N}\}$ is bounded by $\|\mu\|$. Conversely, let $b_0, b_1, \ldots$ be a sequence in K for which $\{|b_j|/j : j \in \mathbb{N}\}$ is bounded. We show that there exists a unique $\mu \in C^1(\mathbb{Z}_p \to K)'$ such that $\mu\binom{x}{j} = b_j$ $(j \in \{0, 1, 2, \ldots\})$. Let $f = \sum_{j=0}^{\infty} a_j \binom{x}{j} \in C^1(\mathbb{Z}_p \to K)$. Since by Theorem 53.5 this series converges with respect to the norm $\|\ \ \|_1$ we are forced to define

$$\mu(f) = \sum_{j=0}^{\infty} a_j b_j$$

By Theorem 53.5 we have $\lim_{j\to\infty} |a_j| j = 0$ so that $\lim_{j\to\infty} |a_j b_j| = \lim_{j\to\infty} (|a_j| j)(|b_j|/j) = 0$, hence the series $\Sigma\, a_j b_j$ converges. Further we have

$$|\mu(f)| \leqslant \sup_j (|a_j| j)(|b_j|/j) \leqslant C \|f\|_1$$

for some constant C independent of f. It follows that $\mu \in C^1(\mathbb{Z}_p \to K)'$.

Remark. The power series corresponding to the Volkenborn integral equals

$$\frac{\log(1+X)}{X} = \sum_{j=0}^{\infty} \frac{(-1)^j}{j+1} X^j$$

(see Proposition 55.3).

If we try to define a convolution multiplication in $C^1(\mathbb{Z}_p \to K)'$ we run up against difficulties. If $f \in C^1(\mathbb{Z}_p \to K)$, $\mu \in C^1(\mathbb{Z}_p \to K)'$ then the function $y \mapsto \int f(x+y)\,d\mu(x)$ is easily seen to be continuous but in general it fails to be C^1 so that $\int\int f(x+y)\,d\mu(x)\,d\nu(y)$ is meaningless for $\nu \in C^1(\mathbb{Z}_p \to K)'$. We meet the same state of things when we take the power series view. If we multiply formally $\Sigma\, b_j X^j$ and $\Sigma\, c_j X^j$ where $\{|b_j|/j : j \in \mathbb{N}\}$ and $\{|c_j|/j : j \in \mathbb{N}\}$ are bounded, then for the coefficients of the product $\Sigma\, d_j X^j$ we do have boundedness of $\{|d_j|/j^2 : j \in \mathbb{N}\}$ but in general not of $\{|d_j|/j : j \in \mathbb{N}\}$. This leads to the thought that the convolution of two elements of $C^1(\mathbb{Z}_p \to K)'$ 'is' an element of $C^2(\mathbb{Z}_p \to K)'$ rather than $C^1(\mathbb{Z}_p \to K)'$. Thus, we proceed as follows.

Recall that $C^n(\mathbb{Z}_p \to K)$ $(n \in \{1, 2, \ldots\})$ is a K-Banach space with respect to the norm $\|\ \ \|_n$ and that this norm is equivalent to $\|\ \ \|_n^{\sim}$ where $\|f\|_n^{\sim} := \sup_{j\geqslant 0} |a_j| j^n$ $(f = \sum_{j=0}^{\infty} a_j \binom{x}{j} \in C^n(\mathbb{Z}_p \to K))$ (see Exercise 54.A).

DEFINITION. Let $n \in \{0, 1, 2, \ldots\}$. A C^n*-distribution* is an element μ of the dual space of $C^n(\mathbb{Z}_p \to K)$. Its *corresponding power series* is $\sum_{j=0}^{\infty} \mu\binom{x}{j} X^j$.

THEOREM. *Let $n, m \in \{0, 1, 2, \ldots\}$.*
(i) The map $\mu \mapsto \sum_{j=0}^{\infty} \mu(\binom{*}{j}) X^j$ is a 1-1 correspondence between the C^n-distributions and the formal power series $\sum_{j=0}^{\infty} b_j X^j$ for which $\{|b_j|/j^n : j \in \mathbb{N}\}$ is bounded.
(ii) Let μ be a C^n-distribution, let ν be a C^m-distribution. Then the product of their corresponding power series correspond to an element of $C^{n+m}(\mathbb{Z}_p \to K)'$.
Proof. With the necessary facts provided by Theorem 54.1 and Exercise 54.A the proof of (i) is a simple adaptation of the proof of the previous theorem. To prove (ii), let $\sum_{j=0}^{\infty} b_j X^j$ correspond to μ, let $\sum_{j=0}^{\infty} c_j X^j$ correspond to ν. For the coefficient d_k ($k \in \{1, 2, \ldots\}$) in the product we have

$$|d_k| = \left| \sum_{j=0}^{k} b_j c_{k-j} \right| \leqslant \max_{0 \leqslant j \leqslant k} |b_j| \, |c_{k-j}|$$

According to (i) there is a constant $M > 0$ such that $|b_j| \leqslant Mj^n$, $|c_j| \leqslant Mj^m$ for all $j \in \{1, 2, \ldots\}$. We may assume $|b_0| \leqslant M$, $|c_0| \leqslant M$. We see that

$$\begin{aligned} |d_k| &\leqslant |b_0 c_k| \vee |b_1 c_{k-1}| \vee \ldots \vee |b_k c_0| \\ &\leqslant M^2 k^m \vee \max_{1 \leqslant j \leqslant k-1} Mj^n M(k-j)^m \vee M^2 k^n \\ &\leqslant M^2 k^{n+m} \end{aligned}$$

so that, again by (i), the product $(\sum_{j=0}^{\infty} b_j X^j)(\sum_{j=0}^{\infty} c_j X^j)$ corresponds to a C^{n+m}-distribution.

DEFINITION. Let $\mu \in C^n(\mathbb{Z}_p \to K)'$, $\nu \in C^m(\mathbb{Z}_p \to K)'$ for some $m, n \in \{0, 1, 2, \ldots\}$. Then the *convolution* $\mu * \nu$ of μ and ν is the element of $C^{n+m}(\mathbb{Z}_p \to K)'$ that corresponds to the product $(\sum_{j=0}^{\infty} \mu(\binom{*}{j}) X^j)(\sum_{j=0}^{\infty} \nu(\binom{*}{j}) X^j)$.

**Exercise.* Show that the above convolution also can be defined directly by the formula

$$(\mu * \nu)(f) = \int\int f(x+y)\, d\mu(x)\, d\nu(y) \quad (f \in C^{n+m}(\mathbb{Z}_p \to K))$$

(Hint. Use the Mahler expansion of f and apply the results of Section 54.)

Exercise. Let $f : \mathbb{Z}_p \times \mathbb{Z}_p \to K$ be a C^2-function in the sense of Section 84 and let μ, ν be C^1-distributions. Show that

(i) $\dfrac{d}{dy} \int f(x, y)\, d\mu(x) = \int \dfrac{\partial f}{\partial y}(x, y)\, d\mu(x)$

(ii) $\iint f(x, y)\,d\mu(x)\,d\nu(y) = \iint f(x, y)\,d\nu(y)\,d\mu(x)$

Finally we define C^∞-distributions and convolutions between them. A K-linear map $\mu : C^\infty(\mathbb{Z}_p \to K) \to K$ is a C^∞*-distribution* if there exists an $n \in \mathbb{N} \cup \{0\}$ and a $C > 0$ such that $|\mu(f)| \leqslant C\|f\|_n$ for all $f \in C^\infty(\mathbb{Z}_p \to K)$. The following theorem is not hard to prove.

THEOREM. *The map* $\mu \mapsto \sum_{j=0}^{\infty} \mu(\binom{*}{j})X^j$ *is a 1-1 correspondence between the* C^∞*-distributions and the formal series* $\sum_{j=0}^{\infty} b_j X^j$ *for which there exists an* $n \in \mathbb{N}$ *such that* $\{|b_j|/j^n : j \in \mathbb{N}\}$ *is bounded. The formula*

$$(\mu * \nu)(f) = \iint f(x+y)\,d\mu(x)\,d\nu(y) \quad (f \in C^\infty(\mathbb{Z}_p \to K))$$

defines a convolution in the space of C^∞*-distributions. Equivalently,* $\mu * \nu$ *can be defined as the* C^∞*-distribution corresponding to the product* $(\sum_{j=0}^{\infty} \mu(\binom{*}{j})X^j)(\sum_{j=0}^{\infty} \nu(\binom{*}{j})X^j)$.

For more about distribution theory see, for example, Y. Amice : *Duals. Proceedings of the conference on p-adic analysis*. Report 7806, Mathematisch Instituut, Nijmegen, the Netherlands (1978) and the references given in that paper.

A.7. A substitution formula for real valued integrals

There exists an extensive theory of real or complex valued functions whose domain is in a non-archimedean valued field. For a good background account, see, for example, M. Taibleson, *Fourier analysis on local fields*. Princeton University Press, Princeton, New Jersey (1975).

We choose one particular subject where C^1-functions $K \to K$ are involved.

In this section K is locally compact. We assume that the valuation is chosen such that $|\pi| = q^{-1}$ where $\pi \in K$, $|\pi| < 1$ generates the value group and where q is the number of elements of the residue class field of K. We need the following fact from classical harmonic analysis. For a proof see, for example, E. Hewitt & K.A. Ross: *Abstract harmonic analysis I*, Springer-Verlag, Berlin-Göttingen-Heidelberg (1963).

THEOREM. (Haar integral) *Let* $C_c(K \to \mathbb{R})$ *be the space of all continuous functions* $f : K \to \mathbb{R}$ *with compact support (i.e. f vanishes outside some compact set). Then there exists a unique* $\mathbb{R}$*-linear function* $\varphi : C_c(K \to \mathbb{R}) \to \mathbb{R}$ *satisfying the following conditions.*

(i) $\varphi(\xi_B) = d(B)$ *for each ball B in K.*

(ii) $\varphi(f_s) = \varphi(f)$ *for all* $s \in K$ *(where* $f_s(x) := f(s+x)$ $(x \in K)$*).*

(iii) *For each compact set* $C \subset K$ *there is a constant* $M_C > 0$ *such that for all* $f \in C_c(K \to \mathbb{R})$ *that vanish outside* C *we have* $|\varphi(f)| \leqslant M_C \|f\|_\infty$ *(where* $\|f\|_\infty := \max\{|f(x)| : x \in K\}$*).*

φ is called the *Haar integral* on $C_c(K \to \mathbb{R})$. We shall write $\int f d\lambda$ instead of $\varphi(f)$ (this notation refers to a measure λ corresponding to φ, but we do not need this measure for our purpose). For an open compact subset U of K and a continuous function $f : U \to \mathbb{R}$ we define

$$\int_U f d\lambda := \int \tilde{f} \xi_U \, d\lambda$$

where

$$\tilde{f}(x) := \begin{cases} f(x) \text{ if } x \in U \\ 0 \text{ if } x \in K \setminus U \end{cases}$$

Our aim in this section is to prove the following theorem.

THEOREM. (Substitution theorem) *Let* U, V *be compact open subsets of* K *and let* $\sigma : U \to V$ *be a* C^1*-homeomorphism,* $\sigma'(x) \neq 0$ *for all* $x \in U$. *Let* $f : V \to \mathbb{R}$ *be continuous. Then*

$$\int_V f d\lambda = \int_U f \circ \sigma \ |\sigma'| d\lambda$$

Proof. The expressions on either side of the equality sign define continuous linear forms on the space $C(V \to \mathbb{R})$ consisting of all continuous real valued functions on V, normed by $\| \ \|_\infty$. It is easy to see that the locally constant functions are dense in $C(V \to \mathbb{R})$ so it suffices to check the formula for $f = \xi_S$ where $S \subset V$, S compact open. Then $T := \sigma^{-1}(S)$ is a clopen subset of U. By Theorem 27.5, for each $a \in T$ there is a ball $B \subset T$ containing a such that $\sigma(B)$ is a ball with diameter $|\sigma'(a)| d(B)$. We may assume that $|\sigma'|$ is constant on B. A compactness argument yields the existence of $a_1, \ldots, a_n \in T$ and balls $B_1, \ldots, B_n$ in T such that $a_j \in B_j$ for each j, $B_1, \ldots, B_n$ is a partition of T, $\sigma(B_j)$ is a ball with diameter $|\sigma'(a_j)| d(B_j)$ for each j, $|\sigma'|$ is constant on each B_j. Then $\sigma(B_1), \ldots, \sigma(B_n)$ is a partition of S and $\int_V f d\lambda = \int_S 1 d\lambda = \Sigma_{i=1}^n \int_S \xi_{\sigma(B_j)} d\lambda = \Sigma_{j=1}^n d(\sigma(B_j)) = \Sigma_{j=1}^n |\sigma'(a_j)| d(B_j) = \int_T |\sigma'| d\lambda = \int_U f \circ \sigma \, |\sigma'| d\lambda$ and we are done.

Remark. The reader who is familiar with the Haar measure λ shall have no trouble in proving the substitution formula for λ-integrable functions $f : V \to \mathbb{R}$.

A. 8. The ultrametric Hahn-Banach theorem

THEOREM. (Ingleton) *Let E be a normed space over a spherically complete field K. Let D be a linear subspace of E and let $f : D \to K$ be a continuous linear function. Then f can be extended to a continuous linear function $\bar{f} : E \to K$ such that $\|\bar{f}\| = \|f\|$.*

Proof. By an application of Zorn's lemma it suffices to consider the case $E = D + [\![a]\!]$ where $a \notin D$. The wanted extension $\bar{f}$ must satisfy $|\bar{f}(\lambda a + d)| \leqslant \|f\| \; \|\lambda a + d\|$ for all $\lambda \in K$, $d \in D$. In other words we have to find an element $w (=\bar{f}(a))$ in K such that

$$|w - f(d)| \leqslant \|f\| \quad \|a - d\| \qquad (d \in D)$$

i.e.

$$w \in \bigcap_{d \in D} B_{f(d)} \; (\|f\| \quad \|a - d\|)$$

To prove that the intersection is nonempty it is, by spherical completeness, enough to prove that for each $d, d' \in D$

$$B_{f(d)} \; (\|f\| \; \|a - d\|) \; \cap B_{f(d')} \; (\|f\| \quad \|a - d'\|)$$

is not empty. But the latter follows from $|f(d) - f(d')| \leqslant \|f\| \quad \|d - d'\| \leqslant \|f\| \max(\|d - a\|, \|a - d'\|)$.

Remarks.

1. Similar techniques have been used in Theorems 76.4 (extension of isometries to isometries), 76.5 (extension of increasing functions to increasing functions), 86.3 (existence of functions, monotone of certain type).
2. If $f \neq 0$ and D is not dense in E an extension of f in the above sense is not unique.
3. Ingleton's theorem becomes a falsity if we replace K by a non-spherically complete field. For more about this, see van Rooij (1978).

A. 9. A field with prescribed residue class field and value group

In Section 11 we announced the following theorem.

THEOREM. *Let F be a field and let Γ be a multiplicative subgroup of* $(0, \infty)$. *Then there exists a (complete) non-archimedean valued field whose residue class field is isomorphic to F and whose value group equals Γ.*

In order to make the general construction easier to understand we first consider a more modest problem, namely to find a *discretely* valued field whose residue class field is F.

A *formal Laurent series over* F is a two-sided sequence $(\ldots, a_{-1}, a_0, a_1, \ldots)$ of elements of F such that $a_{-n} = 0$ for large n. Such an element is often written as $\Sigma\, a_n X^n$ or as $\Sigma_{n \geqslant N}\, a_n X^n$ (if $a_n = 0$ for $n < N$). The formal Laurent series over F form a field $F((X))$ under the operations

$$\Sigma\, a_n X^n + \Sigma\, b_n X^n := \Sigma\, (a_n + b_n) X^n$$

$$(\Sigma\, a_n X^n)(\Sigma\, b_n X^n) := \Sigma\, (\sum_i a_i b_{n-i}) X^n$$

(observe that $\{i : a_i b_{n-i} \neq 0\}$ is finite for each n). The formula

$$|\Sigma\, a_n X^n| := \begin{cases} 0 \text{ if } a_n = 0 \text{ for all } n \\ e^{-\min\{n \,:\, a_n \neq 0\}} \text{ otherwise} \\ (\text{NB } e = 2.71\ldots) \end{cases}$$

defines a valuation on $F((X))$. In the following exercise we prove that $(F((X)), |\ \ |)$ is a complete discretely valued field whose residue class field is (isomorphic to) F.

Exercise. (i) Show that, with the above definitions, $(F((X)), |\ \ |)$ is a non-archimedean valued field whose value group is $\{e^n : n \in \mathbb{Z}\}$.
(ii) A *formal power series over* F is an element $\Sigma\, a_n X^n \in F((X))$ for which $a_n = 0$ for negative n. Prove that the set $F[[X]]$ of all formal power series constitutes a subring of $F((X))$. In fact, show that $F[[X]]$ is the 'closed' unit disc of $F((X))$.
(iii) Now prove successively the following. $F[[X]]$ and $F((X))$ are complete. $F[X]$ is dense in $F[[X]]$ and $F(X)$ is dense in $F((X))$. (Observe that it follows that $\mathbb{R}((X))$ is the completion of the field $(\mathbb{R}(X), |\ \ |_1)$ of Exercise 2.B (with $\rho = e$)). The map $\Sigma a_n X^n \mapsto a_0$ is a homomorphism of $F[[X]]$ onto F whose kernel is $XF[[X]] = \{f \in F((X)) : |f| < 1\}$. The residue class field of $F((X))$ is isomorphic to F.

One may also identify $F((X))$ with the set of all functions $f : \mathbb{Z} \to F$ for which $f(-n) = 0$ for large (positive) n with the pointwise addition

$$(f+g)(n) := f(n) + g(n) \qquad (n \in \mathbb{Z})$$

and the convolution multiplication

$$(f * g)(n) := \sum_{i \in \mathbb{Z}} f(i)\, g(n-i) \qquad (n \in \mathbb{Z})$$

It is this view that we shall adopt in the general proof of the theorem in which, however, we shall admit a larger class of functions f, namely those for which $\{x : f(x) \neq 0\}$ is a well-ordered subset of $\mathbb{R}$ (see below).

A subset X of $\mathbb{R}$ is *well-ordered* if each nonempty subset of X has a smallest element. Examples are finite sets, $\mathbb{N}$, $\{-3, -2, -1, 0, 1, 2, \ldots\}$. Subsets of well-ordered sets are well-ordered. The sets $\mathbb{R}$, $\mathbb{Z}$, $\mathbb{Q}$ are not well-ordered.

Exercise. Show that $\{2^{-n} : n \in \mathbb{N}\}$ is not well-ordered but $\{-2^{-n} : n \in \mathbb{N}\}$ is. Give an example of a well-ordered set having two accumulation points.

**Exercise.* Show that each sequence in $\mathbb{R}$ has a monotone subsequence. Use this fact to show that the following conditions for a subset X of $\mathbb{R}$ are equivalent.

(α) X is well-ordered.

(β) Each sequence in X has an increasing subsequence.

(γ) X 'does not have strictly decreasing sequences' (i.e. there are no $a_1, a_2, \ldots$ in X such that $a_1 > a_2 > a_3 > \ldots$).

Exercise. Let X be a well-ordered set, let $a \in X$. Prove that either a is the largest element of X or there is an element $a' \in X$, the *successor* of a, such that $a' > a$ and $(a, a') \cap X$ is empty. Show that a well-ordered set is countable. (Hint. If a has a successor a', choose a rational number in (a, a').)

**Exercise.* For $X, Y \subset \mathbb{R}, a \in \mathbb{R}$ set $-X := \{-x : x \in X\}$, $a + X := \{a + x : x \in X\}$, $X + Y := \{x + y : x \in X, y \in Y\}$. Use the results of the above exercises to show the following.

(i) Let $a \in R$. If X and Y are well-ordered then so are $a + X$, $X \cup Y$, $X + Y$. The set $X \cap (-Y)$ is finite.

(ii) Let $a_1, a_2, \ldots$ be a sequence in $\mathbb{R}$ such that $\lim_{n\to\infty} a_n = \infty$. If $(-\infty, a_n) \cap X$ is well-ordered for each n, then X is well-ordered.

After this preparatory work we are ready to present a

Proof of the theorem. Let $\log \Gamma := \{\log x : x \in \Gamma\}$ where log is the ordinary real valued logarithm. Then $\log \Gamma$ is an additive subgroup of $\mathbb{R}$. We define

$$K := \{f : \log \Gamma \to F : \operatorname{supp} f \text{ is well-ordered}\}$$

where as usual supp f, the *support* of f, is $\{x \in \log \Gamma : f(x) \neq 0\}$. For $f \in K$, $f \neq 0$, let $r(f)$ be the smallest element of supp f. We shall prove that the formulas

$$(f + g)(x) := f(x) + g(x)$$

$$f * g(x) := \sum_{y \in \log \Gamma} f(y)\, g(x - y)$$

$$|f| := \begin{cases} e^{-r(f)} & (f \neq 0) \\ 0 & (f = 0) \end{cases}$$

define addition, multiplication and a valuation making K into a complete non-archimedean valued field whose residue class field is isomorphic to F and whose value group is Γ. First observe that, due to part (i) of the previous exercise, the sum in the formula for $f * g$ is in fact a finite sum and supp $(f * g)$ is well-ordered. We may conclude that $+$, $*$, $|\ |$ are well defined.

If follows easily that K is a commutative ring with identity ϵ where

$$\epsilon(x) := \begin{cases} 1 \text{ if } x = 0 \\ 0 \text{ if } x \neq 0 \end{cases}$$

We proceed to prove that K is a field, i.e. that a nonzero $f \in K$ has an inverse. First suppose that the smallest element of $X :=$ supp f is 0. Then $f(0) \neq 0$ and $f(x) = 0$ for $x \in \log \Gamma$, $x < 0$. Let $X_1 := X$, $X_2 := X + X, \ldots$ Then by the previous exercise each X_n is well-ordered. Since $0 \in X_n$ we have $X_1 \subset X_2 \subset X_3 \ldots$ Set $X_\infty := \bigcup_{n=1}^{\infty} X_n$. We show that X_∞ is well-ordered. First, choose $a \in \mathbb{R}$, $a > 0$ such that $x \in X$, $x \neq 0$ implies $x > a$. If $x \in X_\infty$ and $x < na$ for some $n \in \mathbb{N}$ then $x \in X_n$. (For example, if $x \in X_{n+1} \setminus X_n$ then $x = x_1 + \ldots + x_{n+1}$ where $x_i \in X$, $x_i \neq 0$, hence $\geqslant a$ for each i so that $x \geqslant (n+1)a$.) Therefore, for all $n \in \mathbb{N}$ we have

$$Y_n := X_\infty \cap (-\infty, na) \subset X_n$$

so that Y_n is well-ordered. By the previous exercise it follows that $\bigcup_n Y_n = X_\infty$ is well-ordered. We now define a function g such that $g * f = \epsilon$. If we choose $g : \log \Gamma \to F$ such that $g(0) = f(0)^{-1}$ and supp $g \subset X_\infty$ then it is true already that supp g is well-ordered so $g \in K$ and $g(0) f(0) = 1$. Further, if $x \in \log \Gamma$, $x \notin X_\infty$ then for each $y \in \log \Gamma$ we have $f(y)\, g(x - y) = 0$. We need only to adjust g in such a way that

$$\sum_y f(y)\, g(x - y) = 0 \qquad (x \in X_\infty, x > 0)$$

In other words

$$f(0) g(x) = - \sum_{y > 0} f(y)\, g(x - y)$$

or (using the definition of a)

$$(*) \qquad f(0)\, g(x) = - \sum_{y \geqslant a} f(y)\, g(x - y) \qquad (x \in X_\infty, x > 0)$$

If $x \in (0, a]$ and $y > a$ then $x - y < 0$, hence $f(y)\, g(x-y) = 0$. If $y = a$ then $f(a) = 0$ so $f(y)\, g(x-y) = 0$. We see that (*) requires that $g = 0$ on $(0, a]$. We can now define g inductively as follows. Set $g := 0$ on $(0, a]$. Suppose we have defined g on $X_\infty \cap (0, na]$ for some $n \in \mathbb{N}$ such that (*) holds for $x \in X_\infty \cap (0, na]$. Then (*) prescribes the values of g on $(na, (n+1)a]$, etc. In this way we obtain an element $g \in K$ for which $g * f = \epsilon$. For an arbitrary nonzero element f of K let $s := r(f)$ and notice that f has an inverse if and only if $x \mapsto f(x + s)$ has one, and the latter function is invertible by the foregoing. Thus, we now have that K is a field.

It is easy to check that if $f, g, f+g$ are nonzero then $r(f+g) \geqslant \min(r(f), r(g))$ and $r(fg) = r(f) + r(g)$. It follows that $|\ \ |$ is a valuation on K. Trivially, the value group is $\{e^{-r} : r \in \log \Gamma\} = \Gamma$. Next, we consider the residue class field of K. We have $B_0(1) = \{f \in K : |f| \leqslant 1\} = \{f \in K : r(f) \geqslant 0\} = \{f \in K : f(x) = 0 \text{ for } x \in (-\infty, 0)\}$ and $B_0(1^-) = \{f \in K : |f| < 1\} = \{f \in K : f(x) = 0 \text{ for } x \in (-\infty, 0]\}$. We see that the map $f \mapsto f(0)$ is a homomorphism of $B_0(1)$ onto F whose kernel is $B_0(1^-)$. So, the residue class field of K is isomorphic to F. Finally we show that K is complete. In fact we prove a little more, viz. if $B_1 \supset B_2 \supset \ldots$ are 'closed' discs in K then $\bigcap_n B_n$ is not empty (i.e. K is 'spherically complete' in the sense of Section 20; from this the completeness of K follows easily). To prove it, let

$$B_n := \{f \in K : |f - f_n| \leqslant r_n\}$$

where $r_1 > r_2 > \ldots$ Then $-\log r_1 < -\log r_2 < \ldots$ Now if $x \in \log \Gamma$, $x < -\log r_n$ and $m \geqslant n$ then $f_m(x) = f_n(x)$ (since $f_m \in B_n$ we have $|f_m - f_n| \leqslant r_n$ so $f_m = f_n$ on $\log \Gamma \cap (-\infty, -\log r_n)$). Define $g : \log \Gamma \to F$ as follows.

$$g(x) := \begin{cases} \lim\limits_{m \to \infty} f_m(x) & \text{if } x < -\log r_n \text{ for some } n \\ 0 & \text{otherwise.} \end{cases}$$

We claim that $g \in K$ and $g \in \bigcap_n B_n$. In fact, first observe that if $x \in \operatorname{supp} g$ then $x < -\log r_n$ for some n so $g(y) = f_m(y)$ for large m and all $y \leqslant x$. Let S be a nonempty subset of supp g. Then for large m, $S \cap (-\infty, -\log r_m)$ is nonempty and is contained in supp f_m. Thus S has a smallest element, supp g is well-ordered, so $g \in K$. Finally, to prove that $g \in B_n$ for each n observe that $|g - f_n| \leqslant r_n$, which follows directly from the definition of g.

Remark. The above result shall be used in the next section.

A.10. Isometrical embedding of an ultrametric space into K

THEOREM. *Each ultrametric space can isometrically be embedded into a non-archimedean valued (spherically complete) field.*

Proof. Let X be an ultrametric space. By Appendix A.9 there exists a spherically complete field K such that the value group of K is equal to all of $(0, \infty)$ and such that the cardinality of X is strictly less than the cardinality of the residue class field of K. We shall prove that there is an isometry of X into K. By Zorn's lemma it suffices to prove that for $Y \subset X$, $a \in X \backslash Y$, an isometry $f: Y \to K$ can be extended to an isometry $\bar{f}: Y \cup \{a\} \to K$. We may assume that Y is closed so that $d(a, Y) > 0$. We distinguish two cases.

(i) a has no best approximation in Y (see Section 21). Then choose x_1, $x_2, \ldots \in Y$ such that $d(a, x_1), d(a, x_2), \ldots$ is a strictly decreasing sequence converging to $d(a, Y)$. For each $n \in \mathbb{N}$ define the ball B_n in K by

$$B_n := B_{f(x_n)} \quad (d(a, x_n))$$

If $z \in B_{n+1}$ then $|z - f(x_{n+1})| \leqslant d(a, x_{n+1})$, so $|z - f(x_n)| \leqslant \max(|z - f(x_{n+1})|, |f(x_{n+1}) - f(x_n)|) \leqslant \max(d(a, x_{n+1}), d(x_n, x_{n+1})) \leqslant \max(d(a, x_{n+1}), d(x_n, a), d(a, x_{n+1})) = d(x_n, a)$, which proves that $z \in B_n$. So we have

$$B_1 \supset B_2 \supset \ldots$$

Extend f by defining $\bar{f}(a)$ to be an arbitrary element of $\bigcap_n B_n$. To prove that $\bar{f}$ is an isometry it suffices to show that $|\bar{f}(a) - f(x)| = d(a, x)$ for each $x \in Y$. Since x is not a best approximation of a in Y there exists $n \in \mathbb{N}$ such that $d(a, x_n) < d(a, x)$, so that $d(x, x_n) = d(x, a)$ and therefore $|f(x) - f(x_n)| = d(x, x_n) = d(x, a)$. As $\bar{f}(a) \in B_n$ we have also $|\bar{f}(a) - f(x_n)| \leqslant d(a, x_n) < d(a, x)$. Hence, $|\bar{f}(a) - f(x)| = \max(|\bar{f}(a) - f(x_n)|, |f(x_n) - f(x)|) = d(a, x)$, and we are done.

(ii) a has a best approximation in Y. Let A_1 be the set of all these approximations, viz.

$$A_1 := \{x \in Y : d(a, x) = d(a, Y)\}$$

Observe that $x, y \in A_1$ implies $d(x, y) \leqslant \max(d(x, a), d(a, y)) = d(a, Y)$. Let A_2 be a maximal subset of A_1 with the property

$$\text{if } x, y \in A_2, x \neq y \text{ then } d(x, y) = d(a, Y)$$

To define $\bar{f}(a)$ we distinguish two cases.

(ii)' A_2 consists of a single point m. Since $|K^{\times}| = (0, \infty)$ we can find $\gamma \in K$ such that $|\gamma - f(m)| = d(a, m)$. Set $\bar{f}(a) := \gamma$.

(ii)'' A_2 contains more than one point. The points of $f(A_2)$ are equidistant and contained in a ball in K of the form

$$B := \{\beta \in K : |\beta - a| \leq d(a, Y)\}$$

Let $B^- := \{\beta \in K : |\beta - a| < d(a, Y)\}$. Then the map

$$A_2 \xrightarrow{f} B \to B/B^- \simeq k$$

(where k is the residue class field of K) is injective, but not surjective since the cardinality of A_2 is strictly less than the cardinality of k. It follows that we can find $\gamma \in B$ such that $d(\gamma, f(A_2)) = d(a, Y)$. Set $\bar{f}(a) := \gamma$.

In both cases (ii)' and (ii)'' we have defined $\bar{f}(a)$ in such a way that

$$|\bar{f}(a) - f(m)| = d(a, m) \qquad (m \in A_2)$$

Now let $x \in A_1$. Then by maximality there is $m \in A_2$ such that $d(x, m) < d(a, m)$, so that $|f(x) - f(m)| < d(a, m) = |\bar{f}(a) - f(m)|$, whence $|\bar{f}(a) - f(x)| = \max(|\bar{f}(a) - f(m)|, |f(m) - f(x)|) = d(a, m) = d(x, a)$. So we have arrived at

$$|\bar{f}(a) - f(x)| = d(a, x) \qquad (x \in A_1)$$

Finally, let $x \in Y \backslash A_1$. Choose any $y \in A_1$. Then $d(a, x) > d(a, y)$, so that $|f(x) - f(y)| = d(x, y) = d(a, x)$ and $|\bar{f}(a) - f(y)| = d(a, y) < d(a, x)$. It follows that $|\bar{f}(a) - f(x)| = \max(|\bar{f}(a) - f(y)|, |f(y) - f(x)|) = d(a, x)$. We have now that

$$|\bar{f}(a) - f(x)| = d(a, x) \qquad (x \in Y)$$

which finishes the proof.

According to the technique of the above proof one needs rather 'big' fields in which to embed an ultrametric space. But sometimes one can afterwards reduce the 'size' of K.

COROLLARY. *A separable ultrametric space can isometrically be embedded into a separable non-archimedean valued field.*

Proof. By the above theorem a (separable) ultrametric space X can be considered as a subset of some non-archimedean valued field L. Let Y be a countable dense subset of X. The field S generated by Y is countable, so its closure $\bar{S}$ is a separable field containing X.

Exercise. Show that a compact ultrametric space can isometrically be embedded into a 'compactly generated field', i.e. a non-archimedean valued field K having a compact subset X such that the smallest closed field containing X is K. Show however that such 'compactly generated fields' are nothing else but separable fields.

Exercise. Let X be an ultrametric space and let K be a discretely valued complete non-archimedean valued field. Suppose that the cardinality of X is strictly less than the cardinality of the residue class field of K. Show that there exists an embedding $\sigma : X \to K$ such that

$$d(x, y) \leqslant |\sigma(x) - \sigma(y)| \leqslant |\pi|^{-1} d(x, y) \qquad (x, y \in X)$$

where $\pi \in K$, $0 < |\pi| < 1$, $|\pi|$ is a generator of the value group. (Hint. Find a metric ρ on X whose nonzero values are in $|K^\times|$ such that $d \leqslant \rho \leqslant |\pi|^{-1} d$. Use the technique of the proof of the above theorem to embed (X, ρ) isometrically into K.)

APPENDIX B

Glossary of terms

We list a few notations, definitions and statements used in the main text. For what still remains unexplained or unproved we refer to the usual textbooks on elementary analysis, topology and algebra.

B.1. Sets

Let Y, Z be sets. If $Z \subset Y$ then $Y \backslash Z := \{y \in Y : y \notin Z\}$. The *product set* $Y \times Z$ equals $\{(y, z) : y \in Y, z \in Z\}$. We write $Y^2 := Y \times Y$, $Y^{n+1} := Y^n \times Y$ ($n \in \mathbb{N}, n \geqslant 2$). Y is *countable* if there exists a surjection $\mathbb{N} \to Y$, otherwise Y is *uncountable. The cardinality of Y is strictly less than the cardinality of Z* if no map $Y \to Z$ is surjective. A *partition* of Y is a covering of Y by mutually disjoint subsets. The classes of an equivalence relation $\sim$ on Y form a partition of Y. The set of these classes is denoted $Y/\sim$. The *quotient map* $\pi : Y \to Y/\sim$ sends each element $x \in Y$ into its class $\pi(x)$. A *(full) set of representatives in Y of $\sim$*, or a *(full) set of representatives in Y modulo $\sim$* is a subset R of Y such that π maps R bijectively onto $Y/\sim$.

Let $Y = (Y, \geqslant)$ be a partially ordered set. A *maximal element* of Y is an element $y \in Y$ such that $z \in Y$, $z \geqslant y$ implies $z = y$. A *majorant* of a subset S of Y is an element $y \in Y$ such that $y \geqslant s$ for all $s \in S$. Zorn's lemma states that if in a nonempty partially ordered set Y each linearly ordered subset has a majorant then Y has at least one maximal element. In a similar way one can define *minimal* elements, *minorant* and formulate a corresponding version of Zorn's lemma.

B.2. Subsets of $\mathbb{R}$

$\mathbb{Z} := \{\ldots, -1, 0, 1, \ldots\}$ is the ring of *integers*. An element $s \in \mathbb{Z}$ is *divisible* by an element $t \in \mathbb{Z}$ (notation $t \mid s$) if there is an integer m such that $s = mt$. The notation $t \nmid s$ stands for 's is not divisible by t'. Two integers s, t are congruent modulo $n \in \mathbb{Z}$ (notation $s \equiv t \pmod{n}$) if $s - t$ is divisible by n.

$\mathbb{N} := \{1, 2, 3, \ldots\}$ is the collection of *positive integers* (*natural numbers*). The *greatest common divisor* of $m, n \in \mathbb{N}$ is the largest $d \in \mathbb{N}$ with the property $d \mid m$, $d \mid n$.

$\mathbb{Q}$ is the field of the *rational numbers*, $\mathbb{R}$ is the field of the *real numbers* with its natural *ordering* $\geqslant$ and *absolute value function* $|\ |$. The maximum of two real numbers a and b is denoted by $\max(a, b)$ or by $a \vee b$, their minimum by $\min(a, b)$ or $a \wedge b$. Similarly we use $\max V$ to indicate the maximum of a subset V of $\mathbb{R}$. The *entire part* $[x]$ of a real number x is defined by $[x] := \max \mathbb{Z} \cap (-\infty, x]$. A subset C of $\mathbb{R}$ is *convex* if $x, y \in C$, $\lambda \in \mathbb{R}$, $0 \leqslant \lambda \leqslant 1$ implies $\lambda x + (1 - \lambda) y \in C$. 'Convex' is identical to 'connected' (see B. 3). The convex subsets of $\mathbb{R}$ are just $\emptyset$, the singleton sets and the intervals. A *real sequence* is a map $\mathbb{N} \to \mathbb{R}$. For such a sequence $n \mapsto a_n$ we use $\overline{\lim}_{n \to \infty} a_n = a$ as an abbrevation for $\lim_{n \to \infty} \sup \{a_k : k \geqslant n\} = a$. Also by definition $\underline{\lim}_{n \to \infty} a_n = a$ means $\lim_{n \to \infty} \inf \{a_k : k \geqslant n\} = a$.

B.3. Metric and topology

A *metric space* $Y = (Y, d)$ is a set Y together with a map $d : Y \times Y \to \mathbb{R}$ such that for all $x, y, z \in Y$ we have $d(x, y) \geqslant 0$, $d(x, y) = 0$ if and only if $x = y$, $d(x, y) = d(y, x)$, $d(x, z) \leqslant d(x, y) + d(y, z)$. d is a *metric* on Y. A subset U of Y is *open* if for each $a \in U$ there is an $\epsilon > 0$ such that $\{x \in Y : d(x, a) < \epsilon\} \subset U$. The collection of all open subsets of Y is the *topology induced by* d. We have (i) $\emptyset$ and Y are open, (ii) finite intersections of open sets are open, (iii) arbitrary unions of open sets are open. A *topology* on a set Z is a class of subsets of Z (called open sets) satisfying (i), (ii), (iii). A set Z together with a topology is a *topological space*. Let Y, Z be topological spaces. The product topology on $Y \times Z$ is the smallest topology containing the collection $\{U \times V : U$ open in Y, V open in $Z\}$.

Throughout the rest of B.3, $Y = (Y, d)$ and $Z = (Z, d)$ are metric spaces. Topological notions refer to the induced topologies. A *neighbourhood* of a point $a \in Y$ is a subset U of Y containing $\{x \in Y : d(x, a) < \epsilon\}$ for some $\epsilon > 0$. The point a is an *isolated point* if $\{a\}$ is a neighbourhood of a, a is an *accumulation point* of a subset S of Y if $S \cap U$ is infinite for every neighbourhood U of a. A subset S of Y is *closed* if its complement is open, *bounded* if

$\sup \{d(x, y) : x, y \in S\} < \infty$. The *closure* $\bar{S}$ of S is the smallest closed set containing S, the *interior* of S is the largest open set contained in S. The *boundary* of S is $\bar{S} \cap \overline{Y \backslash S}$. S is *dense* (in Y) if $\bar{S} = Y$. The space Y is *separable* if Y has a countable dense subset. The topology of Y is *discrete* if each point of Y is an isolated point. Y is *compact* if each covering of Y by means of open sets has a finite subcovering. A sequence $x_1, x_2, \ldots$ in Y is *convergent* to $x \in Y$ if $\lim_{n \to \infty} d(x, x_n) = 0$; it is a *Cauchy sequence* if $\lim_{m, n \to \infty} d(x_m, x_n) = 0$. Y is *complete* if every Cauchy sequence is convergent. Y is compact if and only if each sequence in Y has a convergent subsequence (if and only if each infinite subset of Y has an accumulation point). Compact spaces are complete and separable. Y is *locally compact* if each point of Y has a neighbourhood which – as a metric space – is compact. The topology of Y is *zerodimensional* if for every $a \in Y$ and every neighbourhood U of a there is a set V which is both open and closed such that $a \in V \subset U$.

A map $f : Y \to Z$ is *continuous* if for each open subset $U \subset Z$ its inverse image $f^{-1}(U)$ is open in Y. f is continuous if and only if for each $a \in Y$ and $\epsilon > 0$ there exists a $\delta > 0$ such that $x \in Y$, $d(x, a) < \delta$ implies $d(f(x), f(a)) < \epsilon$ (if and only if for each $a \in Y$ and each sequence $x_1, x_2, \ldots$ in Y converging to a the sequence $f(x_1), f(x_2), \ldots$ converges to $f(a)$). A *homeomorphism* $f : Y \to Z$ is a bijection such that f and f^{-1} are continuous. $f : Y \to Z$ is an *isometry* if $d(f(x), f(y)) = d(x, y)$ for all $x, y \in Y$; f is *uniformly continuous* if for each $\epsilon > 0$ there exists a $\delta > 0$ such that for all $x, y \in Y$ we have $d(x, y) < \delta$ implies $d(f(x), f(y)) < \epsilon$.

Let $\mathbb{R}$ be equipped with the metric $(x, y) \to |x - y|$ $(x, y \in \mathbb{R})$. Y is *connected* if every continuous map $f : Y \to \mathbb{R}$ taking only the values 0 and 1 is constant. Y is connected if and only if $Y = A \cup B$, $A \cap B = \emptyset$, A, B closed and open implies $A = \emptyset$ or $B = \emptyset$. Y is *totally disconnected* if the only subsets of Y that are connected as a metric space are the empty set and the *singleton sets* $\{a\}$ $(a \in Y)$. A zerodimensional space is totally disconnected. A function $f : Y \to \mathbb{R}$ is *upper semicontinuous* if for each $a \in Y$ and $\epsilon > 0$ there exists a $\delta > 0$ such that $d(x, a) < \delta$ implies $f(x) < f(a) + \epsilon$.

B.4. Algebra

We assume the reader to be familiar with the notions group, ring, field, vector space over a field L (L-vector space, L-linear space), dimension (over L) of such a space, L-linear map between L-vector spaces, determinant. Let S be a subset of an L-vector space E. Its *L-linear span* $[\![S]\!]$ is the smallest L-linear subspace of E containing S. If S is a finite set $\{x_1, x_2, \ldots, x_n\}$ or a countable set $\{x_1, x_2, \ldots\}$ we sometimes write $[\![x_1, \ldots, x_n]\!]$ resp $[\![x_1, x_2, \ldots]\!]$ instead of $[\![S]\!]$. From now on in B.4. all groups are abelian and, until further notice, written additively.

Let G_1, G_2 be groups. A (group) *homomorphism* of G_1 into G_2 is a map $f : G_1 \to G_2$ satisfying $f(x) + f(y) = f(x + y)$ for all $x, y \in G_1$. If f is also a bijection then f^{-1} is a homomorphism and f is an *isomorphism*. If also $G_1 = G_2$ then f is an *automorphism*. Let G be a group. A subset H of G is a *subgroup* if $0 \in H$ and $x, y \in H$ implies $x - y \in H$. Then H, with the addition inherited from G, is itself a group. The equivalence relation $\sim$ on G defined by $x \sim y$ if $x - y \in H$ induces a partition of G into *cosets of H*; they have the form $a + H$ $(a \in G)$. The definition $(a + H) + (b + H) := (a + b) + H$ $(a, b \in G)$ makes the collection of these cosets into a group, the *quotient group* G/H, in which the coset $H = 0 + H$ acts as a zero element. The *quotient map* $\pi : a \mapsto a + H$ $(a \in G)$ is a surjective homomorphism. If f is a homomorphism of a group G_1 into a group G_2 then $\mathrm{Ker}\, f := \{x \in G_1 : f(x) = 0\}$ is a subgroup of G_1 and $\mathrm{Im}\, f := \{f(x) : x \in G_1\}$ is a subgroup of G_2 which is isomorphic to $G_1/\mathrm{Ker}\, f$. A group G is *cyclic* if there exists an element $a \in G$ such that the smallest subgroup of G containing a equals G. Then $G = \{na : n \in \mathbb{Z}\}$ where $0a := 0$, $1a := a$ and $(n + 1)a := na + a$, $(-n)a := -na$ $(n \in \mathbb{N})$.

In the sequel rings and fields are commutative with identity 1. Let R_1, R_2 be rings. A (ring) *homomorphism* of R_1 into R_2 is a map $f : R_1 \to R_2$ satisfying $f(1) = 1$, $f(x + y) = f(x) + f(y)$, $f(xy) = f(x) f(y)$ for all $x, y \in R_1$. If f is also a bijection then f^{-1} is a homomorphism and f is an *isomorphism*. If also $R_1 = R_2$ then f is an *automorphism*. Let R be a ring. A subset D of R is a *subring* if D is an additive subgroup of R, $1 \in D$ and $x, y \in D$ implies $xy \in D$. Then D with the operations inherited from R is itself a ring. A subset I of R is an *ideal* if I is an additive subgroup of R and $x \in R$, $y \in I$ implies $xy \in I$. Then if $I \neq R$ the (additive) quotient group R/I can be made into a ring, the *quotient ring* R/I by the multiplication $(a + I)(b + I) := ab + I$ $(a, b \in R)$. $1 + I$ acts as an identity. The quotient map $\pi : R \to R/I$ is a homomorphism. If f is a homomorphism of a ring R_1 into a ring R_2 then $\mathrm{Ker}\, f$ is an ideal in R_1, $\mathrm{Im}\, f$ is a subring of R_2 isomorphic to $R_1/\mathrm{Ker}\, f$. Let a be an element of a ring R. The *principal ideal generated by a*, notation (a), is the smallest ideal in R containing a. Then $(a) = \{ra : r \in R\}$. An ideal I in R is a *maximal ideal* if $I \neq R$ and there are no ideals J for which $I \subset J \subset R$, $J \neq I$, $J \neq R$. I is a maximal ideal if and only if R/I is a field. An *inverse* of an element a of a ring R is an element $b \in R$ for which $ab = 1$. If such an inverse exists it is unique and denoted a^{-1}. The set $\{a \in R : a^{-1} \text{ exists}\}$ is a group under multiplication, the *group of units* of R. The group of units of a field L is $L^\times := \{x \in L : x \neq 0\}$, the *multiplicative group of L*. Let L_1, L_2 be fields and let $f : L_1 \to L_2$ be a (ring) homomorphism. Then f is injective, $\mathrm{Im}\, f$ is a subfield of L_2, i.e. a subring of L_2 in which each nonzero element has an inverse.

An *integral domain* is a ring R for which $x, y \in R$, $xy = 0$ implies $x = 0$ or $y = 0$. Let R be an integral domain and let $S := \{(x, y) \in R \times R : y \neq 0\}$. The relation $\sim$ defined by $(x, y) \sim (x', y')$ if $xy' = x'y$ ($(x, y), (x', y') \in S$) is an equivalence relation in S. Let $\pi : S \to S/\sim$ be the quotient map. The formulas

$$\pi((x, y)) + \pi((x', y')) := \pi((xy' + x'y, yy'))$$
$$\pi((x, y))\, \pi((x', y')) := \pi((xx', yy'))$$

define addition and multiplication operations in $S/\sim$ making it into a field L, the *quotient field* of R. (The zero element of L is $\pi((0,1))$, its identity is $\pi((1, 1))$. Each nonzero element of L can be written as $\pi((x, y))$ where $x, y \in R$, $x \neq 0, y \neq 0$. Its inverse is $\pi((y, x))$.) The map $j : x \mapsto \pi((x, 1))$ is an injective homomorphism of R into L, the smallest subfield of L containing $j(R)$ is L. Instead of $\pi((x,y))$ one usually writes x/y. With this notation we have for $(x, y), (x', y') \in S$

$$x/y = x'/y' \quad \text{if and only if } xy' = x'y$$
$$x/y + x'/y' = (xy' + x'y)/yy'$$
$$(x/y)(x'/y') = xx'/yy'$$
$$j(x) = x/1$$

Let F be a field. The collection of all sequences in F forms a ring under the addition and multiplication defined by the formulas

$$(a_0, a_1, \ldots) + (b_0, b_1, \ldots) = (a_0 + b_0, a_1 + b_1, \ldots)$$
$$(a_0, a_1, \ldots)(b_0, b_1, \ldots) = (a_0 b_0, a_0 b_1 + a_1 b_0, a_0 b_2 + a_1 b_1 + a_2 b_0, \ldots)$$

Instead of $(a_0, a_1, \ldots)$ we shall write $\Sigma_{j=0}^{\infty} a_j X^j$. With this notation we have $\Sigma_{j=0}^{\infty} a_j X^j + \Sigma_{j=0}^{\infty} b_j X^j = \Sigma_{j=0}^{\infty} (a_j + b_j) X^j$, $(\Sigma_{j=0}^{\infty} a_j X^j) \cdot (\Sigma_{j=0}^{\infty} b_j X^j) = \Sigma_{j=0}^{\infty} (\Sigma_{i=0}^{j} a_i b_{j-i}) X^j$. Just the same rules apply for power series. Thus, we define the *ring* $F[[X]]$ *of formal power series over* F to be the set of all *formal power series* $\Sigma_{j=0}^{\infty} a_j X^j$ ($a_j \in F$ for all j) with the operations defined above. $F[[X]]$ is an integral domain. A *polynomial* (*over* F) is a formal power series $f = \Sigma_{j=0}^{\infty} a_j X^j$ for which there exists an n such that $a_j = 0$ for $j > n$. In that case we write $f = \Sigma_{j=0}^{n} a_j X^j$. The polynomials form a subring $F[X]$ of $F[[X]]$, the *ring of polynomials* (in one variable) *over* F. $F[X]$ is an integral domain, its quotient field is the *field* $F(X)$ of *rational functions over* F. The degree $d(f)$ of a nonzero polynomial $f = \Sigma_{j=0}^{n} a_j X^j$ is $\max\{j : a_j \neq 0\}$. The polynomials of degree 0 are the (nonzero) *constant polynomials.* A polynomial $a_0 + a_1 X + \ldots + a_n X^n$ of degree $n \in \{0, 1, 2, \ldots\}$ is *monic* if $a_n = 1$. A nonzero polynomial f is *irreducible* if it cannot be written as a product of two polynomials g, h for which $d(g) < d(f)$, $d(h) < d(f)$, otherwise it is *reducible.* Each nonzero polynomial can be written as a product of irreducible ones. Let f be an irreducible polynomial, $d(f) \geqslant 1$. Then the principal ideal (f) is maximal and $F[X]/(f)$ is a field. The quotient map $\pi : F[X]/(f)$ maps the constant polynomials (which constitute a field isomorphic to F) isomorphically onto a subfield of $F[X]/(f)$. Thus $F[X]/(f)$ can be viewed as a field extending F. Considered as a vector space over F it has dimension $d(f)$.

Let $L \subset M$ be fields. An element $z \in M$ is *algebraic* over L if there is a nonzero polynomial $h = a_0 + a_1 X + \ldots + a_n X^n \in L[X]$ for which $a_0 + a_1 z + \ldots + a_n z^n = 0$. Then z is a *root* of h in M. The smallest subfield of M containing L and $\{z\}$ is denoted by $L(z)$. If z is algebraic then $L(z)$ is finite dimensional as a vector space over L, if not then $L(z)$ is isomorphic to $L(X)$. Let $z \in M$ be algebraic over L. The formula $a_0 + a_1 X + \ldots + a_n X^n \mapsto a_0 + a_1 z + \ldots + a_n z^n$ defines a surjective homomorphism $L[X] \to L(z)$ whose kernel is a maximal ideal of the form (f) where f is irreducible. $L(z)$ and $L[X]/(f)$ are isomorphic. Each polynomial $h \in L[X]$ having z as a root is divisible by f, i.e. $h = fh_1$ where $h_1 \in L[X]$. Among all $g \in L[X]$ for which $(g) = (f)$ there is a unique monic one, the *minimum polynomial* of z over L. M is an *algebraic extension* of L if each element of M is algebraic over L. A field L is *algebraically closed* if every algebraic extension of L is trivial, i.e. equals L. L is algebraically closed if and only if each $f \in L[X]$ has a root in L. An *algebraic closure* of a field L is an algebraically closed, algebraic extension of L. Each field has an algebraic closure. Two algebraic closures of a field L are isomorphic by means of an isomorphism leaving L pointwise fixed.

Let L be a field. The map $\varphi : \mathbb{Z} \to L$ given by $\varphi(n) = n.1$ (where 1 denotes the identity of L) is a ring homomorphism. L has *characteristic* 0 if $\operatorname{Ker} \varphi = (0)$. If $\operatorname{Ker} \varphi \neq (0)$ then it has the form $p\mathbb{Z} = \{ pn : n \in \mathbb{Z} \}$ for some prime number p and L has *characteristic* p. $\mathbb{Q}$ has characteristic 0, $\mathbb{F}_p := \mathbb{Z}/p\mathbb{Z}$ (p prime) has characteristic p.

A *root of unity* in a field L is an element $x \in L$ for which $x^n = 1$ for some $n \in \mathbb{N}$. The roots of unity form a multiplicative group. Each finite subgroup is cyclic. Let $n \in \mathbb{N}$. An nth *root of unity* is an element of $C_n := \{x \in L : x^n = 1\}$. C_n has at most n elements. A *primitive* nth *root of unity* is an element $\theta \in C_n$ such that $C_n = \{\theta, \theta^2, \ldots, \theta^n\}$. Such roots exist if C_n has n elements.

Further reading

Amice, Y. (1975) *Les nombres p-adiques,* Presses Universitaires de France.

Bachman, G. (1964) *Introduction to p-adic numbers and valuation theory,* Academic Press, New York.

Dwork, B.M. (1982) *Lectures on p-adic differential equations,* Springer-Verlag, New York.

Iwasawa, K. (1972) *Lectures on p-adic L-functions,* Princeton University Press.

Koblitz, N. (1977) *p-adic numbers, p-adic analysis and zeta functions,* Springer-Verlag, New York.

Koblitz, N. (1980) *p-adic analysis: a short course on recent work,* London Mathematical Society Lecture Note Series 46, Cambridge University Press.

Mahler, K. (1980) *p-adic numbers and their functions,* Cambridge tracts in mathematics 76, Cambridge University Press.

Monna, A. (1970) *Analyse non-archimédienne,* Springer-Verlag, New York.

van Rooij, A. (1978) *Non-archimedean functional analysis,* Marcel Dekker, Inc., New York.

Notation

The pages indicated in this list are those on which the symbols are first defined

Index

(See also Appendix B)